FLORA OF TROPICAL EAST AFRICA

MALVACEAE

BERNARD VERDCOURT & GEOFFREY MWACHALA[1]

Annual or perennial herbs, shrubs or less often trees, usually with stellate or lepidote indumentum, mucilaginous cells etc. in the parenchymatous tissues and fibrous stems or bark. Leaves alternate, sessile to petiolate, usually stipulate, simple, unlobed to deeply and often palmately divided, the margins entire to serrate; venation often palmate. Flowers regular, hermaphrodite or rarely unisexual, solitary or in cymose inflorescences; epicalyx bracts often present; calyx-lobes (3–)5, sometimes united at base, valvate; petals 5, free but often adnate to the base of the staminal tube, imbricate, contorted or convolute, usually obliquely obovate and with a retuse to emarginate apex. Stamens 5–numerous, the filaments joined into a tube for most of their length; anthers dehiscing by slits; pollen spinulose. Ovary superior with (1–)5–many locules with as many or twice as many style-branches as carpels; style-branches partly free or almost completely to completely connate; placentation axile with 1–numerous ovules per locule. Fruits mostly loculicidal capsules or schizocarpic, less often indehiscent berries or samaras.

A large family with species distributed throughout the world; 121 genera and about 1550 species.

Note from the editors. The circumscription of genera in the tribe Hibisceae (e.g., *Abelmoschus, Hibiscus, Kosteletzkya, Pavonia, Urena*), is based on morphological features that are not corroborated by phylogenetic evidence. Pfeil, Brubaker, Craven & Crisp have published on the phylogeny of this group in Systematic Botany 27 (2): 333–350 (2002) and came to the conclusion that these groups are not monophyletic. They state that nomenclatural upheavals in and around *Hibiscus* are almost inevitable. Specifically, *Abelmoschus* and *Fioria* (in the FTEA treatment treated as *Roifia*) cannot be maintained as separate and will have to sink in *Hibiscus*; or *Hibiscus* will have to be divided into more than ten genera.

In a second article, Pfeil & Crisp discuss 'What to do with Hibiscus' (Australian Systematic Botany 18: 49–60, 2005) and confirm inclusion of *Abelmoschus* and *Fioria* in *Hibiscus,* and add *Kosteletzkya, Pavonia* and *Urena* to *Hibiscus* synonymy.

The editors are grateful to David Mabberley (RBG Kew) and Lyn Craven (Australian National Herbarium, CSIRO) for drawing attention to this recent work and providing references. Lyn Craven also comments: "Now that we have the benefit of DNA technologies we can devise better classifications but there are many groups in which we are not quite there yet, and Malvaceae is one of these (the infrageneric classification of *Hibiscus* in particular will take some considerable work)"

Sadly, Dr Verdcourt is seriously ill at the moment and unable to change his treatment accordingly, or to comment on these developments. I (HB) have therefore decided to let the current FTEA treatment stand, rather than implement as yet incomplete, though scientifically convincing, new findings.

[1] Bernard Verdcourt, Royal Botanic Gardens, Kew: *Abelmoschus, Abutilon, Cienfuegosia, Gossypioides, Malva, Malvastrum, Pavonia, Roifia, Senra, Sida, Sidastrum, Thespesia, Urena, Wissadula,* and cultivated genera

[1] Geoffrey Mwachala, East African Herbarium, National Museums of Kenya, Nairobi: *Hibiscus*
Gossypium by Paul Fryxell (University of Texas, Austin) and Bernard Verdcourt
Kosteletzkya by Orland J. Blanchard, Jr., Herbarium, Florida Museum of Natural History, University of Florida, Gainsville, Florida 32611, U.S.A.

1.	Styles 10, twice as many as carpels; fruit a schizocarp	2
	Styles the same number as carpels	3
2.	Leaves with gland on midrib; epicalyx bracts united in lower half; mericarps with many glochidiate awns	1. **Urena** (p. 4)
	Leaves without gland; epicalyx bracts mostly free; mericarps without glochidiate awns [if awns present, with few spines only]	2. **Pavonia** (p. 8)
3.	Epicalyx absent, or of minute teeth only	4
	Epicalyx present and distinct	8
4.	Fruit a capsule; ovary with 5 carpels and 5 free styles	3. **Hibiscus** (p. 31)
	Fruit a schizocarp of separate mericarps; ovary with 3–many carpels; styles 3–many, free	5
5.	Mericarps and style branches 3–5; carpels contracted near the middle, upper parts spreading	14. **Wissadula** (p. 112)
	Mericarps and style branches 5+; carpels not contracted near middle	6
6.	Mericarps 6–30, opening by an apical slit, with 1–3 ovules per locule	15. **Abutilon** (p. 114)
	Mericarps 5–13, opening laterally, with 1 ovule per locule	7
7.	Flowers solitary or fasciculate; calyx often more than 3 mm, petals more than 4 × 3 mm	16. **Sida** (p. 144)
	Flowers eventually in terminal panicles to 30 cm; calyx 2.5–3 mm long, petals to 5 × 1.5 mm	17. **Sidastrum** (p. 162)
8.	Indument of small scales (lepidote); calyx truncate; fruit fleshy, indehiscent	11. **Thespesia** (p. 101)
	Indument not of scales; calyx dentate or lobed; fruit dry, dehiscent	9
9.	Epicalyx of 3 bracts	10
	Epicalyx of 4–many bracts	14
10.	Epicalyx bracts over 13 mm wide; ovary with 3–5 carpels and 3–5-connate styles; fruit a capsule	11
	Epicalyx bracts less than 3 mm wide; ovary with 3–many carpels; styles 3–many, free; fruit a schizocarp	13
11.	Style with 5 branches	7. **Senra** (p. 89)
	Style mostly united	12
12.	Style clavate, undivided	9. **Gossypium** (p. 93)
	Style united below, 3–5-lobed at apex	10. **Gossypioides** (p. 99)
13.	Corolla white/pink/purple	12. **Malva** (p. 107)
	Corolla yellow	13. **Malvastrum** (p. 110)
14.	Calyx spathaceous, irregularly splitting, falling with corolla; ovary pubescent; style unbranched; fruit oblong	4. **Abelmoschus** (p. 75)
	Calyx dentate or lobed, persistent; ovary glabrous or pubescent; style mostly branched (except sometimes in *Cienfugosia*); fruit globose	15
15.	Indument simple; fruit a schizocarp	5. **Roifia** (p. 79)
	Indument stellate	16
16.	Ovary and fruit 5-locular, each locule with 1 ovule/seed	6. **Kosteletzkya** (p. 81)
	Ovary and fruit 3–5-locular, each locule with 3 or more ovules or seeds	17
17.	Calyx with distinct dark oil glands; style not distinctly branched	8. **Cienfuegosia** (p. 92)
	Calyx without glands; style usually 5-branched	3. **Hibiscus** (p. 31)

Many species are cultivated as ornamentals or for fibres and as vegetables. Those not dealt with in full in the main sequence are mentioned below:

Anoda cristata (*L.*) *Schlecht.* (*A. hastata* Cav.) has been collected in Kenya – whether cultivated or growing as a result of being an impurity in other seeds is not clear. A native of South and central America and southern U.S.A. it has occurred as a weed in experimental plots in South Africa and may well occur as a weed elsewhere in East Africa. It is quite widely naturalised in various parts of the Old World.

Herb to about ± 1 m tall; young stems setose but older glabrescent and with typical anastomosing ribbing characteristic of Malvaceae. Leaves very variable, triangular, hastate or palmately lobed. Flowers solitary in leaf axils or in cymes; calyx accrescent, densely setose outside, reticulately veined, deeply 5-lobed; lobes triangular up to 12 × 9 mm, apiculate. Corolla lilac; petals ± 2 cm long; staminal tube very short. Carpels 9–20, usually with distinct setose dorsal narrow projections up to 6 mm long.

KENYA. Kiambu District: Muguga, E.A.A.F.R.O. Nursery, 20 Oct. 1956, *Verdcourt* 1582A.

Lagunaria patersonius (*Andr.*) *G. Don* (as *patersonii*) subsp. *patersonius*; a native of Norfolk I. and Lord Howe I., widely cultivated in the tropics, particularly as a street tree; has been extensively grown in East Africa [Jex-Blake, Gard. E. Afr. ed. 4: 117 (1957)]. It has been grown in Uganda at Entebbe [Dale, Descr. List Intr. Trees Uganda: 47 (1953)].

Evergreen tree 3–15 m tall (over 20 m in the wild). Leaves subcoriaceous with lamina elliptic, 4–9 × 1.3–5 cm, rounded at the apex, dull green above but distinctively silvery white-scaly beneath. Flowers solitary in the leaf axils; petals pink, 3.5–4.5 cm long. Fruit said to be round, 2–3 cm long in the wild but in Kenya often obovoid and 4 × 2.5 cm.

The specific epithet will be found spelt *patersonii* in virtually all reference works including recent ones such as the European Garden Flora. Craven, Miller & White in Blumea 51: 352 (2006) discuss the nomenclature, and here we follow their advice.

KENYA. Lari Arboretum, 12 Mar. 1953, *Darling* 54 & Nairobi Arboretum, 17 Mar. 1952, *G.R. Williams Sangai* 367 & Nairobi, Mbagathi Road, near Forces Memorial Hospital, 2 Dec. 1987, *Mathei & Ndegwa* 1/87.

Malvaviscus arboreus *Cav.* has been cultivated fairly extensively in East Africa. A native of Mexico to Peru and Brazil.

Shrub 1.5–4 m tall, densely velvety stellate-hairy or in some varieties almost glabrous; leaves narrowly to broadly ovate, 3–12(–23) × 1.5–10(–18) cm, sometimes 3-lobed, rather coarsely crenate-toothed; flowers solitary, usually held ± erect but sometimes pendulous; pedicel 2.5–7 cm long; calyx tubular 1–2 cm long including triangular teeth ± 5 mm long; epicalyx bracts ± 7, linear-spatulate ± 1 cm long; corolla red or pink, never fully opening, the petals 2.5–7 cm long; staminal column included or exserted to 3 cm; fruit a berry; seeds with red fleshy coat. Two varieties occur in East Africa:

var. *arboreus* Cav. (*M. mollis* (Ait.) DC.): KENYA: Nairobi, Karen, 13 May 1966, *Gardner* in EAH 13538.

var. *mexicana* Schltdl. (*M. grandiflorus* Kunth): KENYA. Nairobi Arboretum, 17 Mar. 1952, *G.R. Williams-Sangai* 369 & Karen, 13 May 1966, *Gardner* in EAH 13539. TANZANIA. Lushoto, Jaegertal, 30 Jan. 1971, *Ruffo* 368 & Lushoto, 9 May 1969, *Shabani* 413 & Amani Nursery, 11 Mar. 1973, *Ruffo* 625. Grown on ZANZIBAR and PEMBA fide U.O.P.Z.: 339, fig. (1949).

1. URENA

L., Sp. Pl.: 692 (1753); Gen. Pl., ed. 5: 309 (1754); Gürke in E.J. 16: 361–385 (1892); Borss. Waalk. in Blumea 14: 137–145 (1966)

Perennial or annual herbs or subshrubs with stellate indumentum. Leaves petiolate, subentire to deeply palmately 3–7-lobed, toothed, with often both coarse and finer teeth, palmately 3–9-veined, with a conspicuous gland near base of midrib consisting of a slit ± 1.5 mm long with thickened edges, sometimes also present on veins immediately on either side; stipules linear or setaceous, eventually deciduous. Flowers solitary, fasciculate or merging into terminal pseudoracemes at the ends of the branches. Epicalyx 5-lobed, the lobes ± rigid, narrowly triangular or lanceolate; calyx 5-lobed, the lobes ovate to ovate-lanceolate, keeled. Corolla pink or mauve, mostly with dark centre or rarely white, the petals obovate, densely stellate-pilose outside. Staminal tube equalling the petals. Ovary of 5 free 1-ovulate carpels; styles 10, reflexed, with discoid stigmas. Fruits depressed globose, of 5 indehiscent trigonous mericarps, stellate-pubescent and with dense glochidiate spines dorsally (except in endemic Australian species), striate on the flat contiguous lateral surfaces.

Formerly usually treated as a single species but apart from the pantropical species *U. lobata* with numerous varieties there are certainly three other distinct species in Australia (Craven & Fryxell in Austral. Syst. Bot. 2: 455–460 (1989)). One of these and several Indian species were transferred to *Pavonia* by Hochreutiner in 1901 but the latter may also be better considered as *Urena* again. *Urena* and *Pavonia* are extremely close and the former is the earlier name. It appears there are 4–6 species of true *Urena*.

Urena lobata *L.*, Sp. Pl.: 692 (1753), Mast. in F.T.A. 1: 189 (1868); Gürke in E.J. 16: 370 (1892) & in P.O.A. C: 266 (1895); Hochr. in Ann. Cons. Jard. Bot. Genève 5: 131 (1901); Ulbr. in E.J. 51: 52 (1913); Exell & Mendonça in C.F.A. 1: 155 (1937); T.T.C.L.: 308 (1949); U.O.P.Z.: 479 (1949); Keay, F.W.T.A. ed. 2, 1(2): 341 (1958); Exell in F.Z. 1: 504 (1961); Borss. Waalk. in Blumea 14: 138 (1966); Abedin, Fl. W. Pak. 130: 92, fig. 12 (1979); Blundell, Wild Fl. E. Afr.: 80, fig. 741 (1987); U.K.W.F. ed. 2: 102, t. 29 (1994); Vollesen in Fl. Eth. & Eritr. 2 (2): 224, fig. 82.16.10 (1995). Type: *Herb. Linnaeus* 873.1 (LINN, lecto.)

Erect annual or perennial herb or subshrubby herb 0.5–1.5(–2) m tall; stems often reddish and usually ± unbranched. Lamina oblong or oblanceolate to ovate, round or oblate in outline, 1–12 × 0.5–13 cm, acute at the apex, cordate at base, serrate to coarsely toothed, crenate or subentire, stellate-pubescent with upper surface slightly scabridulous but velvety beneath; petiole 0.2–9(–12) cm long; stipules 2–4 mm long. Terminal pseudoracemes, if present, up to 35 cm long; pedicel 1–3 mm long, becoming 3–7 mm in fruit; calyx and epicalyx bracts 3–6(–10) mm long. Petals 1–1.5(–2.5) cm long. Fruit depressed globose, 6–10 mm wide; mericarps 4–7 mm long, 3–4 mm wide; seeds brown, reniform, 2–3.5 mm wide, glabrous or minutely pubescent.

a. var. **lobata**

Leaves unlobed to deeply palmately lobed, the lobes usually narrowing from base to apex but sometimes oblanceolate, widest near apex but sinuses acute and narrow. Fig. 1, p. 5 & Fig. 2: 1, 2, 5–8, p. 6.

UGANDA. West Nile District: Amua, Dec. 1932, *Eggeling* 895! ('898' corrected according to field book); Bunyoro, Nov. 1862, *Grant* 610!; Teso District: Serere, Dec. 1931, *Chandler* 95!

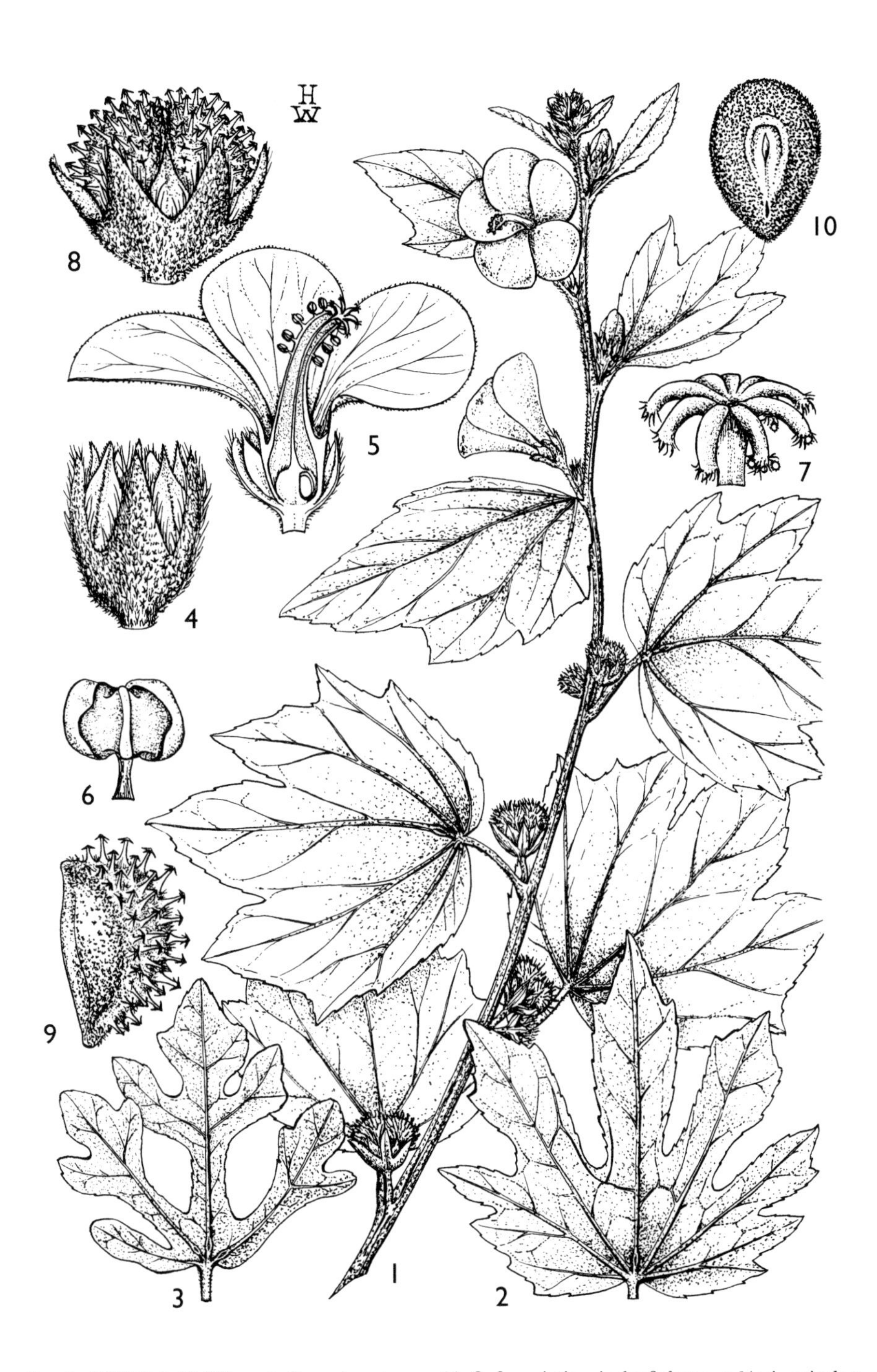

FIG. 1. *URENA LOBATA* — **1**, flowering stem, × ²⁄₃; **2**, **3**, variation in leaf shape, × ²⁄₃; **4**, epicalyx and calyx, × 3; **5**, flower, longitudinal section, × 2; **6**, stamen, × 10; **7**, stigma, × 10; **8**, fruit with epicalyx and calyx, × 3; **9**, mericarp, × 4; **10**, seed, × 4. 1, 4, 8–10 from *Langdale-Brown* 2100; 2 from *Milne-Redhead & Taylor* 9795; 3, 5–7 from *Faulkner* 3044. Drawn by Heather Wood.

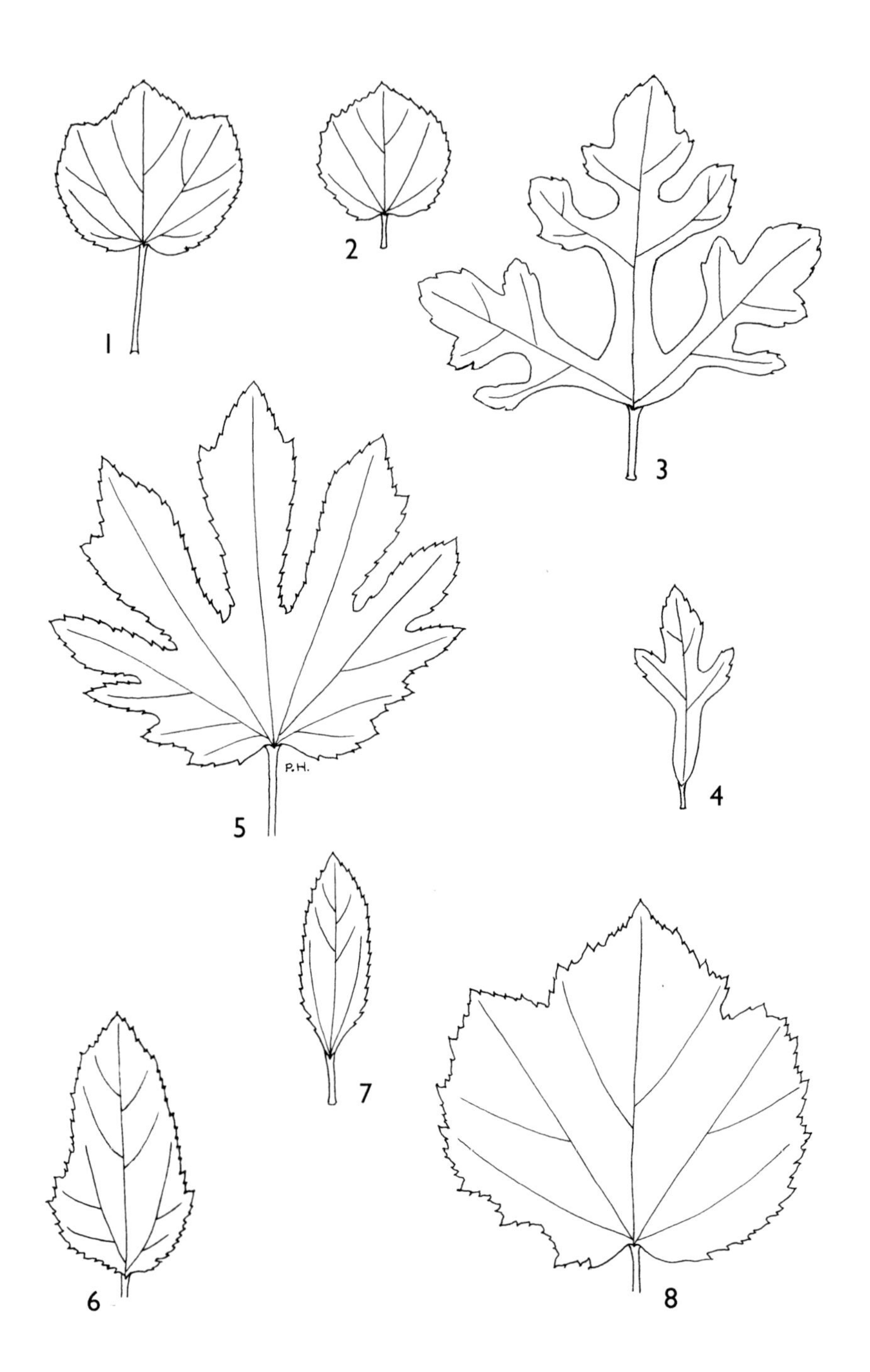

FIG. 2. *URENA LOBATA* — **1–8**, leaf shapes showing variability; 3 & 4 are var. *sinuata*, all others are var. *lobata*. 1–2 from *Last* s.n. anno 1909; 3–4 from *Faulkner* 3044; 5 from *Milne-Redhead & Taylor* 9795; 6 from *Kimani* 132; 7–8 from *Chandler* 380. Drawn by Pat Halliday.

KENYA. Uasin Gishu District: Kipkarren, Sep. 1931, *Brodhurst Hill* 582!; Embu District: S slopes of Mt Kenya near Castle Forest Station, 16 Jan. 1973, *Spjut & Ensor* 3012!; West Kakamega Forest Reserve, 10 July 1960, *Paulo* 526!

TANZANIA. Mwanza District: shore of Speke Gulf near Mwanza, 1 June 1931, *B.D. Burtt* 2466!; Rufiji District: Mafia I., near Kidika, 21 Aug. 1937, *Greenway* 5140!; Songea District: 8 km W of Songea, 24 Apr. 1956, *Milne-Redhead & Taylor* 9795!; Pemba: Chake Chake, Mkoani road, 12 Sep. 1929, *Vaughan* 642!

DISTR. **U** 1–4; **K** 3–5, 6 (fide UKWF), 7; **T** 1, 4, 6–8; **Z**; **P**; widespread in tropical Africa; pantropical

HAB. Grassland, sandy flood plains, roadsides, grassland with scattered *Combretum*, *Terminalia* etc., bushland, termite-mound thicket, less often in semideciduous forest, absent from very dry areas, often near cultivations and villages; sea level–2100 m

SYN. *U. reticulata* Cav., Diss. V1: 335, t. 183, fig. 2 (1788). Type: "Equatorial America" specimen in Herb. Lamarck fide Cav. (P, holo. ? not found in microfiche index)
U. lobata L. var. *reticulata* (Cav.) Gürke in E.J. 16: 376 (1892)

NOTE. A very important fibre plant used for ropes, fishing lines etc. Gürke does not actually have a typical variety which includes the Linnean type. He cites several East and Central African specimens under var. *reticulata.*

b. var. **sinuata** (*L.*) *Hochr.* in Ann. Cons. Jard. Bot. Genève 5: 141 (1901). Lectotype: Herb. Hermann 4: 34, No. 257 (BM-000621193), designated by Fawcett & Rendle in Fl. Jamaica 5 : 127 (1926).

Leaves deeply 3–5-lobed (or sometimes some scarcely or not lobed). The lobes, particularly the central ones, again deeply lobed, the lobules mostly broadest at the middle and sinuses broad and more rounded, leaving a large gap between central and lateral lobes; or occasionally leaves narrow with a pair of lobes above the middle. Fig. 2: 3, 4, p. 6.

KENYA. Kwale District: Bodo, 14 Oct. 1991, *Luke* 2915!

TANZANIA. Tanga District: Sawa, 12 May 1973, *Faulkner* 4780! & Tanga–Pangani road, 4 May 1970, *Archbold* 1231!; Zanzibar: Mkokotoni, 29 May 1962, *Faulkner* 3044! & without locality, 1908, *Last* s.n.!

DISTR. **K** 7; **T** 3; **Z**; Comoro Is., Madagascar and Mascarene Is., also India to Japan and Java, Central and South America and West Indies

HAB. Damp grassland, areas of cultivation, near villages; 0–15 m

SYN. *U. sinuata* L., Sp. Pl.: 692 (1753); Gürke in E.J. 16: 377 (1892) & in P.O.A. C: 266 (1895); Philcox in Rev. Handb. Fl. Ceylon 11: 324 (1997)
U. lobata L. subsp. *sinuata* (L.) Borss. Waalk. in Blumea 14: 142 (1961); Marais, Fl. Mascar. 51: 46 (1987)

NOTE. This taxon has mostly been separated by having narrower epicalyx bracts shorter than the calyx and spreading or slightly reflexed in fruit, but I have found this rather unreliable and it does not always correlate with the leaf character; the latter is my sole reason for separating it. Since there is no geographical reason for using the rank of subspecies and since Borssum Waalkes reduced it to a form on annotation labels for the Flora Malesiana account (unpublished) I have reverted to varietal status which appears more appropriate. The Kenya specimen cited above is somewhat intermediate. Since the species has been so widely cultivated for fibre for centuries it is not possible to be certain of the original range. It is possible neither of the varieties mentioned above is truly native in Africa.

2. PAVONIA*

Cav., Diss. 2, App. 2: [2] (1786) & 3: 132, t. 45 (1787); Ulbr. in E.J. 57: 54–184 (1920–21) & in V.E. 3(2): 380–390 (1921); Vollesen in Fl. Eth. & Eritr. 2(2): 224–236 (1995), *nom. cons.*

Annual or perennial herbs, subshrubs or shrubs, erect or less often procumbent or scrambling; indumentum usually stellate but simple and glandular hairs often present. Leaves unlobed to palmately shallowly to very deeply lobed, often toothed; stipules linear to lanceolate. Flowers solitary in leaf axils or in clusters or heads sometimes ± forming paniculiform, raceme-like or spicate inflorescences; epicalyx bracts 5–16(–19), ovate to linear, free or slightly joined at base; calyx 5-lobed; corolla white, yellow, pink, red or mauve, often with darker centre, longer than calyx. Staminal tube truncate, bearing numerous free filaments mostly in upper part but often some basal ones with much longer filaments. Carpels 5, free, 1-ovulate, styles usually 10, sometimes fewer; stigmas capitate. Fruit a discoid to subglobose schizocarp; mericarps segment-shaped or ± rhomboid or obovoid, glabrous or pubescent, variously reticulate, ribbed or with crests, conical projections or spines or distinctly awned, indehiscent or some thin-walled ones ultimately splitting. Seeds reniform.

About 200 species in tropics or subtropics, the largest number in South America; 25 occur in the Flora area (one introduced and naturalised). Two groups are taxonomically difficult and very different species concepts have been followed by several botanists. I have for the most part followed Vollesen who has extensive field knowledge; Fryxell who began the account for East Africa but was unable to finish it has more conservative concepts explained by notes in the appropriate places.

I have followed Ulbrich's order of species as far as possible. He recognised too many species and infraspecific taxa but his basic arrangement is sound. For an exhaustive account of the genus see Fryxell, Flora Neotropica Monograph 76 (1999) where all aspects of its biology are discussed.

Ripe mericarps are needed for certain identification in some cases. *P. dimorphostemma* (25) from **K** 1 is not keyed out as its fruits are unknown.

1. Mericarps with 3 apical retrorsely barbed awns; epicalyx bracts linear, not widened above base (sect. *Afrotyphalea* Ulbr.) or lanceolate (sect. *Urenoideae* St Hilaire) 2
   Mericarps smooth to reticulate or ridged and with wings or conical or spiny projections but without such awns; epicalyx bracts linear to ovate or almost orbicular 6
2. Introduced South American species with yellow flowers, elliptic unlobed closely serrate leaves and glabrous mericarp bodies (sect. *Urenoideae*) 5. *P. sepioides* (p. 15)
   Native species with pink to purple or white flowers; leaves unlobed to deeply 3–7-lobed; mericarp bodies pubescent, puberulous or glabrous 3

* By B. Verdcourt. P. Fryxell undertook to write up this account but was unable to finish it; I (BV) have used much of his account but since I have in several sections diverged considerably from his specific concepts I have not given him as joint author since it would imply his acceptance of concepts with which he disagrees. Great use has been made of Vollesen's account for Flora of Ethiopia & Eritrea.

3. Retrorsely barbed awns very short, 1–2 mm long; **T** 7 .... 4. *P. stolzii* (p. 14)
   Retrorsely barbed awns up to 8 mm long; widespread 4
4. Stem leaves unlobed with cuneate base; epicalyx bracts 6–8, united for 2–3 mm at base .... 3. *P. kilimandscharica* (p. 14)
   Stem leaves shallowly to deeply palmately 3–7-lobed; epicalyx bracts 8–10, united for 1 mm at base 5
5. Corolla pink to purple with or without a dark centre; flowers mostly in dense subsessile clusters but some solitary and pedicellate; leaves pubescent to very densely velvety all over .... 1. *P. urens* (p. 11)
   Corolla nearly always white with purple centre; flowers much laxer, all ± clearly pedicellate; leaves ± glabrescent or with only short stellate hairs above .... 2. *P. schimperiana* (p. 13)
6. Epicalyx bracts narrowly lanceolate to ovate, elliptic or ± orbicular, distinctly widened above the base; mericarps ± obovoid or rhomboid, ± basally attached, reticulate to strongly keeled and ridged, with conical projections or distinctly spinous (sect. *Afrolebretonia* Ulbr.) 7
   Epicalyx bracts linear or slightly linear-lanceolate, not widened above the base (or rarely narrowly ovate); mericarps with two flat lateral sides and convex back (segmentiform) smooth on flat sides, smooth to very complicatedly spiny or sculptured on curved surface 11
7. Epicalyx bracts stems and leaves densely finely grey tomentose velvety; leaves often with glands and ± viscid; mericarps reticulate, usually ± pubescent; leaves usually not palmately lobed .... 10. *P. gallaensis* (p. 19)
   Epicaclyx bracts etc. not densely finely tomentose but pubescent or hairy with obvious hairs, ± not obscuring the surface; mericarp glabrous [sparsely puberulous in *burchellii*] 8
8. Leaves usually ± palmately lobed; mericarps reticulate with only a low keel and no or faint projections; epicalyx bracts ovate, usually bluntly acute and not acuminate .... 9. *P. burchellii* (p. 18)
   Leaves usually ovate and unlobed; mericarps with narrow spines at apex or with very marked often jagged keel and conical projections or ridges; epicalyx bracts usually more lanceolate or acuminate 9
9. Leaves up to 1–1.5 cm long and wide, ovate or almost round with dense stellate hairs and black dots beneath; stems ± glandular, without long hairs; epicalyx bracts up to 6 × 1.5 mm; **K** 1 .... 7. *P. flavoferruginea* (p. 16)
   Leaves larger 10

10. Young stems glandular with dense short hairs, mericarps with distinct spines; calyx lobes usually distinctly narrowly triangular-lanceolate and acuminate . . . . 7. *P. flavoferruginea* (p. 16)
    Young stems usually with spreading hairs, not glandular; mericarps with distinct often crested keel and 2–3 lateral conical projections from sides; calyx lobes usually more elliptic, shorter and less acuminate . . . . 8. *P. procumbens* (p. 17)
11. Mericarps with 3 conical dorsal spines and/or longitudinal rows of curved or hooked prickles (sect. *Callicarpidium* Ulbr.) 12
    Mericarps not as above 15
12. Mericarps without longitudinal rows of curved or hooked prickles; calyx 6–8 mm long . . . . 12. *P. propinqua* (p. 20)
    Mericarps with longitudinal rows of curved or hooked prickles 13
13. Leaf base (at least of lower leaves) cordate; epicalyx bracts 1.5–4 mm wide, lanceolate to narrowly ovate; mericarps with raised spiny central band and 4 rows of prickles and large lateral spines with small apical prickles . . . . 13. *P. elegans* (p. 20)
    Leaf base truncate to subcordate; epicalyx bracts up to 1.5 mm wide, linear; mericarps without raised central band, with central and 2 lateral spines without apical prickles 14
14. Epicalyx bracts 6–8; calyx 5–8 mm long; petals 10–15 mm long; mericarps dorsally with transverse ridges; leaves broadly elliptic to round, up to 2.8 × 2.7 cm. . . . . 14. *P. cristata* (p. 22)
    Epicalyx bracts 9–12; calyx 10–15 mm long; mericarps without transverse ridges; leaves ovate to elliptic or oblong, up to 9 × 5 cm . . 15. *P. melhanioides* (p. 22)
15. Leaves very deeply divided into 3–5 narrow oblong to oblanceolate lobes (in Flora area); mericarps 3–5 mm long with wings up to 0.5 mm wide or unwinged 16
    Leaves shallowly divided [often almost to the middle in *columella*] or not divided 17
16. Leaves mostly 5-lobed; flowers bright yellow or cream . . . . 17. *P. arenaria* (p. 24)
    Leaves 3-lobed; flower pink or orange . . . . 18. *P. triloba* (p. 26)
17. Mericarps 8–10 mm long, segmentoid, the outer curved surface with strong raised crests and one short projection and two smaller ones at apex; epicalyx bracts 12–16, narrowly linear, up to 1 mm wide, shorter than the 10–16 mm long calyx which has ± ovate 3-veined lobes; corolla yellow (sect. *Afrolopimia* Ulbr.) . . . . . . 6. *P. senegalensis* (p. 15)
    Mericarps shorter and without other characters combined 18
18. Epicalyx bracts 5, linear or narrowly lanceolate 5–8 mm long; leaves distinctly 3–5-lobed; calyx lobes conspicuously 3-veined; mericarps 3 × 2 mm, the back finely reticulate; corolla mauve or pink (sect. *Columella* Ulbr.) . . . . . . 11. *P. columella* (p. 19)
    Epicalyx bracts 7–14, filiform 19

19. Petals pink (rarely white) .... 19. *P. arabica* (p. 26)
    Petals white to yellow 20
20. Leaves entire or with a few irregular teeth near apex 21
    Leaves regularly toothed to near base 23
21. Mericarps with 1–2 mm wide ciliate wings, the back glabrous, conspicuously transversely ribbed .... 22. *P. blepharocarpa* (p. 28)
    Mericarps not winged, the back not or slightly ribbed, glabrous to pilose 22
22. Stems rough, stellate-pubescent to tomentose (but sometimes also glandular); calyx glabrous except for ciliate margins; mericarps with concave transversely ribbed back; petals 7–12 mm long .... 21. *P. ellenbeckii* (p. 27)
    Stems smooth, glandular-pubescent; calyx puberulous all over; mericarps with flat back not ribbed; petals 10–18 mm long .... 20. *P. schweinfurthii* (p. 27)
23. Mericarps not winged; flowers white; leaves up to 8.5 × 4.5 cm, densely velvety pubescent beneath, discolorous, often slightly lobed .. 16. *P. leptocalyx* (p. 23)
    Mericarps winged; flowers white or more usually pale to bright yellow 24
24. Mericarp wing 0.5–2 mm wide, triangular in outline, widest near middle; petals 13–27 mm long, white to pale yellow .... 23. *P. serrata* (p. 28)
    Mericarp wing 2–5 mm wide, semicircular in outline, equally wide from top to bottom; petals 9–16 mm long, bright or lemon yellow 24. *P. kotschyi* (p. 30)

1. **Pavonia urens** *Cav.*, Dissert. 3: 137 (1787) t. 49, fig. 1 (1785); Ulbr. in E.J. 57: 104 (1920) & in V.E. 3 (2): 382 (1921); T.T.C.L.: 306 (1949); Brenan in Mem. N.Y. Bot. Gard. 8: 223 (1953); Meeuse in F.Z. 1(2): 511 (1961); Hauman in F.C.B. 10: 184, t. 17 (1963); U.K.W.F.: 209, fig. on 208 (1974); Cribb & Leedal, Mt. Fl. S Tanz.: 52, t. 6b (1982); Blundell, Wild Fl. E. Afr.: 79, fig. 740 (1987); U.K.W.F. ed. 2: 103, t. 30 (1994); Vollesen in Fl. Eth. & Eritr. 2(2): 228, fig. 82.16.1–5 (1995). Type: Cultivated in Madrid, Herb. Madrid 476081 & 476082 (M, syn.) (microfiche!)*

Shrubby herb or shrub 0.6–2(–4.5)m tall; branches with rather rough ± irritating large stellate and/or simple hairs, rarely glabrescent. Leaves cordate to reniform in outline, 4.5–23 × 5.5–24 cm, shallowly to more deeply 3–7-lobed, the lobes triangular, acute to acuminate or some leaves can be narrowly triangular but cordate at base or lanceolate, margins coarsely toothed, finely pilose to very densely-velvety above with stellate and or simple hairs, densely pilose to very densely velvety beneath; petiole 4–23 cm long. Flowers solitary or in dense clusters, forming spike-like panicles, sessile or pedicel 0–8 mm long; epicalyx bracts 8–10, linear, 5–11 mm long, 0.5–1 mm wide, appressed to the calyx; calyx 6–8 mm long, long-pilose with simple hairs, particularly on the veins. Petals pink to red or purple, darker at base, 1.5–3.5 cm long, staminal column 1–2(–2.6) cm long. Mericarps 5–6 mm long, glabrescent or puberulous, reticulate on the back and with 3 apical retrorsely barbed awns (3–)5–8 mm long.

* In his manuscript Fryxell has suggested P-JU 12349 is the 'holotype' but I prefer to select specimens from which Cavanilles drew up his descriptions.

1. Androecium and styles 2.2–2.6 cm long; leaves with fine indumentum, not densely tomentose .... c. var. *hanangensis*
   Androecium and styles < 2 cm long; leaves with fine indumentum or more often densely tomentose 2
2. Indumentum fine to densely velvety .... a. var. *urens*
   Indumentum on all parts excessively densely velvety ..... b. var. *irakuensis*

var. **urens**; Keay, F.W.T.A. ed. 2, 1(2): 341 (1958)

Indumentum fine to densely velvety; androecium and styles < 2 cm long.

UGANDA. Kigezi District: Kachwekano Farm, May 1949, *Purseglove* 2803!; Mbale District: Bugisho, Buansifwa, 8 Sep. 1932, *A.S. Thomas* 672!; Masaka District: Buddu, *Dawe* 312!
KENYA. Trans-Nzoia District: Cherangani Hills, 6 Dec. 1958, *Symes* 513!; South Nyeri District: Gathiba crossing below Kamweti Forest Station, 10 Nov. 1971, *Robertson* 1627!; Kericho District: W border of Mau Escarpment, about 8 km S of Kericho, near Chegoibe Tea Estate, 4 Feb. 1973, *Spjut & Ensor* 3105!
TANZANIA. Arusha District: Ngurdoto Crater, Saji, 11 Oct. 1965, *Greenway & Kanuri* 11998!; Lushoto District: Mazumbai, 8 July 1966, *Semsei* 4077!; Ufipa District: Sumbawanga, Kito Mt, 21 Apr. 1961, *Richards* 15052!
DISTR. **U** 1–4; **K** 1, 3–6; **T** 1–7; widespread in tropical Africa; Madagascar and Réunion
HAB. *Hagenia* etc. forest margins, riverine and marshy ground forest, *Vernonia* bushland, grassland and bracken patches in forest clearings; 600–2550 m

SYN. *P. schimperiana* A. Rich. var. *hirsuta* Ulbr. in E.J. 57: 109 (1920); T.T.C.L.: 306 (1949). Type: Ethiopia, North Aber Mts, Dschenausa, *Schimper* II.1405 (B†, holo.)
P. schimperiana A. Rich. var. *tomentosa* Ulbr. in E.J. 57: 109 (1920). Type: Ethiopia, near Adua, *Schimper* II, 1026 (B†, holo.); T.T.C.L.: 306 (1949)
P. ruwenzoriensis De Wild., Pl. Bequaert. 1: 500 (1922); Robyns, F.P.N.A 1: 590 (1948). Type: Congo-Kinshasa, Lamia Valley, *Bequaert* 4222 (BR, holo.)

NOTE. There are certainly some intermediates between *P. urens* and *P. schimperiana* and possibly hybrids e.g. *Verdcourt* 1559 (Ngorogoro Crater rim, Sale turn-off, 6 Aug. 1956) which has white flowers and a pink eye, very sparsely pubescent above but the inflorescences more resembling *P. urens*. *Schlieben* 4673 (Kilimanjaro) describes the flower colour as yellow orange but no other field note mentions such a colour.

b. var. **irakuensis** (*Ulbr.*) *Verdc.* **comb. nov.** Type: Tanzania, Mbulu District: Iraku, land of Mama Isara, *Jaeger* 219, 220 (B†, syn.)

All parts excessively densely velvety, the leaves particularly thickly velvety beneath and attaining 23 × 24 cm. Inflorescences very tight velvety clusters up to 3.5 cm diam.

UGANDA. Toro District: without exact locality, Feb. 1902, *Mrs. Tufnell* s.n.!; Kigezi District: Kisoro, 12 Jan. 1938, *Ghesquière* 5691!
KENYA. Trans-Nzoia District: between Suam Sawmills and Suam R., 22 Dec. 1967, *Kahuro* 13!; Ravine District: about 4 km NNE of Londiani, 8 Nov. 1967, *Perdue & Kibuwa* 9014!; Kisumu-Londiani District: Londiani, junction of Eldoret and Kericho roads, 4 Jan. 1967, *Magogo* 15!
TANZANIA. Masai District: near Ngorongoro, 23 June 1970. *B.J. & S. Harris* 4844! & Ngorongoro Crater, W summit above Laroda, 12 Sep. 1932, *B.D. Burtt* 4313! & near Ngorongoro Camp, 5 Aug. 1956, *Verdcourt* 1542!
DISTR. **U** 2; **K** 3, 5; **T** 2; E Congo-Kinshasa
HAB. Burnt hillsides with mist-forest remnants (*Bersama, Cussonia* etc.), grassland, thicket at forest margins, abandoned farmland; 1950–2400 m

SYN. *P. irakuensis* Ulbr. in E.J. 57: 107 (1920); T.T.C.L.: 305 (1949)

NOTE. Brenan annotated *Burtt* 4313 as *Pavonia urens* var. *tomentosa* (Ulbr.) Brenan and it has been usual to look on this taxon as a particularly luxuriant form of that but I think it deserves a name. It is associated with particularly rich volcanic soils and ancient lava.

c. var. **hanangensis** *Verdc.* **var. nov.** a var. *urenti* androecio stylisque 2.2–2.6 cm longis differt. Type: Tanzania, Mbulu District: Mt Hanang, above Katesh, *Polhill & Paulo* 2303 (K!, holo.)

Leaves finely pubescent above, more densely so beneath. Androecium and styles 2.2–2.6 cm long.

TANZANIA. Mbulu District: Mt Hanang, above Katesh, 3 May 1962, *Polhill & Paulo* 2303!
DISTR. **T** 2; not known elsewhere
HAB. Clearings in upland rain forest on dark chocolate volcanic soil; ± 1950 m

NOTE. The petals are pink-mauve with darker veins, dark red at base, 2.5–3 cm long. Staminal column pink, dark red at base. This is also reminiscent of a large-leaved large-flowering specimen of *P. columella* but there are more than 5 epicalyx bracts. Unfortunately there are no young or old fruits on the specimen and these are needed to confirm my placing with absolute certainty.

2. **Pavonia schimperiana** *A. Rich.*, Tent. Fl. Abyss. 1: 52 (1847); Mast. in F.T.A. 1: 192 (1868); Oliv. & Grant in Trans. Linn. Soc. 29: 35, t. 11 (1873); T.T.C.L.: 306 (1949); Vollesen in Fl. Eth. & Eritr. 2 (2): 228, fig. 82.16.6–8 (1995). Type: Ethiopia, Aderbati, *Quartin-Dillon & Petit* s.n. (P, syn.) & Mt Scholoda, *Schimper* 1.53 (P, syn.; BM!, FT, K!, isosyn.)

Shrubby herb or shrub 0.9–3.5(–5)m tall, branches glabrous to puberulous. Leaves ovate-cordate to reniform in outline, 3.5–24 × 2.5–25 cm, shallowly to deeply 3–5-lobed (mostly cut for $^{1}/_{10}$–$^{1}/_{3}$), margins coarsely toothed, ± glabrescent or with only short stellate hairs above, pubescent below with mostly simple hairs; lobes ± triangular, acute or acuminate; petiole 2.5–15(–22) cm long. Flowers solitary or in 2–4-flowered clearly pedunculate cymes often merged to form narrow panicles or pseudoracemes; pedicel mostly clearly evident, 4–13(–20) mm long; epicalyx bracts 8–10, linear, 8–12 mm long; calyx 5–8 mm long, glabrous or puberulous mainly on the veins and often long-ciliate on the margins. Petals white with dark red or purple base, 1.5–2.7 cm long, sometimes pubescent and with dense short glandular hairs; staminal column 1.5–2.5 cm long. Mericarps 5–6 mm long only tardily separating, puberulous and reticulate on the back, with 3 apical retrorsely barbed awns 2–3(–5) mm long.

UGANDA. Ankole District: Igara, Mar. 1939, *Purseglove* 635!; Kigezi District: Bugangali, Kashindo R., 20 Feb. 1953, *Norman* 205!; Mengo District: Mawokota, Feb. 1905, *E. Brown* 140!
KENYA. Trans Nzoia District: NE Elgon, Nov. 1955, *Tweedie* 1374!; Kiambu District: Kiambu, 21 July 1932, *Mainwaring* in *Napier* 2170!; North Kavirondo District: Kakamega Forest, 8 Dec. 1956, *Verdcourt* 1638!
TANZANIA. Bukoba District: Karagwe, Mar. 1862, *Grant* 478!; Mbulu District: between Babati and Galappo, Endanok ravine, 24 Aug. 1951, *Welch* 100!; Mpanda District: Mahali Mts, near Lubugwe R., Kasangazi, 24 July 1958, *Jefford & Juniper* 220!
DISTR. **U** 1?, 2–4; **K** 1–7 (see notes); **T** 1, 2–5, 6?, 7; Ivory Coast to Cameroon, NE Congo-Kinshasa, Rwanda, Burundi, S Sudan, Ethiopia, Angola
HAB. Short grassland, swamp edges, thicket, regenerating forest and forest edges; 1100–2400 m

SYN. *P. schimperiana* A. Rich. var. *glabrescens* Ulbr. in E.J. 57: 108 (1920); T.T.C.L.: 306 (1949); Keay, F.W.T.A. ed. 2, 1(2): 341 (1958); Hauman in F.C.B. 10: 186 (1963). Type: 14 syntypes from East & West Africa including Tanzania, Kyimbila, *Stolz* 311 (B†, syn.; K!, isosyn.) & Uganda, Mengo, *Scott Elliot* 7249 (B†, syn.; BM!, K!, isosyn.)
P. rutshuruensis De Wild., Pl. Bequaert. 1: 499 (1922). Types: Congo-Kinshasa, Rutshuru, *Bequaert* 6055 & Ruwenzori, Butagu Valley, *Bequaert* 3957 (BR, syn.)
P. urens Cav. var. *schimperiana* (A. Rich.) Brenan in Mem. N.Y. Bot. Gard. 8: 223 (1953)
P. urens Cav. var. *glabrescens* (Ulbr.) Brenan in Mem. N.Y. Bot. Gard. 8: 223 (1953)

NOTE. A few specimens which agree exactly in inflorescence and upper leaf indumentum have the flowers described as pink e.g. *Shillito* 59 (Toro), *Purseglove* 624 (Ankole), *Fishlock & Hancock* 207 (Ruwenzori). A specimen *Faden et al.* 816 (Kenya, Teita District: Ngangao Forest, 22 May 1985) has the inflorescence of *P. schimperiana* but flower colour is unknown and the leaves are thinly pilose above.

3. **Pavonia kilimandscharica** *Gürke* in E.J. 19 (Beibl. 47): 40 (1894); Ulbr. in E.J. 57: 114 (1920) and in V.E. 3(2): 385, fig. 182/2 (1921); T.T.C.L.: 305 (1949); Harman in F.C.B. 10: 188 (1963); U.K.W.F. ed. 2: 103 (1994); Vollesen in Fl. Eth. & Eritr. 2(2): 228, fig. 82.16.9 (1995). Type: Tanzania, Kilimanjaro, above Marangu, *Volkens* 1263 (B†, holo.; BM!, K!, iso.)

Shrub 0.9–2.5(–3.5)m tall with branches glabrous or with few stellate hairs. Leaves ovate to elliptic, 2–8.5 × 0.5–4.5(–6) cm, acute to acuminate at the apex, cuneate at base, unlobed or sometimes ± trilobed at base, margins coarsely serrate-crenate, glabrous or with few stellate hairs; petiole 0.2–5 cm long. Flowers solitary in leaf axils sometimes forming terminal raceme-like inflorescences; pedicel 0.3–1.5 cm long; epicalyx bracts 6–8, 4–7(–10) × 1–1.5 mm, spreading, joined at base to form a cup; calyx 7–8 mm long, lobed to about the middle, the lobes ovate, pubescent on the veins and margin ciliate but otherwise ± glabrous or, particularly in eastern Congo-Kinshasa, densely appressed-stellate hairy all over. Petals white, purple with dark centre at base or all pink, 15–25 × 7–8 mm; staminal column purple, 1–1.7 cm long; styles white. Mericarps 5–6 mm long, pubescent and reticulate on back, awns 4–6 mm long.

UGANDA. Toro District: Msandama Hill, 16 Dec. 1925, *Maitland* s.n.! & Bwamba Pass, 22 Sep. 1932, *A.S. Thomas* 687!; Kigezi District: Virunga Mts, pass between Mgahinga and Sabinio, 25 Nov. 1934, *G. Taylor* 2059!

KENYA. Northern Frontier District: Mt Nyiro, June 1936, *M. Jex-Blake* 23!; East Elgon, 30 Nov. 1951, *Irwin* 107!; SW Mt Kenya, Oct. 1932, *Honoré* in F.D. 3048!

TANZANIA. Arusha District: Mt Meru, 6 Nov. 1969, *Richards* 24621!; Kilimanjaro, track above Kilimanjaro Timbers, 23 July 1993, *Grimshaw* 93437!; Rungwe District: Rungwe Mt, 13 Sep. 1932, *Geilinger* 2229!

DISTR. **U** 2; **K** 1, 3–5; 6?; **T** 2, 7; ?Sierra Leone (see Note); Cameroon, Bioko, Congo-Kinshasa, Burundi, Ethiopia

HAB. Bamboo zone, *Ocotea* and *Podocarpus-Juniperus* forest; 1200–2700 m

SYN. *P. kilimandscharica* Gürke var. *triloba* Ulbr. in E.J. 57: 114 (1920). Types: Burundi, Usumbara, Mt Lukona, *Keil* 279 (B†, syn.) & Rwanda, Ninagongo, *Mildbraed* 1322 (B†, syn.)

NOTE. *Glover et al.* 1097 (Masai District: 32 km from Olokurto on road to Elburgon, 14 May 1961, clearing edge of *Cupressus* plantation at 2780 m, differs from *P. kilimandscharica* in the leaves broadest at the truncate or subcordate base, calyx with dense simple hairs and narrower epicalyx lobes not broadened at tip. It may be a distinct taxon or a hybrid with *P. urens*. *Azuma* 677, Kigoma District: 8 km S of Ilagara [Ilagala] has similar leaves but a different facies; the single fruit appears to indicate a ± glabrous calyx.

Gardner 67 (Sierra Leone, Bo, 2 Dec. 1957) appears to be this species, but atypical and not otherwise recorded W of Cameroon.

4. **Pavonia stolzii** *Ulbr. in* E.J. 57: 115 (1920) & in V.E. 3(2): 383, fig. 182/2 & 183 (1921); Brenan in Mem. N.Y. Bot. Gard. 8: 223, 224 ('c.f.') (1953). Type: Tanzania, Njombe District: Ukinga Mts, Tandala, *Stolz* 2202 (B†, holo.; S, iso.)

Annual or perennial 2–3 m tall, foetid (fide Ulbrich, presumably from Stolz field note); branches with very dense rough rigid large stellate hairs. Leaves broadly ovate in outline, 4–20 × 3–19 cm, cordate at base, 3–5(–7)-lobed, densely scabrid-tomentose on both sides, the lobes triangular-ovate, acute, margins coarsely toothed or lobulate; petiole 0.5–13 cm long, densely pilose; stipules linear-lanceolate, densely pilose, very deciduous. Flowers axillary, subsessile, condensed to form dense many-flowered terminal inflorescences; pedicel 2–5 mm long; epicalyx bracts 7, linear-lanceolate, 5 × 1 mm, hispid; calyx campanulate, 6–7 mm long, hispid, lobed to about halfway, the lobes ovate-triangular, acute. Petals rose, ± 15 mm long; staminal column ± 13 mm long. Mericarps ovate-triangular, 5 × 3 mm, glabrous, hardly keeled, laxly reticulate, the sides flat, indistinctly marginate, 3-awned, awns broadly conical projecting 1–2 mm above top of mericarp, decurrent at base with few short retrorse bristles.

TANZANIA. Mbeya District: Kikondo, 21 Oct. 1956, *Richards* 6682! & Mporoto Mts, Aug. 1936, *McGregor* 26!; Njombe District: Ukinga Mts, Tandala, 18 Sep. 1913, *Stolz* 2202
DISTR. **T** 7; Malawi
HAB. Thick vegetation at forest edges; 2100–2350 m

NOTE. In his 'Tentative species enumeration' Fryxell sinks this into *P. urens* presumably following Meeuse (F.Z. 1(2): 511 (1961)), who pointed out *P. stolzii* had much shorter awns, but that there were intermediates such as *Buchanan* 145 from Malawi, Mt Chiradzulu; although Brenan specifically mentions this sheet as *P. stolzii*. Other inadequate sheets with no fruits almost certainly belong here e.g. *Clair-Thompson* 917 (Rungwe District: South slopes of Poroto Mts, Mbeye, 16 Mar. 1932). He mentions irritant stem hairs.

5. **Pavonia sepioides** *Fryxell & Krapovickas* in Fl. Neotropica 76: 221 (1999). Type: Colombia, Antioquia, Medellin, via Santa Helena, *Callejas & Escobar* 3313 (TEX!, holo.; CAS, HUA, NO, NY, TEX, iso.)

Herb 0.6–2 m tall; stems with narrow lines of short hairs. Leaves ovate to lanceolate or elliptic, 5–10(–15) × 1.7–5(–6.5) cm, long-acuminate at the apex, narrowly rounded at base, regularly biserrate, with rather sparse stellate or simple hairs or ± glabrous; petiole 4–15(–40) mm long. Flowers solitary, axillary; pedicel 1.5–6 cm long; epicalyx bracts ± 6, lanceolate to oblanceolate, 7–11 mm long; calyx lobes narrowly ovate, 5–9 mm long, whitish with green veins, ciliate. Corolla yellow, 1.2–2.2 cm long; staminal column 3–7 mm long. Mericarps 6 mm long with 3 sharp rigid spines 6 mm long ± covered with retrorse bristles.

UGANDA. Mengo District: Kampala, Makerere University Hill, 9 Apr. 1970, *Lye* 5183! & Entebbe Botanic Gardens, Nov. 1930, *Snowden* 1814!
DISTR. **U** 4; a South American species occurring in Colombia, Venezuela, Ecuador, Peru and Bolivia
HAB. Roadside ditch; ± 1200 m

NOTE. Snowden says 'possibly introduced' so it was not obviously cultivated; most of the Gardens were wild original cover.

6. **Pavonia senegalensis** (*Cav.*) *Leistner* in Bothalia 10: 74 (1969). Type: Senegal, Senegal R. mouth near Gandiole & Safal Is., *Perrottet* s.n. (P, syn.) & Walo, *Perrotet* s.n. (P, syn.; BM!, isosyn.)

Low herb or shrub with semiprostrate scrambling or ascending branches sometimes forming dense ± round bushes up to 1.25 m tall and 2 m wide; stems rather roughly stellate-tomentose. Leaves ± round to ovate-reniform, 3–8 × 3–10 cm, rounded, truncate or rarely acute at the apex, cordate at base, sometimes angular to shallowly palmately lobed, stellate-hairy becoming glabrescent above, paler and subtomentose beneath, margins ± entire to coarsely crenate or serrate; petiole 0.5–11 cm long, stellate-pubescent; stipules filiform ± 1 cm long. Flowers solitary in the leaf axils, often forming false racemes; epicalyx bracts 12–16, narrowly linear, 7–10 × 0.5–1 mm; calyx campanulate, 1–1.6 cm long, lobed to beyond the middle; lobes ovate-triangular, strongly 3-veined, stellate-tomentose. Petals lemon yellow tinged pink, scarlet to maroon at base, 2.5–4.5 cm long, distinctly veined, fimbriate-ciliate at base. Mericarps 8–10 mm long, the back foveolate-rugose and shortly retrorsely aculeate-hispid, with 3 short protuberances or short blunt spines at the apex; lateral edges narrowly winged and sides flat, smooth and glabrous.

TANZANIA. Kigoma District: Ujiji, Mar. 1939, *Loveridge* 702!; Mpanda District: Lake Katavi, 25 Jan. 1950, *Bullock* 2340! & 10 Feb. 1962, *Richards* 16055!
DISTR. **T** 4; Senegal to Sudan and South to Namibia and South Africa
HAB. Lakeside grassland and sandy shore; ± 1050 m

SYN. *P. hirsuta* Guill. & Perr., Fl. Senegamb. Tent. 1: 51 (1831); Mast. in F.T.A. 1: 191 (1868); Ulbr. in E.J. 51: 56 (1913) & 57: 116 (1920); Exell & Mendonça, C.F.A. 1: 156 (1937); Keay, F.W.T.A. ed. 2, 1(2): 341 (1958); Meeuse in F.Z. 1: 508 (1961); Hauman in F.C.B. 10: 182 (1963): Boulos., Fl. Egypt 2: 105, t 27/1 (2000)

Hibiscus baumii Gürke in Kunene-Samb.-Exped. Baum: 299 (1903). Type: Angola, Bié, between Cutue and Sobi, R. Cuito, *Baum* 760 (B†, holo.; BM!, COI, K!, iso.)

Pavonia zawadae Ulbr. in E.J. 48: 371 (1912) & E.J. 51: 57 (1913) & E.J. 57: 118 (1920). Type: Namibia, Gross Namaland, Arub, *Zawada in Dinter* 1343 (B†, lecto., chosen by Ulbrich (1920))

NOTE. Richards gives a much lower figure for the altitude but from the map Bullock's seems more likely to be correct.

7. **Pavonia flavoferruginea** (*Forssk.*) *Hepper & Wood* in K.B. 38:85 (1983); Hepper & Friis, Pl. Forssk. Fl. Aegypt-Arab: 199 (1994); Wood, Handb. Yemen Fl.: 108 (1997); Thulin, Fl. Somalia 2: 65, fig. 36 A-D, t. 2A (1999). Type: Arabia, North Yemen, Hays ['circa Haes rarius'], Oude, Bulgose, Roboa, *Forsskål* 603 (C, holo., microfiche!)

Erect or straggling shrublet or semi-woody herb 0.9–1.5 m tall; stems pubescent or not and densely glandular. Leaves ovate to reniform, usually unlobed or less often 3-lobed, 1.3–8 × 1–7.5 cm, acute to obtuse at the apex, cordate at base, margins coarsely toothed, pubescent and glandular; petiole 0.5–5(–7.5) cm long. Flowers solitary in the leaf axils; pedicel 1–7 cm long; epicalyx bracts 5–7, narrowly lanceolate or rarely ovate, 6–17 × 1.5–6 mm, obtuse to narrowly acuminate, glandular, ciliolate; calyx 3–5 mm long, glandular, usually ciliolate. Petals cream or yellow to orange with red to purple base, 0.8–2.5 cm long; staminal column 4–15 mm long. Mericarps 4–5 mm long, puberulous, the dorsal keel with a single or double row of spines and lateral ridges with long spines in the upper part.

var. **flavoferruginea**

Leaves usually ovate, up to 8 × 7.5 cm; epicalyx bracts up to 17 × 6 mm.

UGANDA. Karamoja District: Matheniko, Lumeno R., Sep. 1958, *Wilson* 620!

KENYA. Northern Frontier District: Dandu, 6 May 1952, *Gillett* 13091! & 11.2 km S of Kangetet, Lokori, 10 June 1970, *Mathew* 6723!; ? Tana River District: Melka Rupia, May 1957, *J. Adamson* 618!

TANZANIA. Pare District: Ngulu, May 1928, *Haarer* 1381! & Mkomazi Game Reserve, SE of Ndea Hill, 10 June 1996, *Abdallah et al.* 96/158! & Same, Gonja Maore, July 1955, *Semsei* 2118!

DISTR. **U** 1; **K** 1, 2, 4, 7; **T** 2 (atypical), 3; Sudan, Eritrea, Ethiopia, Djibouti, Somalia; Arabia, Pakistan, NW India

HAB. Grassland with scattered trees, bushland on black cotton soil and lava flows, thicket; 30–950 m

SYN. *Hibiscus flavoferrugineus* Forssk., Fl. Aegypt.-Arab CXVII, No. 420 (1775)

Lebretonia glechomifolia A. Rich. in Tent. Fl. Abyss. 1: 54 (1847), as *'glechomaefolia'*. Type: Ethiopia, Chocho, *Quartin Dillon & Petit* s.n. (P, holo.)

Pavonia glechomifolia (A. Rich.) Garcke in Schweinf., Beitrag. Fl. Aethiop. 1: 54 (1867); Mast. in F.T.A. 1: 191 (1968); Abedin, Fl. W. Pak., 130 Malvaceae 96, fig. 22//F-H (1979) (as '*glechomaefolia*'); Vollesen in Fl. Eth. & Eritr. 2 (2): 226, fig. 82.15.1–4 (1995)

P. glechomifolia (A. Rich.) Garcke var. *glabrescens* Ulbr. in E.J. 57:120 (1920). Types: numerous syntypes including 7 from Ethiopia and Kenya, Witu, *F. Thomas* 95 (B!, syn.; K! isosyn.)

P. glechomifolia (A. Rich.) Garcke var. *tomentosa* Ulbr. in E.J. 57: 121 (1920). Types: syntypes from Somalia, Mozambique and Transvaal and Kenya, Nairobi, *F. Thomas* III, 101 (B†, syn.) (I have seen none of these and I have omitted some from the distribution)

P. coxii Tad. & Jacob in J. Ind. Bot. Soc. 5: 11, figs. 1–11 (1926). Type: India, Coimbatore, Agricultural College and neighbourhood, presumably the authors in Madras Herb. 15353 (MH, holo.; K! iso.)

NOTE. *Gillett* 20583 (Kenya, Northern Frontier Province, Garissa–El Lein road ± 28 km from fork off Hagadera road, 10 May 1974) and *Gillett* 21217 (similar area, 46 km from Garissa on

Hagedera road, 29 May 1977) have broader, 6 × 6 mm, ovate epicalyx lobes but the double crest of distinct spines and lateral spines and the typically glandular stems. It occurs in *Acacia-Commiphora-Terminalia orbicularis* bushland at 210–230 m, on black cotton soil.

It is not clear why Vollesen did not employ Forsskål's epithet, whether or not he disagreed with Hepper and Wood's decision. The Malvaceae part of Fl. Ethiopia & Eritrea was prepared long before its publication and this probably explains it.

var. **microphylla** Verdc. **var. nov.** a var. *flavoferruginea* foliis rotundatis parvis ± 1.3 × 1 cm, floribus parvis, bracteis epicalycis 4 × 1 mm differt. Typus: Kenya, Northern Frontier District: Mandera, 30 km from Ramu on Malka Mari road, *Gilbert & Thulin* 1571 (K!, holo.)

Leaves uniformly small, rounded-ovate, 1.3 × 1 cm. Flowers small, the epicalyx bracts only 4 × 1 mm.

KENYA. Northern Frontier District: Mandera, 30 km from Ramu on Melka Mari road, 8 May 1978, *Gilbert & Thulin* 1571! & Furroli, 4 Sep. 1952 *Gillett* 13866!

DISTR. **K** 1; not known elsewhere

HAB. Mixed *Acacia-Commiphora* woodland on limestone rocks, lava hills; 400–1000 m

NOTE. This has the typical stem indumentum of *P. flavoferruginea* but no mature mericarps have been seen. It is certainly very distinctive, but I suspect no more than a variant of *P. flavoferruginea*. Many specimens have been seen with much smaller leaves mixed with normal large ones and intermediates occur but it needs investigation. *Bally* 5650 (Kenya, Mt Kulal, 15 Oct. 1947) may belong here.

8. **Pavonia procumbens** (*Wight & Arn.*) *Walp.*, Repert. 17: 301 (1842); Garcke in Peters Reise Mossamb. Bot. 1: 123 (1861); Abedin, Fl. W. Pak. 130, Malvaceae: 94, fig. 22A–C (1979); Vollesen in Fl. Eth. & Eritr. 2(2): 226, fig. 82.15.5–6 (1995); Thulin, Fl. Somalia 2: 65 (1999). Type: India, not known, *Wallich* 2688 (K-W!, lecto.) (chosen by Borssum Waalkes, 1966)

Woody herb 30–90(–100) cm tall, trailing or scrambling or erect with long prostrate branches; stems pubescent and mostly with ± long spreading hairs and some stellate ones. Leaves cordate, 1.5–10.5 × 1–7.5 cm, acute to acuminate at the apex, unlobed or shallowly 3-lobed, margins coarsely crenate or toothed, stellate-tomentose on both sides and sometimes with some long hairs; petiole 0.5–5.5 cm long. Flowers solitary in the leaf axils; pedicel 1–4.5(–6) cm long, jointed near the apex; epicalyx bracts 5–6, elliptic to narrowly ovate, rarely ± round, (4–)6–12 × 2–7 mm, obtuse to acuminate, usually 3-veined, pubescent and ciliate; calyx 5–8 mm long, densely stellate-pubescent with lobes 3–4 × ± 4 mm, acute or acuminate, sparsely glandular, ciliate. Petals yellow to pale orange or deep apricot mostly with crimson, purple or blackish base, 8–14 × 2–10 mm, hairy at base; staminal column 3–5 mm long. Mericarps 4–5 mm long, puberulous, dorsally longitudinally carinate, the keel with a single or double row of projections or distinct spines and lateral ridges with 3 conical projections.

UGANDA. Karamoja District; Kanamugit, Feb. 1936, *Eggeling* 2924! (needs confirmation, specimen without mericarps)

KENYA. Northern Frontier District: South Turkana, Loriu Plateau, 1 June 1970, *Mathew* 6512!; Baringo District: Chemolingot borehole area, 13 Aug. 1976, *Timberlake* 437!; Masai District; 48 km Nairobi to Ol Orgasalik, 31 May 1958, *Verdcourt et al.* 2194!

TANZANIA. Moshi District: 16–24 km S of Moshi, Apr. 1965, *Beesley* 106!; Pare District: 13 km NW of Mkomazi on Moshi–Korogwe road, 1 Apr. 1969, *Lye* 2495!; Mpwapwa District: 11.2 km S of Gulwe on Kibakwe track, 9 Apr. 1988, *Bidgood et al.* 970!

DISTR. **U** 1?; **K** 1–4, 6, 7; **T** 2, 3, 5; Sudan, Eritrea, Ethiopia, Djibouti, Somalia; Arabia, Pakistan, NW India

HAB. Scattered tree grassland, *Acacia* woodland, *Acacia-Commiphora* scrub and bushland in rocky places, on lava boulders etc.; 30–1500 m

SYN. *Lebretonia procumbens* Wight & Arn., Prodr. Fl. Ind. Or. 1: 47 (1834)

Pavonia ukambanica Ulbr. in E.J. 57: 132 (1920). Type: Kenya, Kitui District: Ukamba, Kitui, *Hildebrandt* 2763* (B†, holo.; BM!, K!, iso.)
P. ctenophora Ulbr. in E.J. 57: 122 (1920). Types: Sudan, Darfur, Gebel Barkin, Surutj, *Pfund* 245 (B†, syn.) & Gebel Chusus, East border of Darfur, *Pfund* 247 (B†, syn.)

9. **Pavonia burchellii** (*DC.*) *R.A. Dyer* in K.B. 1932: 152 (1932); Burtt Davy, Fl. Pl. Transv. 2: 277 (1932); Ardwiss. in Bot. Notis. 1934: 96 (1934); Hauman in F.C.B. 10: 181 (1963); Vollesen in Fl. Eth. & Eritr. 2(2): 225, fig. 82.15.7–8 (1995); Wood, Handbk. Yemen Fl.: 108 (1997); Thulin, Fl. Somalia 2: 63, fig. 36G (1999); Boulos, Fl. Egypt 2: 105, t. 27/3 (2000). Type: South Africa, Cape, Kosi Fontein, *Burchell* 2587 (G, holo.; K!, iso.)

Shrubby herb or herb 0.9–2 m tall, mostly erect but sometimes scrambling; stems puberulous to pubescent and usually with long simple hairs. Leaves broadly ovate and mostly ± 3-lobed but sometimes exactly ovate-cordate and unlobed, 2–14 × 2–10 cm, acute to acuminate at the apex, cordate at base, margins coarsely toothed, pubescent; petiole 0.5–8(–10.5) cm long. Flowers solitary in the leaf axils; pedicel 1–7(–8.5) cm long; epicalyx-bracts 5–6(–8), elliptic, broadly-elliptic or rhomboid-ovate, 5–15 × 2–9 mm, rounded to subacute or acuminate at the apex, slightly joined at base; calyx 4–7 mm long, the lobes ovate, acute, pubescent, glandular and ciliate. Petals bright yellow to orange with or without a purple base, 1.2–2.7 cm long; staminal column 5–15 mm long. Mericarps obovoid, 4–5 mm long, sparsely puberulous, the dorsal keel not strongly raised and lateral ridges usually indistinct, reticulate all over.

UGANDA. Karamoja District: Moroto, Sep. 1956, *J. Wilson* 271!; Toro District: Katwe, 12 Sep. 1941. *A.S. Thomas* 3962!; Mbale District: Mt Elgon, Bomwege, 22 May 1924, *Snowden 882!*
KENYA. Northern Frontier District: Marsabit, above secondary school, 6 Aug. 1968, *Faden* 68/412!; Trans Nzoia District: Kitale, 27 July 1950, *Wiltshire* 31!; Masai District: Ewaso Nyiro–Loliondo road where it crosses the Masan R., Subatia, 14 Dec. 1963, *Verdcourt* 3840!
TANZANIA. Mwanza District: Nascha Is., Igokero, 9 Nov. 1952, *Tanner* 1151!; Masai District: South Serengeti, Naibardad Hill, 22 Dec. 1962, *Newbould* 6439!; Ufipa District: Lake Kwela, 11 Mar. 1959, *Richards* 11149!
DISTR. **U** 1–4; **K** 1, 3–5, 6?, 7; **T** 1–7; Cameroon, Congo-Kinshasa, Rwanda, Sudan, Ethiopia, Eritrea, Somalia, South Africa; Arabia
HAB. Grassland, woodland, (e.g. open *Acacia xanthophloea* to *Juniperus* woodland), forest (even rainforest) margins, scrub, rocky places and cultivations; 750–2250 m

SYN. *Althaea burchellii* DC., Prodr. 1: 438 (1824), *non* Eckl. & Zeyh., Pl. Afr. Aust.: 37 (1835)
Pavonia kraussiana Hochst. in Flora 27: 293 (1844); Ulbr. in E.J. 51: 57 (1913) & E.J. 57: 125 (1920). Type: South Africa, Durban [Port Natal], Umlaas R., *Krauss* s.n. (B†, holo.; K, iso.)
P. macrophylla Harv. in F.C. 1: 169 (1860); Mast. in F.T.A. 190 (1868); Oliv. & Grant in Trans. Linn. Soc. 29: 34 (1872), *nom. illegit.* Types: South Africa, many syntypes including 2 *Drège* s.n. (K!, isosyn.)
P. kraussiana Hochst. subsp. *dictyocarpa* Ulbr. var. *tomentosa* Ulbr. in E.J. 51: 58 (1913) & in E.J. 57: 127 (1920). Types: Amboland, Olukunda, *Schinz* 188, Hereroland, Noassanabia, *Range* 794, Botswana, Massaringani vley, *Seiner* 268; North Rietfontein, Epikuro, *Seiner* 394 (all B†, syn.)
P. kraussiana Hochst. subsp. *dictyocarpa* Ulbr. var. *glandulosa* Ulbr. in E.J. 51: 89 (1913) & in E.J. 57: 128 (1920). Type: Hereroland, Waterburg, *Dinter* 1822 (B†, lecto.) (chosen by Ulbrich, 1920)
P. burchellii DC. var. *glandulosa* (Ulbr.) Heine in Mitt. Bot. München 2: 177 (1956)
P. burchelii DC. var. *tomentosa* (Ulbr.) Heine in Mitt. Bot. München 2: 177 (1956)
P. patens sensu auctt. mult., *non* (Andr.) Chiov.

NOTE. Although the reticulate mericarps without crests or conical protuberances or spines are characteristic of *P. burchellii*, many specimens from NW Kenya show tendencies to crests etc. often correlated with dark purplish stems and sometimes less lobed leaves. This is possibly

* Ulbrich has 6763 in error.

due to some hybridisation with *P. procumbens*, or it might be possible to make subspecies of them. I am convinced, however, it is wrong to have a single species comprising species 7–10 of my account. This is a meaningless jumble contradicting geographical distribution and of no value whatsoever from a practical point of view.

10. **Pavonia gallaensis** *Ulbr.* in E.J. 57: 133, fig. 5/H-P (1920); Vollesen in Fl. Eth. & Eritr. 2 (2): 226, fig. 82.15.9–10 (1995). Type: Ethiopia, Bale, Ginir [Arussi Galla, Ginea], *Ellenbeck* 1961 (B†, holo.)

Shrubby herb, shrub or often semiprostrate or scrambling plant 0.5–1(–1.5) m; all parts densely stellate-tomentose and with glandular hairs, often sticky (Vollesen says rarely glandular) and with long simple hairs. Leaves ovate-cordate, 1.5–5.5 × 1.5–4(–5) cm, unlobed, margins crenate to strongly toothed; acute to rounded at apex, ± cordate at base, densely stellate-tomentose and usually viscid in East Africa; petiole 0.5–3.5 cm long. Flowers solitary in leaf-axils; pedicel 0.5–2.5(–4) cm long; epicalyx bracts 5–7(–9), elliptic to broadly elliptic, sometimes becoming narrowly lanceolate, 4–8 × 2–4 mm, sometimes with a conspicuous tuft of long hairs at base; calyx lobes 4–9 × 2.5–4 mm, densely tomentose. Petals lemon yellow to orange-yellow without a darker base in the Flora area but sometimes with one further north, (7–)8–12 × ± 6 mm. Mericarps obovoid, 4–5 mm long, reticulate.

UGANDA. Karamoja District: between Kaabong and Kotido, July 1966, *M.R. Smith* s.n.! & Kangole, 22 May 1940, *A.S. Thomas* 3461! & E Matheniko, Mar. 1959, *Wilson* 716!
KENYA. Northern Frontier District: Marsabit airstrip, 15 Aug. 1968, *Faden* 68/678!; Nakuru District: Lake Elmenteita, 16 June 1951, *Bogdan* 3045!; Machakos District: Athi R. on main Nairobi–Mombasa road, 30 May 1958, *Verdcourt & Napper* 2180!
TANZANIA. Masai District: Kakessio, 7 April 1961, *Newbould* 5809! & South Serengeti, Naibardad Hill, 22 Dec. 1962, *Newbould* 6441!; Pare District: Kiruru, May 1928, *Haarer* 1338!
DISTR. **U** 1; **K** 1, 3, 4, 6; **T** 1–3; Ethiopia
HAB. Grassland, *Tarchonanthus-Ormocarpum* bushland, swamp edges, lakeside, scrub, *Cynodon* plains; 700–1800 m

NOTE. There is no doubt that this is a distinct species and several collectors have pointed this out e.g. *Gillett* 18936 (Kenya, Meru District: 24 km S of Isiolo on road from Meru, 25 Aug. 1969) states corolla lemon yellow with no dark centre, leaves viscid – clearly not same species as 18916). It has been suggested on one cover of material that it might be a hybrid of *P. gallaensis* and *P. burchellii* but it appears typical *P. gallaensis* to me. The glands are often hidden by the very dense indumentum.

11. **Pavonia columella** *Cav.*, Diss. 3: 138, t. 48, fig. 3 (1787); Baker f. in J.L.S. 40: 27 (1911); Ulbr. in E.J. 57: 135 (1920); Brenan in Mem. N.Y. Bot. Gard. 8: 224 (1953); Meeuse in F.Z. 1: 510. t. 96 (1961). Type: Réunion, Plaine de Caffres, *Commerson* s.n. (P-JU no. 12348, holo.; P, iso.) (seen by Brenan)

Erect or somewhat spreading annual or shrub or subshrub (0.6–)1–2 m tall, usually densely shortly pubescent but sometimes sparsely so, the stems with simple, stellate and minute glandular hairs. Leaves ovate to ± rounded in outline but often narrower above, ovate-lanceolate, 2.5–12 × (1–)2–15 cm, usually 3–5-lobed often almost to the middle, acute at the apex, ± cordate at base, pubescent or subhispid mainly on the veins above, pubescent to ± tomentose beneath; stipules filiform, 2–3 mm long; petiole 1.8–6(–11) cm long. Flowers solitary in upper axils and on short lateral branches, often forming a narrow terminal leafy pseudo-panicle; pedicel 0.5–1.5 cm long; epicalyx bracts 5(–6), linear or narrowly lanceolate, 5–8(–9) × ± 0.7 mm, pubescent; calyx cupuliform-campanulate, ± 7 mm long, lobed to about the middle, pubescent and ciliate, lobes triangular, acute to apiculate, conspicuously 3-veined. Petals mauve, pink or rose, orange-red or yellow at base, 1.9–2.5 cm long; staminal column 10–17 mm long, glabrous. Mericarps ± 3 mm long, ± glabrous, back reticulate with slightly raised ridges.

TANZANIA. Njombe District: Milo, 4 Oct. 1978, *Archbold* 2425!; Songea District: Miyau, 23 May 1956, *Milne-Redhead & Taylor* 10379A! & Lupembe Hill, 20 May 1956, *Milne-Redhead & Taylor* 10379!
DISTR. **T** 7, 8; Zambia, Malawi, Mozambique, Zimbabwe, Swaziland, South Africa; Réunion, Madagascar (introduced)
HAB. Waste ground, secondary bushland; 1600–2300 m

SYN. *P. meyeri* Mast., F.T.A. 1: 191 (1968), *nom. nov.*; Dyer in K.B. 1932: 153 (1952). Type: South Africa, Natal, *Gueinzius* s.n. (K!, lecto., chosen here)

NOTE. Recorded from Uganda in F.Z. but we have seen no specimens. Fryxell (in MS) gives the petals as 10–12 mm and Meeuse as 2–2.5 cm. I have seen only the three specimens cited.

12. **Pavonia propinqua** *Garcke* in Schweinf., Beitr. Fl. Aethiop. 1: 55 (1867); Mast. in F.T.A. 1: 191 (1868); Ulbr. in E.J. 57: 141 (1920); U.K.W.F. ed. 2: 103, t. 29 (1994); Vollesen in Fl. Eth. & Eritr. 2(2): 230, fig. 82.17.1–3 (1995); Thulin, Fl. Somalia 2: 67, fig. 38/A-C (1999). Type: Ethiopia, Tigray, Goelleb, *Schimper* 134 (B†, holo., FT, K!, P, iso.) & *Schimper* in *Hohenacker* 2137 (BM!, K! P, iso.)*

Spreading shrublet 0.3–1.2 m tall; stems pubescent. Leaves narrowly ovate to ovate or elliptic, sometimes linear-oblong, 2.5–7(–9.5) × 0.3–4(–6.5) cm, rounded or subacute at the apex, rounded to truncate at base, unlobed but coarsely toothed, pubescent; petiole 0.3–1.5(–4) cm long. Flowers solitary in the leaf axils; pedicel 0.3–1.5(–3) cm long; epicalyx bracts 7–12, 8–15 mm long, with long strigose bulbous-based hairs; calyx 6–8 mm long, densely pubescent to tomentose-strigose with margins ciliate. Petals pale lemon yellow, yellowish cream, white or darker yellow with orange-red tinge at base or reddish outside, often turning pink or crimson on fading or drying, 1.5–2.5 cm long; staminal column 6–9 mm long. Mericarps 5–6 mm long, glabrous to sparsely puberulous, with 2 lateral and one apical conical spines, reticulate on back. Fig. 3, p. 21.

UGANDA. Karamoja District: 77 km S of Moroto, 13 Sep. 1956, *Hardy & Bally* 10819! & Kangole, 22 May 1950, *A.S. Thomas* 3485! & between Rupa and Nakiloro, 12 June 1970, *Katende & Lye* 429B!
KENYA. Northern Frontier District: 21 km from Marsabit towards Jaldessa, 16 May 1970, *Magogo* 1378!; Machakos District: Sultan Hamud, 5 June 1958, *Irwin* 404!, Masai District: Ol Tukai to Namanga 34 and 54 km, 15 Dec. 1959, *Verdcourt* 2584!
TANZANIA. Masai District: on Nairobi road at foot of Mt Longido, 31 Dec. 1968, *Richards* 23571!; Pare District: Gonja, Jan. 1950, *Bally* 7692! & Kiruru, May 1927, *Haarer* 490! & Mkomazi Game Reserve, SE of Ndea Hill, 10 June 1996, *Abdallah et al.* 96/149!
DISTR. **U** 1; **K** 1, 3, 4, 6, 7; **T** 2, 3; Sudan, Ethiopia, Somalia; Pakistan and NW India
HAB. Seasonally flooded *Acacia* etc. bushland on black cotton soil, grassland on sandy flats, overgrazed wet grassland, tufa hills; 5–1400 m

SYN. *P. grewioides* Boiss., Fl. Orient. 1: 837 (1867); Reidl, Flora Iranica 120, Malvaceae: 35 (1976); Abedin, Fl. W. Pak., No. 130 Malvaceae: 100, fig. 24/A-D (1979); Paul & Nayar, Fasc. Fl. India 19: 192, fig. 42 (1988). Type: Ethiopia, 1200 m, *Schimper* 2137 (K!, lecto.) (chosen by Abedin on the sheet; in Fl. Pak. he states *Schimper* s.n., but data agree)

13. **Pavonia elegans** *Garcke* in Jahrb. Königl. Bot. Gart. 2: 332 (1883); Ulbr. in E.J. 57: 142 (1920); T.T.C.L.: 305 (1949); U.K.W.F. ed. 2: 103 (1994); Vollesen in Fl. Eth. & Eritr. 2(2): 230, fig. 82.17.6 (1995); Thulin, Fl. Somalia 2: 67, fig. 38 D, E. (1999). Types: Kenya, Teita District: Ndara, *Hildebrandt* 2396 (B†, syn.) & Kwale/Kilifi District: Duruma, Tschamtei, Maji ya Chumvi, *Hildebrandt* 2324 (B†, syn.)

* Vollesen is almost certain these are all part of the same *Schimper* collection but they are technically syntypes. Abedin labelled 2137 at K as lectotype but it was not published in Fl. Pakistan. Garcke did not mention Schimper's number but Ulbrich did.

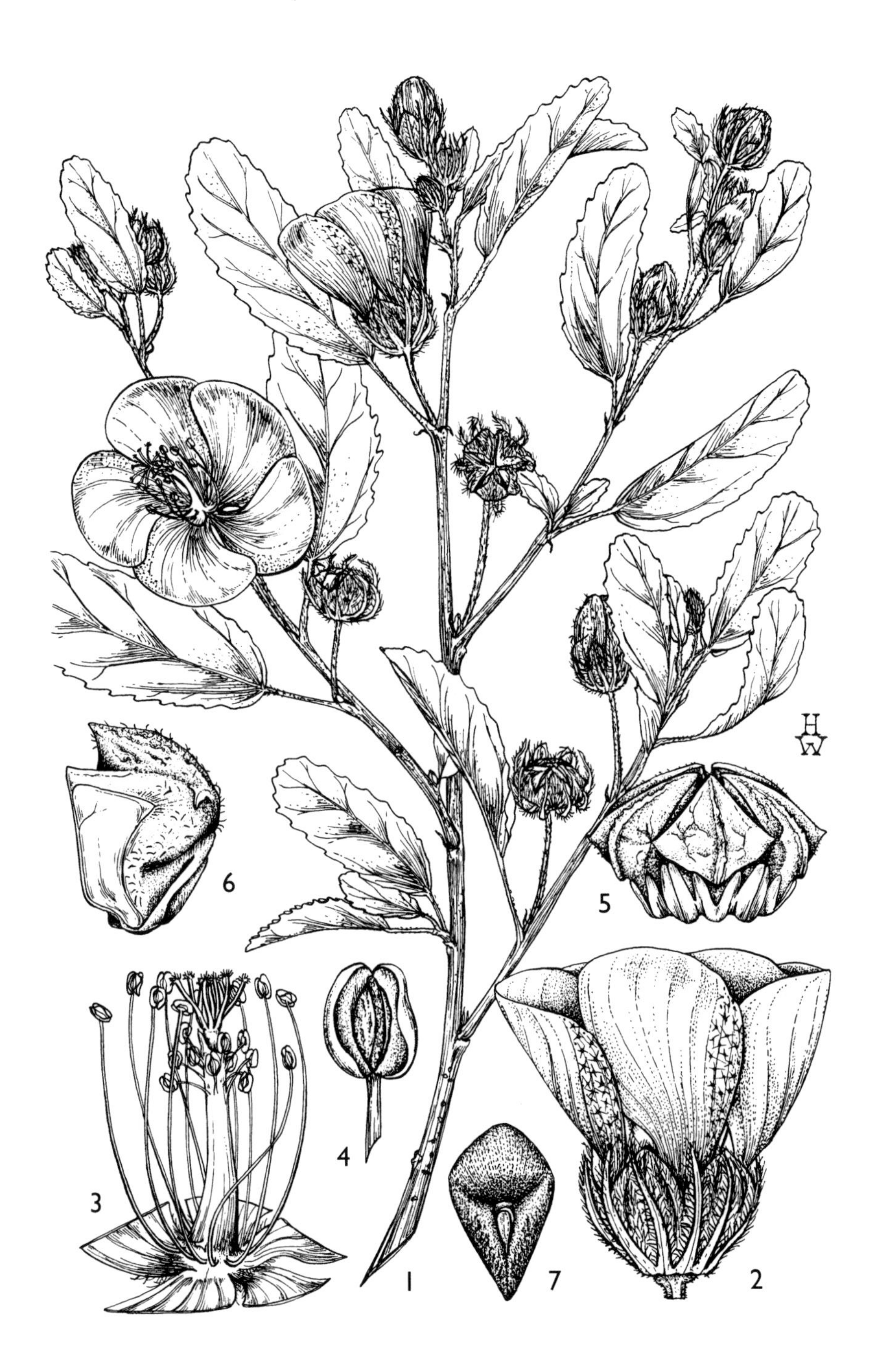

FIG. 3. *PAVONIA PROPINQUA* — **1**, flowering branch, × 1; **2**, flower, × 2; **3**, part of opened flower, × 4; **4**, anther, × 15; **5**, fruit with calyx removed, × 3; **6**, mericarp, side view, × 4; **7**, seed, front view, × 4. 1 from *Bally* 1987; 2–4 from *Gerlinger* 4918; 5–7 from *Bally* 1987. Drawn by Heather Wood.

Shrubby herb or shrub 0.6–1.2 m tall; stems petiole and pedicel viscid with glandular hairs and scattered long simple hairs. Leaves broadly ovate (or upper ones sometimes oblong-lanceolate), 1.2–7 × 1.2–7 cm, subacute to rounded at the apex, cordate at base, margins coarsely toothed, tomentose; petiole 1–8 cm long. Flowers solitary in the leaf axils; pedicel 0.5–2(–3.5) cm long; epicalyx bracts 5–7, lanceolate to narrowly ovate or ovate-triangular, 6–10 × 1.5–4 mm, pubescent and ciliate with long simple hairs; calyx 6–10 mm long, tomentose and yellowish glandular; with simple marginal hairs. Petals pale yellow, turning pink on drying, 1.2–2 cm long; staminal column 7–12 mm long. Mericarps ± 6 mm long, glabrous to sparsely puberulous with a central raised prickly band and 2 rows of hooked prickles on both sides; lateral spines bulbous; also with several small apical spines.

KENYA. Northern Frontier District: Faio, 18 May 1952, *Gillett* 13227!; Embu District: Gathigiriri, 8 Jan. 1968, *Kabuye* 117!; Kwale District: Mackinnon Road, 12 January 1964, *Verdcourt* 3903!
TANZANIA. Pare District: NW spur of N Pare Mts, above Kifaru Estate, 19 May 1968, *Bigger* 1849! & Mkomazi, 17 Sep. 1965, *Semsei* 3986! & same, half way up Koko Hill, 2 Apr. 1972, *Wingfield* 1986!
DISTR. **K** 1, 3, 4, 6, 7; **T** 3; Ethiopia, Somalia
HAB. Mixed dry evergreen bushland, *Acacia-Terminalia* and *Adansonia* woodland, *Acacia-Commiphora* bushland, grassland and cultivations often on black cotton soil; 13–1550 m

14. **Pavonia cristata** *Gürke* in Bull. Herb. Boiss. 3: 407 (1895); Vollesen in Fl. Eth. & Eritr. 2(2): 230, fig. 82.17.5 (1995). Type: Ethiopia, Warandab, *Keller* 68 (Z, holo.)

Shrubby herb or shrub to 75 cm tall; stems pubescent to tomentellous. Leaves broadly elliptic to round, 1–2.8 × 1.2–2.7 cm, truncate to retuse at the apex, cuneate to rounded at base, unlobed ± entire to coarsely toothed, pubescent to tomentellous; petiole 0.3–1.5(–2) cm long. Flowers solitary in the leaf axils; pedicel 0.3–1.5(–2) cm long; epicalyx bracts 6–8, linear, 6–10 × 0.5 mm without long strigose hairs; calyx 5–8 mm long, tomentellous with or without simple hairs on margins. Petals sulphur- or pale yellow, 1–1.5 cm long; staminal column 4–6 mm long. Mericarps ± 6 mm long, sparsely puberulous, with 3 concial spines and with 3 rows of curved to hooked prickles and transverse ribs between them.

KENYA. Northern Frontier District: between Banessa and Ramu, 23 May 1952, *Gillett* 13284!; Teita District: Ndara Ranch to Kajeri, 26 Dec. 1991, *Robertson* 6547!
DISTR. **K** 1, 7; Ethiopia, Somalia (fide *Vollesen* but not mentioned by Thulin; there is a *James & Thrupp* specimen at K)
HAB. On limestone; ± 600 m

NOTE. *Haarer* 1956 (Kilosa District: Tembo, 750 m, Jan. 1931) keys to this species but in the absence of fruits and diverse geography I have not referred it to *P. cristata*. No other material has been seen from **T** 6. It ± agrees with the description above but may have several stems from a woody rootstock; 20 cm tall; epicalyx bracts ± 9; calyx lobes broadly ovate, 9 × 5.5 mm, acute, 3-ribbed; corolla colour unknown, petals 1.5 cm long; pedicel 2.5 cm long, without further material in fruit its staus will remain uncertain.

15. **Pavonia melhanioides** *Thulin* in Nordic J. Bot. 5 (2): 153, fig. 1, 2, (1985); Vollesen in Fl. Eth. & Eritr. 2(2): 230, fig. 82.17.4 (1995). Type: Ethiopia, 1–3 km on track towards Genale R. from turning ± 20 km NW of Neghele, *Thulin et al.* 3488 (UPS, holo.; ETH, FT, K!, iso.)

Perennial herb 30–50 cm tall with erect to procumbent annual stems from a woody rootstock; stems tomentose. Leaves ovate to elliptic or oblong, 2–9 × 1.5–5 cm, rounded to retuse at the apex, rounded at base, unlobed, margins coarsely toothed, tomentose; petiole 0.5–3.5 cm long. Flowers solitary in the leaf axils; epicalyx bracts 9–12, linear, (6–)9–15 × 0.7 mm, with long strigose bulbous-based hairs; calyx 10–15 mm long, tomentose-strigose and with simple marginal hairs. Petals pale lemon-yellow, reddish when fading, 1.5–2.2 cm long; staminal column 5–7 mm long. Mericarps 7–8 mm long, sparsely puberulous, with three conical apines and three rows of curved to hooked prickles but lacking transverse ribs.

Kenya. Northern Frontier District: Moyale, 16 Apr. 1952, *Gillett* 12803! & same locality, 4 Oct. 1952, *Gillett* 14000!
Distr. **K** 1; Ethiopia
Hab. Montane scrub with *Cussonia, Ficus, Acacia* etc; 1050–1150 m

16. **Pavonia leptocalyx** (*Sond.*) *Ulbr.* in E.J. 57: 151 (1920); Meeuse in F.Z. 1: 505 (1961); Vollesen, Opera Bot. 59: 36 (1980). Type: South Africa, Natal, near Durban [Port Natal], *Gueinzius* 424 (?MEL, holo.) (unnumbered *Gueinzius* sheet at B† was probably an iso.)

Annual herb or erect or spreading subshrub 20–90 cm tall; stems brownish stellate-pubescent to -tomentose often with longer stellate or simple hairs and often glandular-viscid. Leaves ovate to ± round or sometimes triangular-hastate, angular or somewhat trilobed, 1–8.5 × 0.5–4.5 cm, subacute to rounded at the apex, rounded to ± cordate at base, margins coarsely crenate-serrate, ± glabrous to stellate-pubescent above, mostly more densely stellate-tomentose beneath, ± discolourous; petiole 1–6 cm long. Flowers solitary in the leaf axils, mostly towards the tips of branches and twigs; pedicel 1–4 cm long; epicalyx bracts 8–12, filiform or narrowly linear, 9–12 mm long, setose; calyx 4–7 mm long, ± campanulate, lobed to below the middle, pubescent, stellate-setulose or ± glabrous. Petals white, 16–18 mm long; staminal tube 12 mm long. Mericarps triquetrous, 3–4 mm long, ± keeled dorsally and with 3 weak transverse ridges, not winged, glabrous or finely stellate-pubescent.

Kenya. Kwale District: Maluganji Forest, 14 Nov. 1989, *Robertson & Luke* 5995!; Kilifi District: Arabuko, Mar. 1930, *R.M Graham* 805 in F.D. 2353!; Lamu District: Ijara–Mkowe, km 67, 1 Oct. 1957, *Greenway* 9261!
Tanzania. Pangani District: Madanga, Mkuzikatani, Kibubu, 24 July 1957, *Tanner* 3634!; Uzaramo District: Dar es Salaam, Mar. 1868, *Kirk* s.n.! & 19 km from Dar es Salaam on Morogoro Road, 19 May 1968, *Batty* 85!
Distr. **K** 7; **T** 3, 6, 8; Somalia, Mozambique, Zimbabwe, South Africa
Hab. Marshy grassland, mixed coastal bushland and thicket on sand behind mangroves, *Julbernardia-Hymenaea-Cynometra-Scorodophloeus* forest; 5–300 m

Syn. *Hibiscus leptocalyx* Sond. in Linnaea 23:7 (1850); Harv. in F.C.1: 175 (1860)
Pavonia odorata sensu Garcke in Peters, Reise Mossam. Bot 1: 123 (1861); Mast. in F.T.A. 1: 193 (1868) pro parte, *non* Willd.
P. discolor Ulbr. in E.J. 57. 148 (1920). Types: Zanzibar, Kondutschi, *Stuhlmann* 7995 & Mbweni, *Stuhlmann* 8001 (B†, syn.), Tanzania, Kikumelela, *Busse* 2969 (B†, syn.) & Lindi, Ruala–Mtua, *Braun* H52 (B†, syn.) & Mozambique, Cabaceiro Grande, *Prelado Barroso* 25 (COI, syn.) & Querimba, *Peters* 57 (K!, syn.) 66 & 68 (B†, syn.)
P. mollissima Ulbr. in E.J. 57: 150 (1920); T.T.C.L.: 305 (1949); Thulin, Fl. Somalia 2: 70 (1999). Type: Kenya, Mombasa, *Hildebrandt* 1928 (B†, holo.; BM!, K! iso.)
P. fruticulosa Ulbr. in E.J. 57: 152 (1920); T.T.C.L.: 305 (1949). Type: Tanzania, Tanga, *Holst* 4029 (B†, syn. & Mozambique, Maputo [Lourenço Marques], *Schlechter* 11708 (B†, syn.; BOL, GRA, K!, L, PRE, isosyn.)

Note. Meeuse considers *P. mollissima* to be a larger-leaved luxuriant form of *P. leptocalyx* but Thulin keeps it separate and gives none of the above names in synonymy. Material from Tanga District certainly matches specimens from Natal. Both Abedin and Sivarajan and Pradeep continue to record *P. odorata* Willd. from Africa.

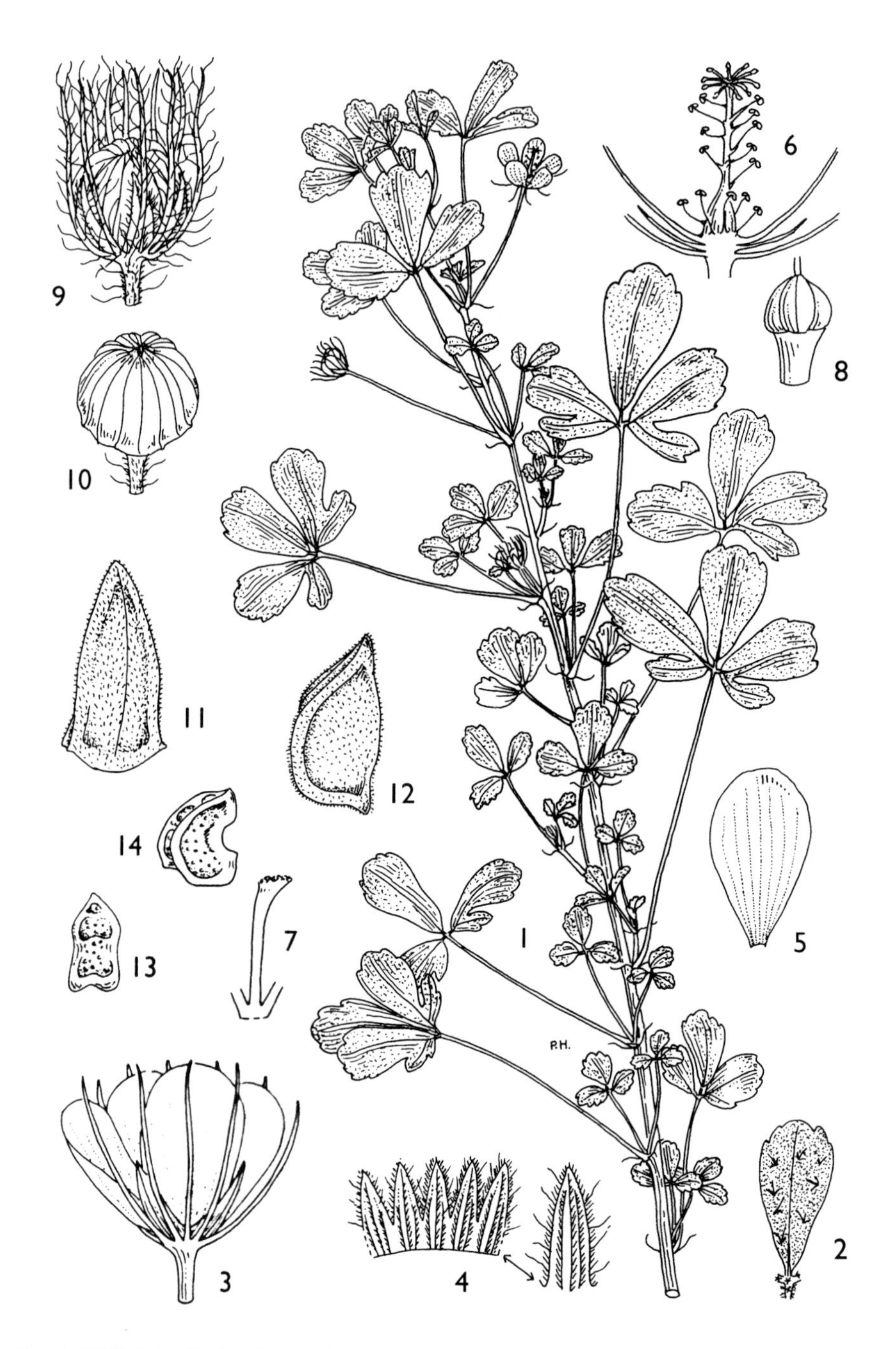

FIG. 4. *PAVONIA ARENARIA* — **1**, habit, × ⅔; **2**, leaf lobe, × 3; **3**, flower, × 3; **4**, calyx opened, × 3; **5**, petal, × 3; **6**, stamens, style & ovary, × 3; **7**, single style branch, much enlarged; **8**, ovary, × 6; **9**, fruit, × 3; **10**, fruit, calices removed, × 3; **11–12**, single valve, front and profile, × 6; **13–14**, seed, back view and profile, × 12. 1 from *Tweedie* 1851; 2–14 from *Kirrika* 72. Drawn by Pat Halliday.

17. **Pavonia arenaria** (*Murr.*) *Roth*, Catalecta Botanica fascicle 2: 80 (1800); Verdc. in K.B. 62(4): 635 (2007). Type: grown from seed sent from Copenhagen by Erik Viborg (ubi?, holo.)

Annual or perennial herb 0.1–1.2 m tall; branches pubescent with simple or stellate hairs, sometimes glandular and viscid and with some long simple hairs. Leaves broadly ovate to round or reniform in outline, 0.5–2.7 × 0.7–3.2 cm, deeply or rarely shallowly (3–)5-lobed, the lobes narrowly oblanceolate, the outer pair ± joined at base, up to 2.5 × 1 cm long, acute or rounded at the apex, margins coarsely toothed in upper part, strigose-pubescent above with simple hairs; petiole 0.5–4 cm long. Flowers solitary in the leaf axils; pedicel 1–5 cm long, articulated near the apex; epicalyx bracts 9–11, filiform, 6–10 mm long, ciliate; calyx 2–4 mm long, glabrous or pubescent with simple hairs particularly on veins. Petals yellow or white sometimes tinged pink on fading or outside flora area sometimes pink, 6–11 mm long; staminal column 3–5 mm long. Mericarps 3–4 mm long with wings up to 0.5 mm wide, puberulous, back with central ridge and transversely reticulate. Fig. 4, p. 24.

SYN. *Hibiscus arenarius* Murr. in Comment. Soc. Reg. Sci. Gottingensis 7: 8, t. 3 (1786)
P. zeylanica sensu Cav., Monad. Classis dissert. 3: 134 (1787) pro parte t. 48, fig. 3 et descript. sed basionym (*Hibiscus zeylonicus* L.) excl.

var. **microphylla** (*Ulbr.*) *Verdc.* **comb. nov**. Types: Senegal, Wallo and Cayor, *Prieur* s.n., near Richard Tol, *Lelièvre* s.n. (K!, isosyn.); without locality, *Richard* s.n.; Tanzania, Pare Mts, *von Trotha* 268; without locality, *Jaeger* 105; Kenya, Northern Frontier/Tana R. District: Tana, Malka Korokoro, *F. Thomas* 96 (BM!, K!, isosyn.); Somalia, Lugh, Ganane R., *Riva* 1153; Ethiopia, Djebel Haquim, *Ellenbeck* 1144; Merehan, *Robecchi-Bricchetti* 456 (all B†, syn.)

Flowers usually bright yellow, occasionally cream, sometimes pinkish outside and when drying.

UGANDA. Karamoja District: Moroto, Kasumeri Estate, Aug. 1972, *Wilson* 2119!
KENYA. Northern Frontier District: 60 km SW of Mandera on El Wak road, 26 May 1952, *Gillett* 13320!; Masai District: Olorgasaille Mt, 10 June 1956, *Bally* 10574!; Tana River District: Tana River National Primate Reserve, Baomo Village, 13 Mar. 1990, *Kabuye et al.* TPR 219!
TANZANIA. Lushoto District: Mkomazi Game Reserve, Kamakota Hill, 11 June 1996, *Abdallah et al.* 96/161! & SW Umba Steppe, Kivingo, 30 Dec. 1929, *Greenway* 1997!
DISTR. **U** 1; **K** 1, 2? 4, 6, 7; **T** 3; W Africa to Eritrea, Ethiopia, Djibouti, Somalia
HAB. *Commiphora-Acacia* scrub and woodland, often in rocky places, bushland on flood plains, rocky slopes with dense grass cover; 0–1500 m

SYN. *P. zeylanica* sensu Vollesen in Fl. Eth. & Eritr. 2(2): 233, fig. 82.18.9–10 (1995); Thulin, Fl. Somalia 2: 67, fig. 40, C, D (1999)
P. zeylanica (L.) Cav. var. *microphylla* Ulbr. in E.J. 58: 154 (1920)
P. zeylanica (L.) Cav. var. *subquinqueloba* Ulbr. in E.J. 58: 154 (1920). Types: Ethiopia, Gageros, Tacasé, *Schimper* 31, Webi, *Ellenbeck* 1144; Tanzania? no locality, *Herb. Amani* 2305; Kenya, Teita District: Ndara, *Hildebrandt* 2405 (all B†, syn.)

NOTE. Trimen (J.L.S. 24: 146 (1887)) demonstrated that the Hermann type of *Hibiscus zeylonicus* L. (original spelling) was in fact not the species long referred to as *Pavonia zeylanica* (L.) Cav. (spelling used by Cavanilles), but *Pavonia odorata* Willd. All modern writers on the genus have ignored this but *Hibiscus zeylonicus* L. is a synonym of the latter and the former needs another name; that of Murray being available. Vollesen and Thulin in their respective accounts considered that the African and Indian plants of *Pavonia zeylanica* auctt. belonged to different species and used the name *Pavonia zeylanica* Cav. 1787 excluding the basionym. Cavanilles himself had doubted his plant was the same as that of Linnaeus and put a '?'. The matter is dealt with fully in my article in K.B. 62: 633–635 (2007).

Fryxell has annotated the Kew sheet of *Thomas* 96 (which is one of many syntypes of var. *microphylla* Ulbr. as *P. pirottae* (Terrac.) Chiov., but there are no ripe mericarps and the other foliage characters mentioned by Vollesen are variable. I do not see how this sheet differs from *P. arenaria*. *P. pirottae* has mericarps with 2–3 mm wide wings.

18. **Pavonia triloba** *Guill. & Perr.*, Fl. Senegal 1: 50 (1831); F.P.S. 2: 39, fig. 15 (1952); Vollesen in Fl. Eth. & Eritr. 2(2): 232, fig. 82.18.7–8 (1995); Wood, Handb. Yemen Fl.: 108 (1997); Boulos, Fl. Egypt 2: 107 (2000). Type: Senegal, Walo, near Dagana, *Perrottet* 70 (P, holo.; BM!, iso.)

Annual or biennial herb or shrub 0.3–1.5 m tall; branches ± glandular-pubescent and with longer simple hairs and scattered stellate hairs. Leaves ovate, elliptic or round in outline, shallowly to deeply 3-lobed (or sometimes unlobed but grossly toothed), 0.5–4.5 × 0.5–6 cm; lobes ovate to elliptic, entire or central one 3–5-toothed at apex and lateral ones usually entire or notched, pubescent on both sides; petiole 0.5–3(–5) cm long. Flowers solitary in the leaf axils; epicalyx bracts filiform, 8–11, 6–10 × 0.2 mm, setose; calyx 3–5 mm long, divided to below middle, the lobes 1–1.5 mm wide, puberulous to pubescent. Petals pink or orange, 8–15 × 5 mm, staminal column 5–9 mm long, the basal stamens 3–6 mm long, upper ones 1–3 mm long. Mericarps 4–5 mm long, unwinged or with 0.5–1 mm wide wings, back slightly reticulate, pubescent to pilose and conspicuously ciliate; sides puberulous and pilose with scattered hairs.

KENYA. Tana River and Kitui District: boundary: 103 km from Garissa on main Thika road, 15 Dec. 1977, *Stannard & Gilbert* 1100!

DISTR. **K** 4/7; Senegal, Sudan, Ethiopia (also probably in Somalia); Egypt, Arabia, NW India and Pakistan

HAB. Inselberg with *Commiphora–Delonix elata* woodland with dense bush layer on silty sand; ± 1200 m

SYN. *P. triloba* Webb, Fragm. Fl. Aeth.-Aegypt.: 43 (1854); Ulbr. in E.J. 57: 157 (1920). Type: Sudan, Kordofan, Mt Arasch-Cool, *Kotschy* 395 (FI, holo.; K!, iso.)

NOTE. This is very close to true *P. zeylanica* (L.) Cav. and has been united with it by Masters; it is not mentioned by FWTA or Abedin. Material from West Africa and India is sparse and poor with inadequate notes. Plants from Sind are I think identical with Sudanese material. F.P.S. gives flower colour as rose or orange and Guillemin & Perrottet's original description as aurantiaca; in eastern Africa it is given as pink. Much more detailed fieldwork and collection of good material with adequate notes are required to sort out this complex. For this Flora account it is certainly necessary to treat the cited specimen as different from species 17. Webb does *not* cite Guillemin & Perrottet and his name must be treated as separate with a different type.

19. **Pavonia arabica** *Boiss.*, Fl. Orient. 1: 837 (1867); Mast. in F.T.A. 1: 193 (1868); Ulbr. in E.J. 57: 160 (1920); Abedin, Fl. Pakistan: 130 Malvaceae: 98, fig. 24/E-G (1979); U.K.W.F. ed. 2: 102 (1994); Vollesen, Fl. Eth. 2 (2): 232, fig. 82.18.5–6 (1995); Wood, Handb. Yemen Fl.: 108 (1997); Thulin, Fl. Somalia 2:68, fig. 40, A, (1999); Boulos, Fl. Egypt 2: 107, t. 27/4 (2000). Type: Arabia, Mt Sedder [Waddi Fatme], *Schimper* 889 (G, lecto.; K! BM!, P, isolecto.)

Erect annual or usually perennial herb, 0.2–1 m tall, sometimes much branched; stems etc. pubescent to tomentose with long simple and/or stellate hairs, sometimes glandular. Leaves ovate or elliptic, 1.5–6 × 1–3.7 cm, subacute to rounded or emarginate at the apex, rounded to cordate at base, entire or toothed near the apex, aromatic and sticky above, petiole 0.5–5 cm long. Flowers solitary in leaf axils or in small cymes; pedicel 0.2–3(–4.5) cm long, articulated above the middle; epicalyx bracts (7–)9–14, 5–17 mm long; calyx 3–6 mm long, puberulous, pubescent or tomentose, sometimes glandular. Petals pink or rarely white, 7–15 mm long. Mericarps 3–4.5 mm long, unwinged or with wing less than 0.3 mm wide, tomentose to lanate on the back, subglabrous to densely puberulous on the sides.

UGANDA. Karamoja District: Matheniko, Rupia, Sep. 1958, *J. Wilson* 584! & Amudat, June 1960, *Tweedie* 2020! & Moroto to Moroto R., May 1948, *Eggeling* 5790!

KENYA. Northern Frontier District: 12 km S of El Wak on road to Wajir, 11 May 1978, *Gilbert & Thulin* 1659!; Machakos District: near Kyulu, km 211 from Mombasa on Nairobi road, 10 Jan. 1964, *Verdcourt* 3833!; Teita District: Tsavo National Park East, Apr. 1965, *Hucks* 44!

TANZANIA. Pare District: Mkomazi Game Reserve, between Kavatera and Kisiman, 4 Jan. 1996, *Abdallah & Mboya* 3935! & same reserve, SE of Ndea Hill, 27 Apr. 1995, *Abdallah & Vollesen* 95/20! & Nyumba ya Mungu, 15 Nov. 1970, *Batty* 1119!

DISTR. **U** 1; **K** 1–4, 7; **T** 3; Sudan, Ethiopia, Djibouti, Somalia; Arabia, Socotra, Pakistan, NW India

HAB. *Sterculia–Givotia–Sesamothamnus* bushland, *Combretum* and *Acacia-Commiphora-Delonix* woodland on dry stony slopes, gravelly soils and lava plains; 400–1500 m

SYN. *P. erlangeri* Ulbr. in E.J. 57: 164 (1921); Vollesen in Fl. Eth. & Eritr. 2(2): 232, Fig. 82.18.1–4 (1995). Type: Ethiopia, Taro Gumbi, *Ellenbeck* 2094 (B†, holo.)

NOTE. Vollesen maintains *P. erlangeri* as a separate species on account of its larger flowers and less hairy mericarp sides. I have, however, found difficulty in distinguishing them and have followed Thulin in uniting them.

20. **Pavonia schweinfurthii** *Ulbr.* in E.J. 57: 165 (1921); Vollesen in Fl. Eth. & Eritr. 2(2): 234, fig. 82.18.17–18 (1995): Type: Eritrea, middle Lawa [Lava] valley, *Schweinfurth* 1685 (B†, holo.)

Perennial or shrubby densely viscid herb up to 60 cm tall; stems, petiole and pedicel glandular-pubescent and with simple curled hairs. Leaves narrowly ovate or ovate to elliptic, 1–4(–5) × 0.4–2.5 cm, acute to rounded at the apex, narrowly cordate at base, pubescent to tomentose; petiole 0.5–5 cm long. Flowers solitary in the leaf axils; pedicel 2–3 cm long geniculate towards apex; epicalyx bracts 8–12, 6–13 mm long; calyx 3–4(–5) mm long, puberulous and glandular and with long hairs. Petals pale yellow or white, 1–1.8 cm long; staminal column 7–12 mm long. Mericarps 1–4 mm long, not winged, back flat, reticulate, densely pilose; sides rugose, with scattered hairs.

KENYA. Northern Frontier District: Dandu, 3 May 1952, *Gillett* 13025! & 18 June 1952, *Gillett* 13455!

DISTR. **K** 1; Eritrea, Ethiopia; Oman

HAB. *Commiphora* bushland on gravel soil, cracks in granite rocks; 750–850 m

21. **Pavonia ellenbeckii** *Gürke* in E.J. 33: 378 (1904); Ulbr. in E.J. 57: 168 (1921) & V.E. 3(2): 388, f. 182, 14 (1921); Vollesen in Opera Bot. 59: 36 (1980) & in Fl. Eth. & Eritr. 2(2): 233, fig. 82.18.19–20 (1995); Thulin, Fl. Somalia 2:70, fig. 40 G, H (1999). Types: Ethiopia 'Dagage Gobell', Gobelle R., *Ellenbeck* 1030 (B†, syn.) & Kenya, Boran, Korakora, *Ellenbeck* 2232 (B†, syn.)

Annual or perennial herb 0.15–1 m tall; stems rough, pubescent to tomentose, sometimes glandular. Leaves narrowly ovate to oblong, 1.3–5(–7) × 0.6–2(–3.5) cm, rounded to subacute at the apex, rounded to cordate at base, entire, hispid pubescent, often with brownish hairs below and along the margin; petiole 0.5–4.5 cm long. Flowers solitary in leaf axils or in small axillary cymes; pedicel 1–2.2 cm long; epicalyx bracts 10–13, filiform, 8–12 mm long, pubescent with long hairs; calyx 3–4 mm long, glabrous except for ciliate margins. Petals white, 7–12 mm long; staminal column 2–5 mm long. Mericarps 3–4 mm long, not winged, back concave, spongy with longitudinal ridge and transversely reticulate, densely puberulous, sides puberulous.

UGANDA. Karamoja District: Kangole, 22 May 1940, *A.S. Thomas* 3486! & about 5 km W of Kotelo, Sep. 1958, *Wilson* 613! & Amudat, 11 June 1959, *Symes* 533!

KENYA. Northern Frontier District: Dandu, 6 May 1952, *Gillett* 13090!; Kitui District: Kibwezi–Kitui road, km 11 after Athi River, 22 Apr. 1969, *Napper & Kanuri* 2035!; Kwale District: Tanga–Mombasa road about 1.6 km from (Tanzania) border, 14 Aug. 1953, *Drummond & Hensley* 3761!

TANZANIA. Pare District: Mkomazi Game Reserve, Ngurunga Dam to Marimboshi Hill, 8 June 1996, *Abdallah et al.* 96!110!; Rufiji District: 10 km NW of Mtemere, 10 June 1976, *Vollesen* in MRC 3858!

DISTR. **U** 1; **K** 1, 4, 7; **T** 3, 6; Ethiopia, Somalia
HAB. Wooded grassland with *Terminalia spinosa, Acacia, Commiphora* etc., also overgrazed wet ground and clay grassland, often on black cotton soil; 30–1350 m

NOTE. *Tweedie* 2941 (also from Amudat, Nov. 1964) has a mixture of rough stem hairs and dense glands. The relationships of *P. ellenbeckii* and *P. schweinfurthii* need further study in the field.

22. **Pavonia blepharicarpa** *N. Brummitt & Vollesen* in K.B. 61: 1 (2006). Type: Kenya, Northern Frontier District: 40 km from Garissa on road to Hagadera, *Brenan, Gillett & Kanuri* 14770 (K!, holo., EA!, K!, iso.)

Annual or perennial herb often densely branched from base, 15–30 cm tall; stems, petiole and pedicel glandular-puberulous to -pubescent and with long simple hairs. Leaves ovate to elliptic or round, 0.7–3.7 × 0.6–3.5 cm, acute to rounded at apex, rounded to subcordate at base, entire or with few large teeth near the apex, pubescent; petiole 1–4.5 cm long. Flowers axillary and solitary; epicalyx bracts filliform, 9–13, 0.8–1.5 cm long; calyx 3–5 mm long, ± glabrous to pubescent, ciliate. Petals white to very pale yellow, sometimes becoming pink or edged with pink, 0.8–1.5 cm long; staminal column 4–7 mm long. Mericarps 5–6.5 mm long with wings 1–2 mm wide, ciliate, back glabrous, strongly transversely ribbed, sides ± pilose.

UGANDA. Karamoja District: near Namunono R., Aug, 1962, *Wilson* 1305!
KENYA. Northern Frontier District: Archer's Post, Nov. 1965, *Makin* 244!; Masai District: Amboseli, 22 km from Ol Tukai towards Namanga, 15 Dec. 1959, *Verdcourt* 2580!; Tana River District: Hola, 29 Dec. 1972, *Robertson* 1815!
DISTR. **U** 1; **K** 1, 4, 6, 7; Ethiopia
HAB. *Acacia-Commiphora-Boswellia-Salvadora* associations, also *Combretum-Commiphora-Sterculia-Lannea*-woodland, *Terminalia-Euphorbia* wooded grassland; 100–550 m

SYN. *P. sp. A*; U.K.W.F. ed. 1: 206 (1974)
P. hildebrandtii sensu Agnew in U.K.W.F. ed. 2: 102 (1994), *non* Gürke
P. sp. = *Corradi* 3527; Vollesen in Fl. Eth. & Eritr. 2(2): 234 (1995)

23. **Pavonia serrata** *Franch.*, Sertum Somalense (in Révoil, Faune des Pays Çomalis): 18 (1882); Ulbr. in E.J. 57: 171 (1921) & in V.E. 3(2): 390 (1921); Thulin, Fl. Somalia 2:70, fig. 40 I, J (1999). Type: North Somalia, Ouarsangeli, *Révoil* 22 (P, holo.)

Perennial (rarely annual) herb or shrublet up to 1.5 m tall; stems etc. pubescent or tomentose with spreading stellate ± yellowish hairs, sometimes glandular. Leaves narrowly to broadly ovate, 3(–4.5) × 2–2.5 cm, acute to obtuse at the apex, ± deeply cordate at base, margins toothed; petiole 0.3–3(–5) cm long. Flowers solitary in the leaf axils; pedicel 1–6 cm long, pubescent, jointed towards apex; epicalyx bracts filiform, (8–)10–13, 1–2 cm long, ciliate; calyx 3.5–7 mm long, glabrous to pubescent, sometimes glandular, ciliate. Petals white to pale yellow, 1.3–2.7 cm long; staminal column 6–10 mm long. Mericarps triangular to ± round in outline, 5–8 mm long, with wings about 0.5 mm wide at base and apex but 2(–2.5) mm wide in the middle, slightly reticulate and pubescent on the back; sides finely puberulous.

KENYA. Northern Frontier District: Dandu, 3 May 1952, *Gillett* 13026!
DISTR. **K** 1; Ethiopia, Djibouti, Somalia; Yemen
HAB. *Commiphora-Acacia* with scattered trees *Delonix, Terminalia, Gyrocarpus* etc.; ± 750 m

SYN. *P. hildebrandtii* Gürke in E.J. 48: 371 (1912); Ulbr. in E.J. 57: 170 (1921) & V.E. 3 (2): 390, fig. 182/15 (1921); Hepper & Friis, Pl. Forssk. Fl. Aegypt.-Arab.: 199 (1994); Vollesen in Fl. Eth. & Eritr. 2(2): 234, fig. 82.18.13–14 (1995); Wood, Handb. Yemen Fl.: 108 (1997). Types: Ethiopia, Ogaden, Milmil, *Ruspoli & Riva* 1067 (FT, syn.) & North Somalia, Ahl Mts, *Hildebrandt* 834f (B†, syn.). Note: *Hildebrandt* 1372, not an original syntype, is cited in Ulbrich 1921 (loc. cit.) and present at BM!

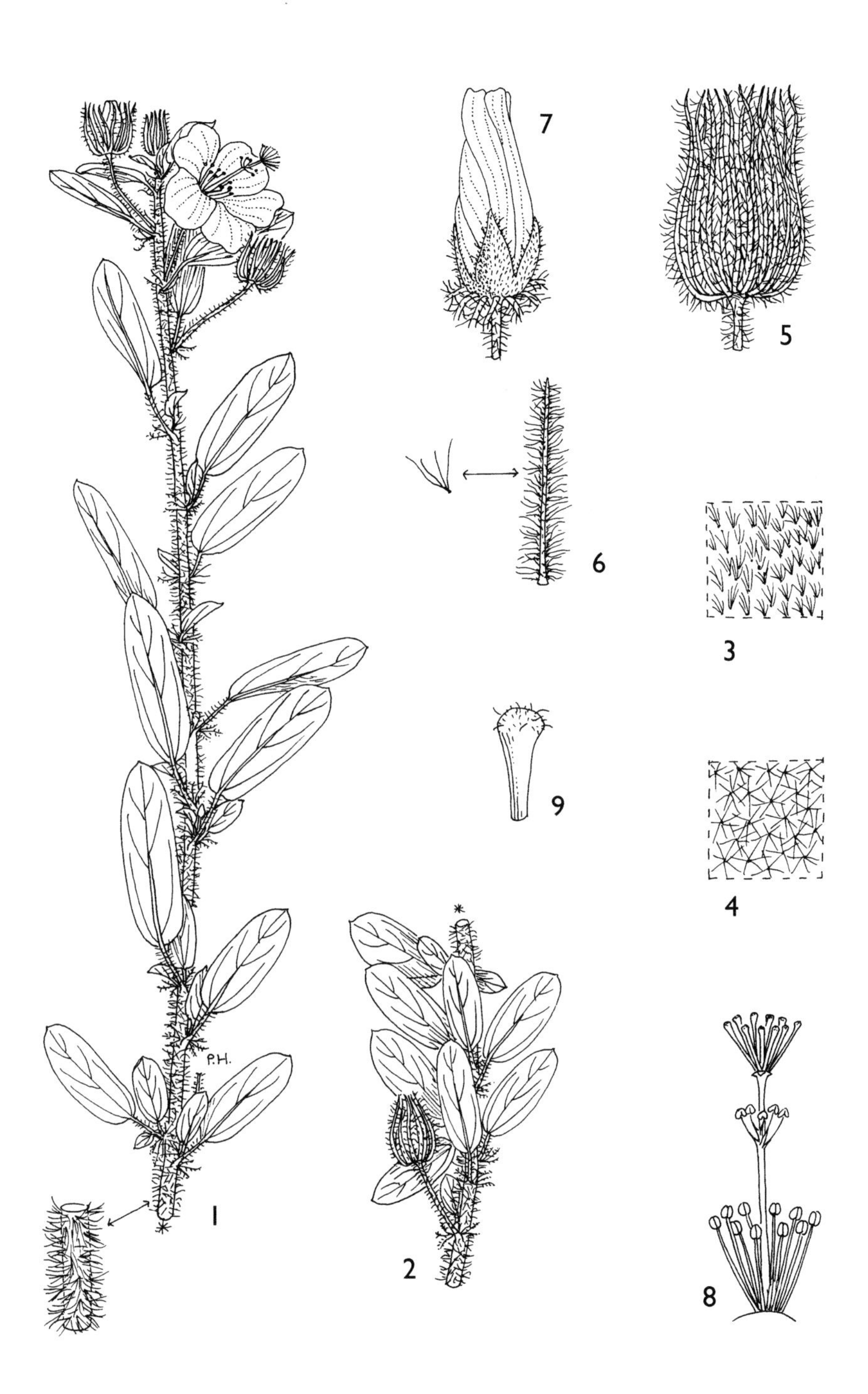

FIG. 5. *PAVONIA DIMORPHOSTEMON* — **1–2**, flowering stem, × $^{2}/_{3}$; **3–4**, leaf, respectively upper and lower surface detail, much enlarged; **5**, epicalyx, × 2; **6**, epicalyx lobe, × 2; **7**, bud, × 2; **8**, stamens and style, × 3; **9**, stigma, much enlarged. All from *Gilbert & Thulin* 1475. Drawn by Pat Halliday.

24. **Pavonia kotschyi** *Webb*, Fragment. Florula. Aethiop.-Aegypt.: 43 (1854); Mast. in F.T.A. 1: 192 (1868); Ulbr. in E.J. 57: 171 (1921) & in V.E. 3(2): 390, fig. 182/16 (1921); F.P.S. 2: 346 (1952); Keay, F.W.T.A. ed. 2, 1(2): 341 (1958); Collenette, Wildfl. Saudi Arabia: 560 (1999); Vollesen in Fl. Eth. & Eritr. 2(2): 234, fig. 82.18.15–16 (1995); Wood, Handb. Yemen Fl.: 108 (1997); Thulin, Fl. Somalia 2:72, fig. 40, K, L (1999); Boulos, Fl. Egypt 2: 105, t. 27/2 (2000). Type: Sudan, Nubia, Abu Gerad, *Kotschy* 12 (FT, holo.; BM!, EA!, K!, iso.)

Erect woody perennial herb or shrublet up to 60 cm tall; stems etc. densely pubescent or tomentose with short stellate hairs. Leaves oblong or elliptic to broadly ovate or ± round, 1.5–3 × 0.8–2.5 cm, subacute to truncate at the apex, cuneate to rounded at base, margins coarsely toothed, with short and long stellate hairs, petiole 1–3 cm long. Flowers solitary in the leaf axils; pedicel slender, 0.4–2.5 cm long, jointed above middle; epicalyx bracts filiform, 8–11, (0.6–)0.8–1.8 cm long, with short stellate hairs and longer bristles; calyx 4–8 mm long, puberulous to densely pubescent. Petals pale and bright yellow or rarely white, 0.9–1.6 cm long; staminal column 4–10 mm long. Mericarps papery when mature, 5–12 mm long with wings 2–5 mm wide, puberulous to pubescent on the back; sides glabrous or with long curled hairs.

KENYA. Northern Frontier District: 1 km S of Tarbaj, 17 Dec. 1971, *Bally & Smith* 14666! & 43 km on road from Wajir to El Wak, 29 Apr. 1978, *Gilbert & Thulin* 1169A! & Malka Korokor, 26 Mar. 1896, *F. Thomas* 97

DISTR. **K** 1, ?7; Mali, Nigeria, Sudan, Ethiopia and Somalia; Arabia, SE Egypt

HAB. *Acacia-Commiphora-Delonix* bushland, *Commiphora-Boswellia-Acacia-Cordia* woodland; 400–500 m

25. **Pavonia dimorphostemon** *Verdc.* **sp. nov.** in sectione *Craspedocarpidio* ponenda inter species folia parva integra, bracteaeque epicalycis lineares etiam flores solitarios albos et filamenta staminum dimorpha habentes, simul collectione characterum id est caulibus dense breviter pilosis foliis stricte oblongis utrinque dense tomentosis basi conspicue trinervis, corollas apertis 3 cm latis differt. Type: Kenya, Northern Frontier District: Mandera, Darecha at the turning, 4 km S of Banissa, *Gilbert & Thulin* 1475 (K!, holo.; EA, UPS, iso.)

Erect shrubby herb to 1 m tall, stems roughly woolly with dense pale yellowish spreading stellate hairs. Leaves narrowly oblong, 1–4 × 0.6–1.1 cm, rounded at base and apex, slightly apiculate at apex, margins entire, with silky long-armed stellate hairs above and more appressed denser stellate hairs above, conspicuously 3-veined at base; petiole 4–10 mm long. Flowers solitary in the leaf axils; pedicel up to 2.2 cm long; epicalyx bracts 14–16, linear, ± 14 mm long, densely spreading-ciliate; calyx pale, ± 5 mm long, densely spreading-ciliate, divided for 3.5 mm into ovate-lanceolate acuminate lobes, the veins pale but darker than rest. Petals creamy white, obovate, ± 1.8 × 1.2 cm; staminal tube 1.3 cm long with 9 basal stamens having filaments ± 5 mm long and 5–6 upper ones near apex of tube with filaments 1–1.5 mm long; style arms 7–10, the stigmatic tips with hairs. Fruit unknown. Fig. 5, p. 29.

KENYA. Northern Frontier District: Mandera, Darecha at the turning 4 km S of Banissa, 5 May 1978, *Gilbert & Thulin* 1475!

DISTR. **K** 1; not known elsewhere

HAB. *Commiphora-Sterculia-Acacia* woodland on shallow stony soil with much exposed bare rock (not limestone); 900 m

NOTE. I have not hesitated to describe this very distinct plant, unfortunately known from only one gathering and without fruits. Fryxell recognised it as an undescribed plant and drew attention to the dimorphic stamens but many species show this character.

3. HIBISCUS

L., Sp. Pl.: 693 (1753); Gen. Pl. ed. 5: 310 (1754); Miller, Gard. Bot. Abr. ed. 4 (1754); Hochr. in Ann. Conserv. Jard. Bot. Geneve, 4: 23 (1900); Ulbr. in Engl. Pflanzenw. Afr. 3, 2: 391 (1921); Bayer & Kubitzki in Kubitzki, Fam. & Gen. Vasc. Pl. 5: 284 (2003)

Ketmia Burm. in Thes. Zeyl.: 133 (1737)

Annual or perennial herbs, sometimes arising from woody rootstocks or underground stems, shrublets, shrubs or (rarely) small trees; indument stellate, rarely also simple. Leaves petiolate, simple, lobed or digitately compound, stipulate, the stipules usually subulate to linear. Flowers usually solitary and axillary but often forming terminal, racemose or corymbose inflorescences by reduction of the upper leaves; peduncle usually articulated. Epicalyx of 5–20 bracts, free or adnate to the base of the calyx, occasionally absent. Calyx 5-lobed, rarely 5- or 10-toothed, joined at base or more rarely almost to the apex, persistent. Corolla often yellow with a dark red centre or red, pink, purple or white. Staminal tube with five minute teeth at the apex; free parts of filaments variable in length, sometimes whorled. Ovary 4–5-locular; loculi 3–many-ovulate; stigmas 5. Fruit a loculicidal capsule, not separating from the receptacle. Seeds 3-more per locule, reniform.

200 species in warm temperate and tropical regions. 53 species are found in the wild in East Africa, and 6 introduced species commonly cultivated as ornamentals. Variation within many species is wide with some apparently intergrading in southern Africa, the centre of diversity for this genus. Further study of live material in the field will enable improvement of this account.

Several species of *Hibiscus* have been introduced in cultivation, either from other parts of the world or from the wild.

KEY TO THE COMMONLY CULTIVATED SPECIES
(key to native species on p. 33)

1. Calyx fleshy or woody, each sepal lobe with three veins, two along the edges and one down the middle, the ones at the edge fusing at the sinus with those of the next lobe 2
   Calyx not as above, lobes variously veined or the veins indistinct 3
2. Epicalyx bracts forked at the tip .... *Hibiscus acetosella*
   Epicalyx bracts not forked at the tip .... *Hibiscus sabdariffa*
3. Underside of leaves and young stems stellate-hispid to stellate-pubescent 4
   Leaves and young stems glabrous 5
4 Young stems and underside of leaves stellate-tomentose; leaves 3-lobed; peduncles > 6 cm long; epicalyx bracts lanceolate .... *Hibiscus mutabilis*
   Young stems and underside of leaves stellate-hispid; leaves unlobed; peduncles ≤ 5 cm long; epicalyx bracts linear to subulate, < 2mm wide .... *Hibiscus scottii*
5. Peduncles at least 2.5 × as long as calyx; staminal column included .... *Hibiscus syriacus*
   Peduncles ± equal to calyx length; staminal column exserted 6
6. Flower pendent; corolla fimbriate (see p. 60) .... *Hibiscus schizopetalus*
   Flower erect; corolla entire .... *Hibiscus rosa-sinensis*

Hibiscus acetosella *Welw.*

Herb or subshrub to 2 m tall; stems and leaves purple, or at least purple-tinged in our area, rarely green, stems glabrous to sparsely pubescent. Leaves unlobed to shallowly to moderately 3–5-lobed. Flowers axilllary, solitary, dark pink or purple, rarely yellow.

Frequently cultivated, also ruderal and naturalised in abandoned fields and settlements.

UGANDA. Mengo District: Kampala Government plantation, June 1929, *Liebenberg* 838!
KENYA. Kiambu District: Muguga, 3 Mar. 1963, *Verdcourt* 3596!
TANZANIA. Lushoto District: Amani, 8 Aug. 1963, *Omari in Richards* 18155

Hibiscus mutabilis *L.*

Shrub or tree to 5 m tall; stems and underside of leaves stellate-tomentose. Leaves cordate, shallowly 3–7-lobed, discolorous. Flowers white when freshly open, turning through pink to dark red before withering.
Originally from China, cultivated at Amani and other places.

TANZANIA. Lushoto District: Amani nursery, 31 Jan. 1949, *Greenway* 8321!

Hibiscus rosa-sinensis *L.*

Shrub or tree to 5 m; stems and leaves glabrous. Leaves simple, unlobed, usually green, ocassionaly reddish. Flowers white, yellow, red with a conspicuously exserted staminal column; double-flowered varieties lack the exserted staminal column.

Originally from tropical Asia. This is the most common *Hibiscus* in cultivation, and is also used as a street tree and as a hedging plant.

UGANDA. Mengo district: Masaka, 5 May 1972, *Lye* 6841!
KENYA. Kiambu District: Muguga, 8 June 1962, *Gichuru* 13!; Nairobi District: Kenya Technical Teachers College, 26 July 1985, *Mutanga & Nyakundi* 66!
TANZANIA. Mwanza District: Ukiriguru, 20 Dec. 1966, *Juma* 4978!; Lushoto District: 3 km E of Lushoto town, 26 Dec. 1966, *Semsei* 4176; **Z** & **P** fide U.O.P.Z.: 296 (1949)

Hibiscus sabdariffa *L.*

Annual herb to 4.5 m tall; stems and foliage green, green-red, or red, glabrous to sparsely pubescent, occasionally aculeate. Leaves entire to shallowly or deeply 3–7-lobed. Flowers solitary in the leaf axils, calyx green, green-red or red, fleshy to leathery; corolla pale yellow, with or without a dark red centre.

Widely cultivated. The most common form of this species is the form with fleshy edible calyces that are used for drinks and jams, and are the source of 'Hibiscus tea' of commerce.

UGANDA. Teso District: Serere, Dec. 1931, *Chandler* 76!
TANZANIA. Ngara District: Rugan̄zo Bugufi, 10 July 1960, *Tanner* 5033!; **Z** & **P** fide U.O.P.Z.: 297 (1949)

Hibiscus scottii *Balf.*

Shrub to 2 m tall; stems stellate-hispid. Leaves simple, ovate with a cuneate to truncate base. Flowers yellow with a dark-red centre.
Rare in cultivation.

KENYA. Nairobi district: Hort. P. Greensmith, Langata, 2 Dec. 1975, *Gillett* 20909!

Hibiscus syriacus *L.*

Shrub or tree to 4 m; stems and leaves glabrous. Leaves unlobed to 3(–5)-lobed. Flowers blue, pink or purple.

Mostly cultivated in highlands gardens; occasional in Nairobi gardens.

KENYA. Nairobi District: City Park, 6 Jan. 1972, *City Park Sup.* in EA 15040!

KEY TO THE NATIVE SPECIES OF *HIBISCUS* IN EAST AFRICA

1. Epicalyx bracts at least 2 mm long 2
   Epicalyx bracts not more than 1 mm long, or absent 53
2. Epicalyx cup-like, with up to 10 teeth; coastal shrub (sect. *Azanza*) .... 1. *H. tiliaceus* (p. 37)
   Epicalyx not cup-like, the bracts separate for most of their length; plant of various habitats 3
3. Calyx fleshy (in flower) or woody (in fruit), each calyx lobe with 3 veins - 2 along the edges and one down the middle, the ones at the edges fusing at the sinus with those of the next lobe (sect. 2. *Furcaria*) ....
   Calyx not as above, lobes variously veined, or the veins indistinct 16
4. Bracts of epicalyx forked 5
   Bracts of epicalyx not forked 9
5. Stipules auriculate and/or amplexicaul, 7–10 mm wide .... 2. *H. surattensis* (p. 38)
   Stipules not auriculate or amplexicaul, spatulate, linear or subulate, 0.2–5 mm wide 6
6. Plants erect, annual; leaves membranous, usually obtuse, rarely cordate at base, up to 18 cm long and wide; peduncle up to 8 mm long 7
   Plants erect or scandent, annual or perennial; leaves membranous or not, usually (sub)cordate; peduncle 1–12 cm long 8
7. Stem prickles stellate; leaf lamina unlobed or shallowly 3-5-lobed; stipules narrowly linear . 3. *H. mastersianus* (p. 38)
   Stem prickles simple; leaf lamina 3–5-palmatipartite; stipules lanceolate .... 4. *H. noldeae* (p. 39)
8. Leaf stellate-tomentose below with prickles of the main veins; leaf sinuses to 2 cm deep; leaf lobes at least as broad as long; stems and petiole hispid .... 5. *H. rostellatus* (p. 39)
   Leaf hispid or stellate-tomentose below; sinuses 2–4 cm deep; leaf lobes longer than broad; stems and petiole glabrous except often with a thin longitudinal line of pubescence .... 6. *H. holstii* (p. 40)
9. Calyx without glands on the median veins of the sepals 10
   Calyx with a conspicuous gland on the median vein of each sepal 11

10. Leaves 3–5-palmatipartite; plant of fallow fields at 80–1100 m .... 11. *H. mechowii* (p. 44)
    Leaves unlobed or shallowly lobed; plants of forest at 1550–3000 m .... 13. *H. berberidifolius* (p. 46)
11. Stem glabrous (except when very young) or with a single line of pubescence 12
    Stem with several lines of pubescence or pubescent all over 14
12. Leaves circular or broadly ovate, 1–3.5 × 0.8–4.5 cm .... 10. *H. sparseaculeatus* (p. 44)
    Leaves usually much larger 13
13. Annual herb; stem with upcurved prickles; staminal tube 2–4 cm .... 7. *H. cannabinus* (p. 41)
    Shrub; stem with recurved prickles; staminal tube 4.5–7 cm .... 9. *H. greenwayi* (p. 42)
14. Filaments 2–3 mm; staminal column 2–4 cm; stem pubescent .... 13. *H. berberidifolius* (p. 46)
    Filaments 0.3–0.5 cm; staminal column 1.5–2 cm; stem stellate-pubescent 15
15. Calyx 1–1.5 cm; fruit 12–15 mm long, appressed-pubescent .... 8. *H. reekmansii* (p. 42)
    Calyx 2–3 cm; fruit 20 mm long, setose .... 12. *H. diversifolius* (p. 45)
16. Capsule (semi)spherical; seeds lanate; petals less than 20 mm in length (sect. *Bombycella*) 17
    Capsule variously shaped but rarely spherical; seeds never lanate; petals over 20 mm in length 30
17. Corolla white, pink or mauve 18
    Corolla red to maroon 21
18. Corolla 2–6 mm long, white, fading pink; staminal tube to 5 mm long .... 14. *H. micranthus* (p. 47)
    Corolla 10–25 mm long, variously coloured; staminal tube 7 mm or more in length 19
19. Calyx 5–7 mm long; epicalyx 2–3 mm long ... 15. *H. meyeri* (p. 47)
    Calyx 8–15 mm long; epicalyx 4–20 mm long 20
20. Staminodes present at base of the staminal column; plant brown-hairy all over .... 16. *H. fuscus* (p. 49)
    Staminodes absent; plant without brown hairs 17. *H. flavifolius* (p. 50)
21. Plant a suffrutex from a tuberous root-stock 22
    Plant a herb or shrub, not a suffrutex from a tuberous root-stock 23
22. Epicalyx 2–3 mm long, bracts not reaching the calyx sinus; petals 15–20 mm long .... 18. *H. rhodanthus* (p. 51)
    Epicalyx 4–5 mm long, bracts reaching beyond the calyx sinus; petals 10–13 mm long .... 19. *H. somalensis* (p. 51)
23. Epicalyx more than half the length of the calyx 24
    Epicalyx half the length of the calyx, or shorter 27
24. Pedicels and calyx brown-hispid, or brown pubescent 25
    Pedicel and calyx not brown-hispid, not brown pubescent .... 20. *H. aponeurus* (p. 52)

25. Corolla 7–12 mm long; veins on the underside of the leaves prominently raised; plants from **K** 1 . . . . 21. *H. crassinervius* (p. 52)
    Corolla more than 12 mm long, veins on the underside of the leaves not prominently raised; plants from various places 26
26. Epicalyx bracts 0.6–1 mm wide, shorter than the calyx . . . . 22. *H. shirensis* (p. 53)
    Epicalyx bracts 0.2–0.3 mm wide, ± as long as the calyx . . . . 23. *H. debeersti* (p. 53)
27. Staminal column 5–10 mm long, at least some leaves 3-or 5-lobed 28
    Staminal column 9–14 mm long; leaves unlobed 29
28. Stems stellate- or simple-hispid, hairs 1.5–2 mm long; seeds glabrous or minutely pubescent . . 25. *H. allenii* (p. 54)
    Stems shortly appressed stellate-hispid, hairs to 0.5 mm; seeds lanate . . . . 26. *H. zanzibaricus* (p. 55)
29. Leaves (broadly) ovate to rotund with a rounded or obtuse apex; epicalyx and calyx densely stellate hispid-pubescent . . . . 24. *H. hildebrandtii* (p. 54)
    Leaves ovate with an acute apex; epicalyx ciliate, calyx pilose outside . . . . 27. *H. richardsiae* (p. 55)
30. Capsule winged or prominently angled, or awned, disintegrating at maturity and falling off the receptacle (sect. *Pterocarpus*) 31
    Capsule neither winged nor angled, not disintegrating at maturity, persistent on receptacle 33
31. Leaves deeply 3–5-palmatipartite; capsule valves aristate; plant a prostrate herb . . . . . . 28. *H. palmatus* (p. 56)
    Leaves 3–5-lobed or not lobed; capsule valves winged, aristate; plant erect 32
32. Epicalyx bracts usually reaching the sinus of calyx; flowers yellow, erect . . . . 29. *H. vitifolius* (p. 56)
    Epicalyx bracts rarely reaching the calyx sinus; flowers pink, pendent . . . . 30. *H. kabuyeana* (p. 60)
33. Staminal tube 40–50 mm long, longer than the petals, conspicuously exserted from the petals (sect. *Lilibiscus*) . . . . 31. *H. schizopetalus* (p. 60)
    Staminal tube not exceeding the petals, not conspicuously exserted, less than 50 mm long (to 65 mm in *H. macranthus*) 34
34. Epicalyx bracts 5 (5–6 in *H. ludwigii*), linear to ovate (Sect. *Calyphylli*) 35
    Epicalyx bracts 6–12, variously shaped 43
35. Epicalyx bracts widest at base, or at the middle 36
    Epicalyx bracts linear 42
36. Epicalyx bracts widest at or near the middle 37
    Epicalyx bracts widest at base 41
37. Leaves concolorous, or only weakly discolorous 38
    Leaves discolorous 39

38. Calyx with prominent rib-like median vein, not reticulately veined, shorter than the capsule; capsule densely stellate-setose, not enclosed in the calyx .... 32. *H. calyphyllus* (p. 61)
    Calyx reticulately veined, obscurely 3–5-veined; capsule glabrous or puberulous, enclosed in the accrescent calyx .... 33. *H. ovalifolius* (p. 63)
39. Calyx 2.5–3 cm long, lobes triangular, reticulately veined; corolla 6 cm long; capsule 3.6-4 cm long .... 34. *H. platycalyx* (p. 63)
    Calyx up to 2 cm long, lobes ovate to triangular, 3-veined, not reticulately veined; corolla 3.5–8 cm long; capsule up to 2.5 cm long 40
40. Epicalyx bracts adnate halfway to sinus between calyx lobes; seeds 4.5–5.5 mm long, white-brown tomentose .... 35. *H. seineri* (p. 64)
    Epicalyx bracts adnate to calyx up to sinus; seeds 3-4 mm long, white-brown sericeous .. 36. *H. faulknerae* (p. 64)
41. Epicalyx bracts in flower shorter than the calyx; corolla 4–6.5 cm long; staminal column and styles shorter than the petals .... 37. *H. ludwigii* (p. 65)
    Epicalyx bracts in flower longer than calyx; corolla 6-10 cm long; staminal column and styles ± equal to the petals .... 38. *H. macranthus* (p. 65)
42. Leaves glabrous or pubescent, unlobed or rarely 3-lobed; stems glabrous or pubescent; capsule sub-globose; seeds glabrous or with scattered stellate tubercles .... 39. *H. dongolensis* (p. 66)
    Leaves densely pubescent, 3-lobed, rarely unlobed; stems pubescent; capsule ellipsoid; seeds with pectinate scale-like hairs. ...... 40. *H. lunarifolius* (p. 66)
43. Epicalyx bracts spatulate; peduncle in flower less than 8 mm long (Sect. *Panduriformes*) .. 41. *H. panduriformis* (p. 67)
    Epicalyx bracts not spatulate; peduncle in flower over 10 mm long 44
44. Calyx inflated, lobes greatly enlarged in fruit (Sect. *Trionum*) .... 42. *H. trionum* (p. 68)
    Calyx not inflated, lobes not enlarged in fruit (Sect. *Ketmia*) 45
45. Most petioles 3–10 cm long; capsule round, shorter than the calyx; plant an annual herb 46
    Petioles less than 3 cm long; capsule various; plant a suffrutex 51
46. Capsule beak conspicuous, at least one third as long as capsule body; most petioles over 7 cm long 47
    Capsule beak inconspicuous, less than quarter as long as capsule body; petiole up to 7 cm long 48
47. Leaves 3–5-lobed, bearing flat calcareous concretions on the underside; plants of southern Tanzania .... 43. *H. masasiana* (p. 68)
    Leaves unlobed, without calcareous concretions on the underside; plants of central and northern Kenya .... 44. *H. rhabdotospermus* (p. 69)

48. Pedicel 6–14 mm long; epicalyx bracts 20–25 mm long; leaves 3–5-foliolate .... 45. *H. caesius* (p. 69)
    Pedicels 20–100 mm long; epicalyx bracts 2–12 mm long; leaves at most lobed 49
49. Pedicels 20–60 mm long; petals 6–25 mm long, white, mauve or pink; leaves lacking calcareous concretions on the underside 50
    Pedicels 40–80 mm long; petals 50–70 mm long, yellow with maroon spot at base; leaves with calcareous concretions below .... 47. *H. physaloides* (p. 70)
50. Leaves densely pubescent; epicalyx bracts 5–12 mm long; corolla 12–25 mm long .... 16. *H. fuscus* (p. 49)
    Leaves glabrous or sparsely hairy; epicalyx bracts 2–3 mm long; corolla 6–10 mm long .... 46. *H. obtusilobus* (p. 70)
51. Most flowers in a terminal corymb; style branches densely pilose .... 48. *H. corymbosus* (p. 71)
    All flowers solitary in leaf axils 52
52. Plant with white, simple (or two-armed stellate) hairs; capsule puberulous, pilose along the sutures .... 49. *H. articulatus* (p. 71)
    Plant yellow-stellate-hairy all over, the hairs mostly 4-armed; capsule pilose all over ..... 50. *H. aethiopicus* (p. 73)
53. Petals 3–4 cm long, red, fimbriate (sect. *Lilibiscus*) 31. *H. schizopetalus* (p. 60)
    Petals up to 2 cm long, white or yellow, entire (Sect. *Solandra*) 54
54. Flowers white, in a terminal raceme; leaf margin serrate-dentate; epicalyx lacking .... 51. *H. lobatus* (p. 74)
    Flowers yellow, solitary in the leaf axils; leaf margins entire to dentate; rudimentary epicalyx bracts to 0.5 mm present .... 52. *H. sidiformis* (p. 74)

1. **Hibiscus tiliaceus** *L.*, Sp. Pl. 694 (1753); U.O.P.Z.: 298 (1949); Keay, F.W.T.A. ed. 2, 1(2): 345 (1958); E.P.A.: 568 (1959); Exell in F.Z. 1: 435 (1961); Hauman in F.C.B. 10: 134 (1963); K.T.S.L.: 172 (1994). Lectotype: India, Herb. Hermann 3: 51, No. 259 (BM-000594693)

Tree to 9 m tall; stems glabrous or sparsely to densely stellate-pubescent, becoming brown-grey with prominent lenticels. Leaves ovate to orbicular, 6–22 × 6–23 cm, unlobed or rarely 3-lobed, apex acuminate or rounded, base cordate, margin entire to obscurely dentate, glabrous or sparsely stellate-pubescent above, glabrous or stellate-tomentose below; petiole 6–15(–17) cm long, sparsely to densely stellate-tomentose; stipules lanceolate to narrowly ovate, 1.7–3.5 × 0.8–1.5 cm, caducuous. Flowers in terminal 3–6-flowered cymes; pedicel 0.5–2 cm long, articulated at base, densely stellate-pubescent, angular in the upper part; epicalyx cuplike, to 7 mm long, adnate to calyx, with ten teeth to 2 mm long, densely stellate-tomentose; calyx lobes lanceolate, 11–18 × 5–7 mm, fused in the lower $^1/_3$, each with prominent median vein on which a gland (nectary) occurs, acute, densely tomentose. Petals yellow, 4–6.5 cm long, the basal dark red, 1–2 cm, densely stellate-tomentose on the outside, sparsely pubescent or glabrous within; staminal tube 2–2.5 cm long, filaments 2 mm long; exserted part of style 5–8 mm long, pubescent. Capsule ovoid-ellipsoid, 1.6–2.2 cm long, valves acuminate, densely grey-tomentose; seeds brown, reniform, 4 × 2 mm, tuberculate.

KENYA. Kwale District: Ngoa, Vanga, Nov. 1929, *Graham* 738; Kilifi District: Kilifi, 21 Mar. 1945, *Jeffery* 142!; Lamu District: Kigangwe, Sep. 1929, *Abdulla* 2197!
TANZANIA. Tanga District: Tanga, 4 Nov. 1929, *Greenway* 1840!; Rufiji District: Salale, 2 July 1932, *Schlieben* 2548!; Zanzibar: 26 Aug. 1951, *Taylor* in *Williams* 88!
DISTR. **K** 7; **T** 3, 6; **Z**; **P**; pan-tropical

HAB. Sandy seashores and mangrove swamps; 0–20 m

SYN. *Talipariti tiliaceum* (L.) Fryxell, Contr. Univ. Michigan Herb. 23: 258 (2001)

2. **Hibiscus surattensis** *L.*, Sp. Pl. 2: 696 (1753); Harv. in Fl. Cap. 1: 177 (1860); Garcke in Peters, Reise Moss. Bot. 1: 127 (1861); Mast in F.T.A. 1: 201 (1868); Hochr. in Ann. Conserv. Jard. Bot. Geneve, 4:110 (1900); Baker f. in J.L.S. 40: 27 (1911); Ulbr. in Engl. Pflanzenw. Afr. 3, 2: 400, fig. 187 A (1921); U.O.P.Z.: 298 (1949); Keay, F.W.T.A. ed. 2, 1(2): 346 (1950); Exell in F.Z. 1: 438, t. 89/4 (1961); Hauman in F.C.B. 10: 121 (1963); Edmonds, Distr. *Hibiscus* Sect. *Furcaria* Trop. E. Afr.: 15 (1991); U.K.W.F. ed. 2: 99 (1994); Vollesen in Fl. Eth. & Eritr. 2(2): 198 (1995); Wilson, Bull. Nat. Hist. Mus. Lond. (Bot.) 29(1): 51 (1999). Type: not designated (see note below).

Annual herb, prostrate or climbing or scrambling up to 3 m; stems hispid and/or pubescent and sparsely to rather densely aculeate. Leaves broadly ovate to suborbicular in outline, 4–11 cm long, 3–10 cm, palmately 3–5-lobed (sometimes not lobed or only shallowly so) to palmatipartite, apex acute or acuminate, base truncate or shallowly cordate, margin coarsely toothed, sparsely hispid on both surfaces and aculeate on the veins beneath; petiole 3–15 cm long, sparsely pilose, aculeate, the prickles on a swollen base, recurved, with an adaxial line of pubescence; stipules auriculate, ± amplexicaul, 15–20 × 7–10 mm, sparsely pilose. Flowers solitary in the leaf axils; pedicel up to 6–7 cm long, hispid above the articulation; epicalyx of 8–9 bracts 10–15 mm long, forked with the inner branch linear and inflexed and the outer branch spatulate; calyx lobes ovate-lanceolate, 12–25 × 5–10 mm, usually somewhat rigid with long sharp apices, hispid and aculeate on the median and marginal veins. Corolla yellow with purple patch at base, 3–5 cm long, glabrous; staminal tube 12–15 mm long, filaments 1.5–2.5 mm long; exserted part of styles 7–9 mm long, glabrous. Capsule globose-ovoid, 12–15 mm long, 10–12 mm diameter, apex acute, densely appressed-pilose; seeds grey-brown, subreniform, 3–3.5 mm × 2.5 mm, with scattered appressed ovate scales.

UGANDA. Bunyoro District: Buhaguzi, Sep. 1939, *Sangster* 561; Kigezi District: Ishasha Gorge, May 1950, *Purseglove* 3430!; Bugishu District: Budadiri, Jan. 1932, *Chandler* 446!
KENYA. South Nyeri District: 8 km N of Murang'a [Fort Hall], 7 Aug. 1959, *Bogdan* 4844!; Meru District: Chuka Kithutuni, 2 Nov. 1968, *Kimani* 113!; Teita District: Voi, 8 May 1931, *Napier* 991!
TANZANIA. Lushoto District: Gereza East, Mnyusi, Sep. 1954, *Semsei* 1796!; Kigoma District: Utahya, 2 Aug. 1958, *Newbould & Jefford* 1297!; Mpanda District: Silambula, Mahali Mts, 17 Sep. 1958, *Newbould & Jefford* 2040!; Zanzibar: km 27, Chwaka, 21 June 1960, *Faulkner* 2607!
DISTR. **U** 2–4; **K** 3, 4, 7; **T** 1, 3–8; **Z**; **P**; widespread in tropical Africa; India; southeast Asia
HAB. Grassland, bushland and forest; 0–1700 m

SYN. *Furcaria surattensis* (L.) Kostel., Allg. Med.-Pharm. Flora, 5: 1856 (1836)
Hibiscus hypoglossus Harv., Fl. Cap. 1: 177 (1860), *nomen nudum*
H. surattensis L. var. *genuinus* Hochr. in Ann. Conserv. Jard. Bot. Geneve, 4: 111 (1900). Type as for species
H. surattensis L. var. *villosus* Hochr. in Ann. Conserv. Jard. Bot. Geneve, 4: 112 (1900). Type: Malawi, *Whyte* s.n. (K!, holo.)

NOTE. Borssum Waalkes in Blumea 14: 58 (1966), and others have treated 875.29 (LINN) (K!, microfiche) as the type, but the sheet lacks the relevant Species Plantarum number (i.e. "12"), and was a post-1753 addition to the collection, and is not original material for the name.

3. **Hibiscus mastersianus** *Hiern*, Cat. Afr. Pl. Welw. 1: 71 (1896); Milne-Redh. in K.B. 1935: 272 (1935); Exell in F.Z. 1: 439 (1961); Edmonds, Distr. *Hibiscus* Sect. *Furcaria* Trop. E. Afr.: 19 (1991); Wilson, Bull. Nat. Hist. Mus. Lond. (Bot.) 29(1): 71 (1999). Lectotype: Mozambique, Lupata, *Kirk* s.n. (K!, lecto., chosen by Milne-Redhead (1935))

Erect annual herb up to 2 m tall; stems with short, sometimes sparse, usually stellate prickles with a swollen base, and one or more longitudinal lines of crisped pubescence. Leaves suborbicular to narrowly oblong, 8–18 × 6–18 cm, not lobed or shallowly 3–5-palmatilobed, lobes acute to rounded, apex acute or acuminate, base rounded to truncate or shallowly cordate, margin irregularly serrate, stellate-pubescent on both surfaces; 5–7-veined, midrib with a longitudinal fissure 2–3 mm long near the base; petiole 4–7(–18) mm long, pubescent; stipules linear to filiform, 4–8 × 0.2 mm, hispid. Flowers solitary, axillary; pedicel 4–8 mm long, hispid; epicalyx of 9–10 bracts, linear, 8–10 × 1 mm, forked, hispid; calyx lobes lanceolate, 10–20 × 2–3 mm, acuminate, hispid or aculeolate on the margins. Corolla yellow to orange with purple or maroon spot at base, 3–5 cm long, sparsely puberulous; staminal tube 12–16 mm long; filaments 0.5–1 mm long; exserted part of styles 1–2 mm long. Capsule ovoid-acuminate, 13–15 mm long, 10 mm diameter, densely appressed-setose; seeds 4 × 2 mm, densely covered in triangular scales.

KENYA. Kitui District: Mutomo Hill Plant Sanctuary, 26 May 1968, *Gillett* 18606! & Ndui, Voo Location, 12 Aug. 1968, *Kimani* 39! & 7 km from Nuu to Endau, 13 Feb 2002, *Kirika et al.* NMK 315!

TANZANIA. Shinyanga District: Huruhuru, 17 May 1931, *Burtt* 2454K!; Iringa District: Msembi–Mbagi junction track at 9.4 km, 20 Apr. 1970, *Greenway & Kanuri* 14393!; Mpanda District: Kongwa, 5 Apr. 1950, *Anderson* 658!

DISTR. **K** 4; **T** 1, 3–7; Congo-Kinshasa, Rwanda, Zambia, Mozambique, Zimbabwe, Botswana

HAB. Grassland, woodland and thicket: 500–1300 m

SYN. *H. furcatus* sensu Mast. in F.T.A. 1: 201 (1868), *non* Roxb. Type as for *Hibiscus mastersianus* Hiern

4. **Hibiscus noldeae** *Baker f.* in J.B. 77: 20 (1939); Hauman in F.C.B. 10: 119 (1963); Wilson in Bull. Nat. Hist. Mus. Lond. (Bot.) 29(1): 59 (1999). Type: Angola, Malange, Quela, Apr. 1938, *Nolde* 713 (BM!, holo.)

Annual herb, erect to 1 m high; stems with recurved red prickles. Leaves 3–5-palmatipartite, 5–8 cm long; lobes between 4–6 cm long and often membranaceous, apices acute, base truncate, margin serrate, veins beneath aculeolate, sparsely hispid above and below; stipules lanceolate, 7–10 × 2–3 mm; petiole 3–6 cm long, aculeolate with a line of pubescence. Flowers solitary in the leaf axils; pedicel ± 3 mm long, sparsely hispid; epicalyx bracts forked, ± 15 mm long, patently pilose; sepals ± 17 mm long, acuminate, sparsely hispid. Corolla 5 cm long, glabrous; staminal tube 1.8 cm long; filaments 1.5–2 mm long. Capsule 10–15 mm long, 12 mm diameter; seeds dark brown, wedge-shaped, 3 × 2 mm, sparsely covered with ovate scales.

UGANDA. Busoga District: Lolui Island, 13 May 1964, *Jackson* 62!; Masaka District: Bugalla Island, Sesse, 7 Oct. 1958, *Symes* 470!

TANZANIA. Lushoto District: Mlinga peak, 28 Aug. 1942, *Greenway* 6630!

DISTR. **U** 3, 4; **T** 3; Sierra Leone to Ethiopia and South to Angola

HAB. Woodland; 1400–1600 m

SYN. *H. eetveldeanus* De Wild. & T. Durand var. *asperatus* De Wild. in B.J.B.B. 3: 279 (1911). Type: Congo-Kinshasa, Kasai Province, Katola, Apr. 1908, *Sapin* s.n. (BR, holo.)

5. **Hibiscus rostellatus** *Guill. & Perr.* in Fl. Seneg. 1: 55 (1831); Mast. in F.T.A. 1: 201 (1868); F.P.S. 2: 26 (1952); Hauman in F.C.B. 10: 117 (1963); Edmonds, Distr. *Hibiscus* Sect. *Furcaria* Trop. E. Afr.: 20 (1991); Wilson in Bull. Nat. Hist. Mus. Lond. (Bot.) 29(1): 53 (1999). Type: Senegal, *Perrottet* s.n. (P, lecto; BM!, G, isolecto.), chosen by Wilson (1999)

Erect or semiscandent undershrub, 2–3 m high; stems hispid with yellow hairs to 3 mm long; prickles small, to 2 mm long, from swollen base, recurved. Leaves palmately 5-lobed, 8–10 × 9–11 cm, lobes deltoid, apex acute or acuminate, base cordate or truncate, margin crenate-serrate, with gland on the midrib below, near the base, tomentose; petiole 4–10 cm long, hispid and aculeate, the prickles recurved to 1 mm, and with an adaxial line of pubescence; stipules linear, 6–9 × 0.5–1.5 mm, glabrous or hispid. Flowers solitary in the leaf axils; pedicel 4–12 cm long, as long as or longer than the adjacent petiole, articulated near the top; epicalyx of 7–10 linear bracts, 17 × 1–3.5 mm, forked, the outer branch spatulate, hispid; calyx lobes lanceolate, united in the bottom half, shortly stellate-hispid along the veins; without a nectary on the median vein, 15–20 × 7–10 mm. Corolla pink or yellow with dark centre, 4–6 cm long, glabrous; staminal tube 2–2.5 cm long; filaments 2–3 mm long; exserted part of style 4–8 mm long, glabrous or sparsely villous. Capsule ovoid, 13 mm long, 20 mm diameter; seeds subreniform, brown, 4 × 3 mm, brown, with pectinate scales on a background of concentric rings.

UGANDA. Bunyoro District: Budongo Forest, 25 Oct. 1971, *Synnott* 711!; Mengo District: Entebbe, Oct. 1922, *Maitland* 166! & Kampala, Apr. 1966, *Smith* 587!

TANZANIA. Mwanza District: Ukerewe Island, 1930, *Conrads* s.n.; Mpanda District: Kasoje, Kungwe Mt, 16 July 1959, *Newbould & Harley* 4367! & Mpanda–Uvinza road, 28 May 2000, *Bidgood et al.* 4476!

DISTR. **U** 2, 4; **T** 1, 4; West and Central Africa, South to Zambia and Angola

HAB. Edge of wet forest; 1000–1300 m

6. **Hibiscus holstii** *Mwachala* **sp. nov.** *Hibiscus rostellato* Guill. & Perr. similis sed caule glabro, foliis plusminus glabriis et lobi longioribus quam latioribus differt. Typus: Tanzania, Lushoto District: Usambaras, *Holst* 3198 (K!, holo.)

Scandent shrub; stems 3–10 m long, aculeate with prickles recurved and swollen at base, otherwise glabrous but for longitudinal lines or areas of pubescence. Leaf lamina ovate to broadly suborbicular, 3–12 × 2–15 cm, 3-lobed, rarely not lobed, lobes acute, apex acute base truncate or cordate, margins irregularly serrate, glabrous to glabrescent with prickles on the veins of the under surface, sinuses acute, midrib (and often the two adjacent veins) with a longitudinal fissure-like gland up to 3 mm long near the base; petiole 3–11 cm, glabrous but for a line of pubescence adaxially; stipules linear or filiform, 6–8 mm long, aculeate. Flowers solitary in the leaf axils; pedicel 1–6 cm long, articulated in the top quarter, shortly pubescent; epicalyx of 7–8 bracts 7–10 mm long, forked, the outer branch somewhat reflexed, shortly stellate-pubescent with scattered setae; calyx lobes lanceolate, 13–22 × 6–10 mm, acute, fused for half their length, hispid on the midrib and margins. Corolla 5–7 cm long, pubescent outside, glabrous inside; staminal tube 14–25 mm long; filaments 0.5–1 mm long; exserted part of styles 0.6–1 cm long, glabrous. Capsule spherical or elliptic, 20 mm long, 18 mm diameter, shorter than the calyx, densely appressed setose; seeds subreniform, brown, 4 × 3 mm, with pectinate scales on a background of concentric rings.

KENYA. Kilifi District: between Dzitsoni & Jaribuni, 21 Feb. 1989, *Luke & Robertson* 1666! & Cha Simba, 14 Aug. 1989, *Luke & Robertson* 1858! & Kombeni River, Kaya Rabai, 22 May 1990, *Luke & Robertson* 2266!

TANZANIA. Tanga District: Pongwe, Maweni, 5 Oct. 1966, *Faulkner* 3667! & Steinbruch Forest Reserve near Maweni, 31 Dec. 1969, *Botany Students in* DSM 1398!; Morogoro District: Kingolwira, Mkumbe Hill, July 1935, *Burtt* 5170!

DISTR. **K** 7; **T** 3, 6; not known elsewhere

HAB. Bush and forest on limestone; 70–500 m

SYN. *H. sp. aff. rostellatus* Guill. & Perr. of K.T.S.L.: 172 (1994)

NOTE. Close to *Hibiscus rostellatus* Guill. & Perr. from which it differs in its glabrous stem, glabrous or glabrescent leaves and shape of its leaf lobes, which are longer than broad.

7. **Hibiscus cannabinus** *L.*, Syst. Nat. ed. 10, 2: 1149 (1759); Harv. in Fl. Cap. 1: 176 (1860); Mast. in F.T.A. 1: 204 (1868); Hochr. in Ann. Conserv. Jard. Bot. Geneve, 4. 114 (1900); Baker f. in J.L.S. 40: 27 (1911); Engl., Pflanzenw. Afr. 3, 2: 400, fig 188 (1921); F.P.S. 2: 28 fig. 11 (1952); Exell in F.Z. 1 (2): 441 (1961); Hauman in F.C.B. 10: 108 (1963); Blundell, Wild Fl. E. Afr.: 76 (1987); Edmonds, Distr. *Hibiscus* Sect. *Furcaria* Trop. E. Afr.: 15 (1991); U.K.W.F.: 99 (1994); Vollesen in Fl. Eth. & Eritr. 2(2): 198 (1995); Thulin in Fl. Somalia 2: 44, fig. 25a–f (1999); Wilson in Bull. Nat. Hist. Mus. Lond. (Bot.) 29(1): 66 (1999). Neotype: *"Alcea Bengalensis spinosissima"* in Commelin, Hort. Med. Amstelod. Pl. Rar., 1: 35, t. 18, 1697, designated by Wijnands in *Bot. Commelins*: 144 (1983)

Annual herb up to 3 m tall; stems aculeate with small sparse prickles usually pointing upwards, otherwise nearly glabrous or with a longitudinal line of crisped pubescence changing its radial position at each node. Leaves narrowly ovate, ovate or suborbicular in outline, 3–17 × 3–20 cm, unlobed to 3–7-palmatisect to palmatilobed, apex acute, base broadly cuneate to shallowly cordate, margin serrate or dentate, rarely subentire, scaberulous or almost glabrous with a few minute prickles on the veins, usually with a prominent gland on the undersurface near the base of the midrib; petiole 6–22 cm long, prickly with a line of pubescence, like the stem; stipules narrowly linear to filiform, 4–5 mm long, caducous. Flowers solitary in the leaf axils, pedicel 2-6 mm long, articulated at base, aculeate or hispid; epicalyx of 7–8 linear bracts 5–10 × 1–1.5 mm, joined to calyx for about 2 mm at base; calyx lobes 10–20 × 3–5 mm, long-acuminate (sometimes subcaudate) joined for up to a third of their length from the base, aculeate or hispid outside especially near the margin, margin sometimes with a woolly tomentum, usually with a prominent gland 1.5–2 mm in diameter on midrib. Corolla deep purple, pale yellow or white, 3–8 cm long, whitish or greyish pubescent outside, glabrous within; staminal tube 2–4 cm long; filaments 1-3 mm long; exserted part of style 3–6 mm long, glabrous. Capsule ovoid-acuminate, 13–18 mm long, 9–12 mm in diameter, appressed setose; seeds irregularly subreniform, 3–3.5 × 1.5–2.5 mm, sparsely to densely covered with truncate scales.

UGANDA. Karamoja District: W of Apedet, 20 July 1957, *Hudson* 305!; Bunyoro District: Budongo forest, 3 Dec. 1938, *Loveridge* 178; Teso District: Serere, Oct. 1932, *Chandler* 961!

KENYA. Northern Frontier District: Moyale, 16 Aug. 1952, *Gillett* 13723!; Trans-Nzoia District: Kitale, 18 Sep. 1956, Bogdan 4304!; Narok District: Aitong, 3 June 1961, *Glover et al.* 1825!

TANZANIA. Arusha District: Olmotonyi, 1 June 1963, *Verdcourt* 3647A!; Kigoma District: Kungwe Mt, W of Kapalagulu near the Mugombasi R., 18 July 1959, *Newbould & Harley* 9005!; Morogoro District: Ruvu, 28 Apr. 1969, *Batty* 480!; Zanzibar: Vitongoge, 14 Oct. 1929, *Vaughan* 808!

DISTR. **U** 1–4; **K** 1–7; **T** 1–8; **Z**; most of Africa and extending to India

HAB. Grassland, bushland, wooded grassland and forest margins; 0–2000 m

SYN. *H. verrucosus* Guill. & Perr. var. *punctatus* A. Rich., Tent. Fl. Abyss. 1: 59 (1847). Type: Ethiopia, Chire, *Petit* s.n. (?P, holo.)

H. asper Hook. f., Niger Fl.: 228 (1849). Type: Sierra Leone, *Turner* s.n. (K!, holo.)

H. cordofanus Turcz. in Bull. Soc. Nat. Mosc. 31: 193 (1858). Type: Ethiopia, Kordofan, *Kotschy* 65 (K, holo.; W, iso.)

H. cannabinus L. var. *chevalieri* Hochr. in Ann. Conserv. Jard. Bot. Geneve 5: 125 (1901). Types: Mali, Koulikoro, middle Niger, 6–14 Oct. 1899, *Chevalier* s.n., and Mali, Sindou, *Chevalier* 856 (BR, syn.)

H. vanderystii De Wild. in B.J.B.B. 5: 35 (1915). Type: Angola, Kwango, *Vanderyst* 1377 (BR, holo.)

H. henriquesii P. Lima in Broteria Ser. Bot. 19: 138 (1921)

H. malangensis Baker f. in J.B. 77: 22 (1939). Type: Angola, Malange, River Cuango, near Xa Sengue, *Exell & Mendonça* 274 (BM, holo.)

NOTE. Following a suggestion by Borssum Waalkes (in *Blumea* 14: 63. 1966), Wijnands treated a Commelin plate as the lectotype. However, this is not cited in the protologue (though it was added in the later account in *Sp. Pl.*, ed. 2, 2: 979. 1763). It is therefore not original material for the name but, in the absence of any original material at all (sheet 875.27 (LINN) is original material for *Hibiscus sabdariffa* L.), Wijnands' statement is treated as correctable to a neotypification (Art. 9.8).

Several strains of *Hibiscus cannabinus* were introduced in the Amani plantations by the Germans in the early part of the twentieth century. Some of these were later introduced in other parts of the Flora area and have become naturalized. These are generally not aculeate or very sparsely so, and may have unlobed or palmatilobed leaves. Wild *Hibiscus cannabinus* appears to get less aculeate the further South from Kenya one travels.

Tanner 4303 (K) is probably an undescribed species, but the material available is too poor for a formal diagnosis.

8. **Hibiscus reekmansii** *F.D. Wilson* in Bull. Nat. Hist. Mus. Lond. (Bot.) 29(1): 64 (1999). Type: Burundi, Muramvya Province, Mpotsa road, *Reekmans* 7993 (BR, holo.)

Shrub to 2 m tall; stems aculeate, the aculei pointing up or down the stem, densely stellate-pubescent or with lines of stellate pubescence. Leaves ovate to obovate, 0.6–6 × 0.2–7 cm, unlobed to deeply 5–7-lobed, apex acute or acuminate, base cuneate, truncate or cordate, margin serrate to irregularly serrate, aculeolate below and above, the hairs stellate or simple; petiole 0.4–5 cm long, densely stellate-pubescent and aculeolate; stipules filiform or linear, 2–6 mm long, tomentose. Flowers solitary, axillary, or with 2–3 nodes clustered together near the stem apex, some flowers with a subtending linear flattened pink bract resembling the stipules; pedicel 2–6 mm long, articulated at base with simple, appressed hairs to 2 mm long; epicalyx of 6–8 linear bracts 6–10 mm long, flattened, with hairs like those of the pedicel; calyx 1–1.5 cm long, lobes triangular-acute to lanceolate-acuminate, gland small. Petals bright yellow to primrose yellow with a small purple centre, obovate, 3–4 × 2–2.5 cm, finely stellate-pubescent outside, glabrous inside; staminal column 16 mm long; filaments 0.5 mm long; style branches 5 mm long, densely appressed-pubescent. Capsule 12–15 mm long, 10–13 mm in diameter, densely appressed-pubescent; seeds subreniform, dark brown, 3 × 2 mm with minute protruberances.

TANZANIA. Iringa District: Mbosi, May 1935, *Horsbrugh-Porter* s.n.!
DISTR. **T** 7; eastern Congo-Kinshasa, Rwanda, Burundi
HABITAT. Grassland, rocky areas, fields and pastures; 1800–2400 m

9. **Hibiscus greenwayi** *Baker f.* in J.B. 75: 100 (1937); E.P.A.: 562 (1959); Edmonds, Distr. *Hibiscus* Sect. *Furcaria* Trop. E. Afr.: 18 (1991); Wilson in Bull. Nat. Hist. Mus. Lond. (Bot.) 29(1): 62 (1999). Type: Tanzania, Lushoto District: Mnazi, Umba Steppe, 12 Jan. 1930, *Greenway* 2034 (EA!, holo.; BM, K!, iso.)

Shrub 2–5 m high; stems glabrous, except the young parts which may be finely pubescent with stout and recurved prickles, 4 mm long, 3 mm wide at base, from a shield-shaped base, scattered, sometimes grouped in 4's or 5's; occasionally very sparse. Leaves 3–5-lobed, 4–9 × 3.5–12 cm, the lobes obovate, rounded at the apices, those at the stem apex sometimes not lobed, leaf base cordate, leaf margin entire, undulate or crenate-serrate; gland on midrib below near the base; stellate-hispid above, stellate-tomentose below; petiole 5–15 mm long, pubescent; stipules subulate, ± 5 mm long, stellate-hispid. Flowers solitary in the leaf axils; pedicel 5–10 mm, articulated at base, pubescent with scattered prickles; epicalyx bracts 6–10, rigid, linear, 6–15 × 0.5–2 mm, entire, stellate-tomentose; calyx lobes lanceolate, 17–24 × 4–6 mm, with a nectary on the midrib, glabrous, hispid or pubescent with stiff hairs along midrib and margins. Corolla yellow, 6–8 cm long, sparsely pubescent outside, glabrous inside; staminal tube dark red, 4.5–7 cm long; filaments 1–2 mm long; exserted parts of style 2–6 mm long, glabrous. Capsule conical, 12–22 mm long, 10–13 mm in diameter always shorter than calyx, densely setose; seeds pyramidal to wedge shaped, 3–5 × 2–3 mm, covered in linear scales. Fig. 6, p. 43.

UGANDA. Mengo District: Mabira Forest, 18 Feb 1918, *Dummer* 3908 (BM, fide Wilson)

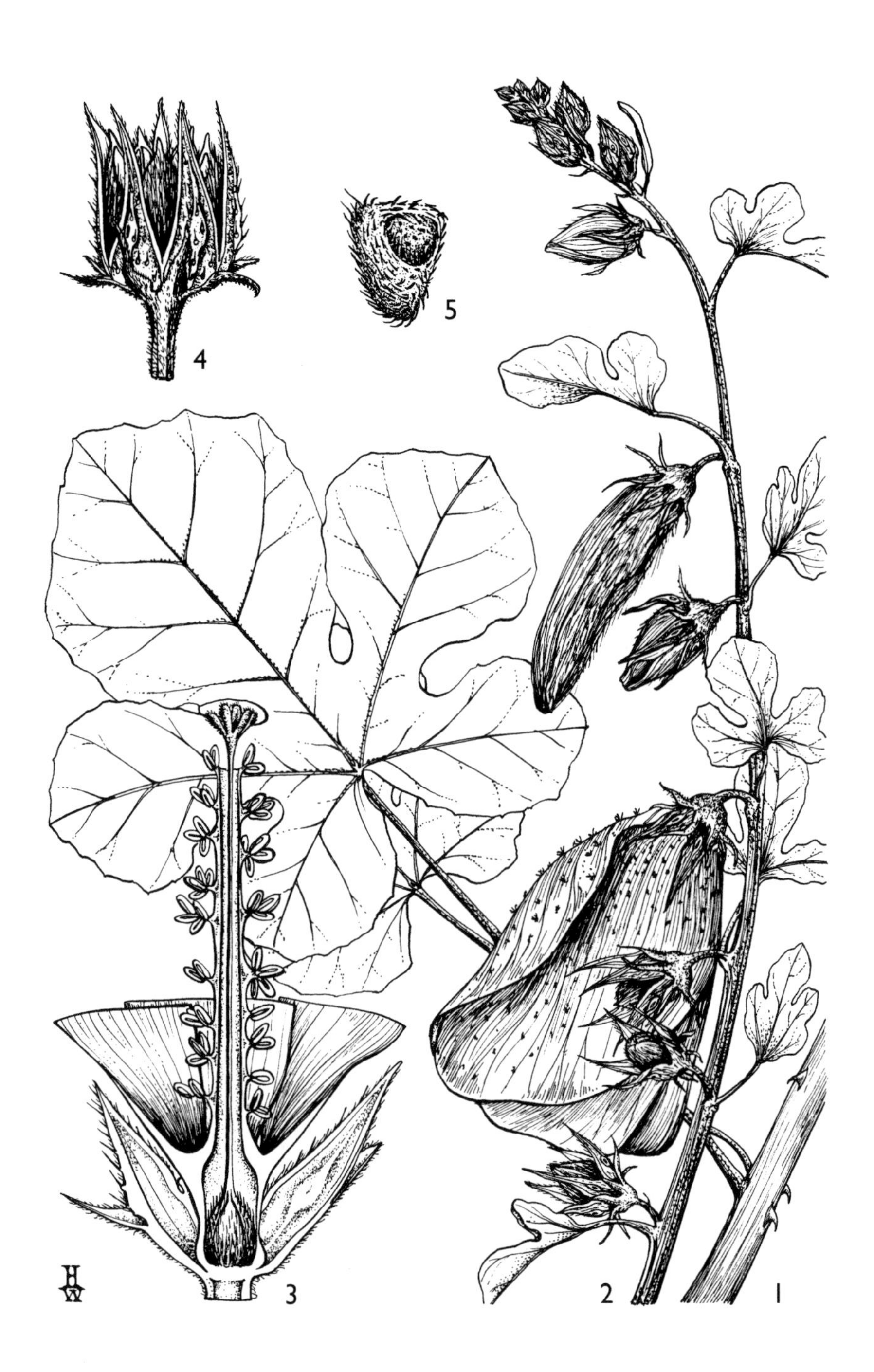

FIG. 6. *HIBISCUS GREENWAYII* — **1**, stem with leaves, × $^{2}/_{3}$; **2**, flowering stem, × $^{2}/_{3}$; **3**, longitudinal section of flower, × $1^{1}/_{2}$; **4**, fruit, × 1; **5**, seed, × 4. 1 from *Polhill & Paulo* 835; 2 from *Verdcourt* 3904; 3–5 from *Semsei* 2121. Drawn by Heather Wood.

KENYA. Kitui District: Mutomo, 24 Apr. 1969, *Napper & Kanuri* 2070!; Kilifi District: Marafa, 22 Nov. 1961, *Polhill & Paulo* 835!; Kwale District: Mackinnon Road, 80 km from Mombasa, 12 Jan. 1964, *Verdcourt* 3904!

TANZANIA. Lushoto District: Usambaras, Aug. 1893, *Holst* 3204!; Pare District: Mkomazi Game Reserve, Kisima Hill, 30 Apr. 1995, *Abdallah & Vollesen* 95/96!; Kigoma District: Buhanda Escarpment, Malagarasi River, July 1926, *Grant* s.n.!

DISTR. **U** 4; **K** 4, 7; **T** 3, 4; Ethiopia

HAB. *Acacia-Commiphora* bushland; 0–1400 m

10. **Hibiscus sparseaculeatus** *Baker f.* in J.B. 76: 22 (1938); E.P.A.: 568 (1959); Edmonds, Distr. *Hibiscus* Sect. *Furcaria* Trop. E. Afr.: 19 (1991); Vollesen in Fl. Eth. & Eritr. 2 (2): 196 (1995); Thulin in Fl. Somalia 2: 46, fig. 25g (1999); Wilson in Bull. Nat. Hist. Mus. Lond. (Bot.) 29(1): 62 (1999). Type: Somalia, Sheik Pass, *Freemantle* s.n. (BM!, holo.)

Shrub 1.2–5 m high; stems glabrous, pale brown, with a waxy appearance; prickles stout, 2–4 mm long, from shield-like base, sometimes 3-pronged or recurved. Leaves circular to broadly ovate, 10–35 × 8–45 mm, sometimes slightly trilobed, apex obtuse to truncate, base obtuse to slightly cordate, margin entire, serrate or shallowly dentate, midrib often with prominent extra floral gland on underside at base, subtomentose to subscabrid; petiole 6–35 mm long, subtomentose; stipules lanceolate, 2.5–3.5 × 0.5–1 mm, sparsely hispid. Flowers solitary in the leaf axils, pendulous; pedicel 7–10 mm long, articulated at base, tomentose and/or hispid; epicalyx of 7 or 8 narrowly triangular bracts, 5–12 × 0.3–0.5 mm, spreading, pubescent. Sepals narrowly triangular, 7–8 × 1–4 mm, each with a nectary midway along its central vein. Corolla yellow with brown or purple patch at base inside, 30–65 mm long, sparsely pubescent outside, glabrous inside; staminal column 27–40 mm long; filaments 1 mm long; exserted part of style 4–7 mm long, glabrous or puberulous. Capsule 15–17 mm long, 12–13 mm in diameter, setose; seeds 4–5-angular, 4–5 × 3 mm, with appressed ovate, truncate scales.

KENYA. Northern Frontier District: Furroli Mt, 15 Sept. 1952, *Gillett* 13874!; Kitui District: Kangonde–Embu road, 8 May 1960, *Napper* 1657!; Teita District: SE of Manda Hill, NE Murka area, 24 July 1969, *Gilbert* 4073!

TANZANIA. Mbulu District: N end of Lake Eyasi, 23 July 1957, *Bally* B11588!

DISTR. **K** 1, 3, 4, 7; **T** 2, 3; Ethiopia, Somalia

HAB. Rocky slopes in deciduous woodland; 1000–1900 m

SYN. *H. greenwayi* Cufod., E.P.A.: 562 (1959), *non* Baker f. (1937) *pro parte quoad* specimen *Bally* 9128

H. greenwayi Baker f. var. *megensis* J-P. Lebrun in Adansonia 2, 15: 379 (1976). Type: Kenya, Northern Frontier District: Nyiro, *Haylett* 12 (K!, holo.)

11. **Hibiscus mechowii** *Garcke* in Linnaea 43: 121 (1881); C.F.A. 1, 1: 169 (1937); Exell in F.Z. 1(2): 443 (1961); Hauman in F.C.B. 10: 113 (1963); Edmonds, Distr. *Hibiscus* Sect. *Furcaria* Trop. E. Afr.: 19 (1991); Wilson in Bull. Nat. Hist. Mus. Lond. (Bot.) 29(1): 61 (1999). Type: Angola, Cuanza Norte, *Mechow* 105 (BM!, illustration)

Annual or perennial herb up to 3 m tall, stems with small rather sparse prickles and sometimes with an additional longitudinal line of crisped pubescence changing its radial position at each node. Leaves obovate to broadly obovate in outline, 7.5–15 × 6.5–13 cm, 3–5-palmatipartite, the lobes very narrowly elliptic, apex acute, cuneate to obtuse at base, margins serrate, sparsely pilose or hispid on both surfaces, midrib with longitudinal fissure-like gland 2–3 mm long near the base; petiole 3.5–12 cm long, hairy like the young stems; stipules linear, 7–12 × 1–1.5 mm. Flowers solitary in the leaf axils; pedicel 8–25 mm long, articulated just below the middle, densely setose or setose-aculeate above the articulation, less densely so and more pubescent beneath it; epicalyx of 7–8 linear bracts 12–20 × 1–1.5 mm, hispid and/or aculeolate, free. Sepals triangular, 2–3.5 cm long, 4–6 mm wide, pilose-hispid and aculeolate especially on the margins and midrib, joined at base. Corolla yellow with maroon base, 3–4 cm long, sparsely stellate-pubescent; staminal tube 20 mm long; filaments 0.5 mm long; exserted parts of style ± 1 mm long. Capsule ovoid, 16–17 mm long, 12–13 mm in diameter, appressed-setose; seeds subreniform, 3.5–4 × 2–3 mm with concentric rings of minute scales made of the fused bases of long white hairs.

TANZANIA. Shinyanga District: Shinyanga, Nov. 1938, *Koritschoner* 2051!; Mpanda District: 5 km on Iruwira–Inyonga road, 3 June 2000, *Bidgood* et al. 4594!; Kigoma District: Lukoma, 29 May 1975, *Kahurananga et al.* 2688!

DISTR. **T** 1, 4, 5; Central African Republic, Congo-Kinshasa, Angola, Mozambique, Zimbabwe, Botswana, Namibia

HAB. Old pasture and fallow fields; 800–1100 m

SYN. *H. lancibracteatus* De Wild. & T. Durand in B.S.B.B. 38(2): 25 (1899). Type: Congo-Kinshasa, Mbandaka [Coquilhatville], *Dewevre* 752 (BR, holo.)

H. cannabinus sensu Hochr. in Ann. Cons. Jard. Geneve 4: 114 (1900) *non* L., *pro parte*

12. **Hibiscus diversifolius** *Jacq.*, Collectanea Botanicum 2: 307 (1789); Ic. Pl. Rar. 3: t. 551 (1792); Edwards, Bot. Reg. 5: t. 381 (1819); Harv. in Fl. Cap. 1: 171 (1860); Mast. in F.T.A. 1: 198 (1868); Hochr. in Ann. Conserv. Jard. Bot. Geneve, 4 (1900); Baker f. in J.L.S. 40: 27 (1914); Eyles in Trans. Roy. Soc. S. Afr. 5: 415 (1916); Engl., Pflanzenw. Afr. 3, 2: 402, fig. 187 E-G (1921); C.F.A. 1, 1: 173 (1937); Garcia in Bol. Soc. Brot. Ser. 2, 20: 40 (1946); Mendonça & Torre, Contr. Conhec. Fl. Moçamb. 1: 14 (1950); F.P.S. 2: 24, fig. 10 (1952); Brenan in Mem. N.Y. Bot. Gard. 8, 3: 225 (1953); Exell in F.Z. 1: 443 (1961); Edmonds, Distr. *Hibiscus* Sect. *Furcaria* Trop. E. Afr.: 17 (1991); Vollesen in Fl. Eth. & Eritr. 2(2): 196 (1995); Wilson in Bull. Nat. Hist. Mus. Lond. (Bot.) 29(1): 63 (1999). Neotype: Ic. Pl. Rar. 3, t. 551! (1792), chosen by Fryxell, 1988

Small tree, shrub or scrambling perennial herb up to 10 m tall; stems stellate-tomentose or densely stellate-pubescent and aculeolate. Leaves suborbicular in outline, 4–6 × 5–16 cm, ± distinctly 3–7-palmatilobed or -palmatipartite, apex acute or rounded, base truncate to cordate, usually with a linear or elliptic fissure or a suborbicular to elliptic gland with a central fissure on the lower surface near the base of the midrib, margin irregularly serrate or crenate serrate, stellate-pubescent above, stellate-tomentellous beneath; petiole 2–12 cm long, hairy like the young stems; stipules linear to filiform, 3–4 mm long, hispid. Flowers solitary, axillary or in terminal racemes; pedicel 5–7 mm long, densely hispid, articulated near the base; epicalyx of 7–8 free, linear bracts 8–12 × 1–2 mm; calyx lobes lanceolate, 2–3 cm long, 3–6 mm wide, densely setose, fused in the basal quarter, with an elliptic gland on the median vein. Corolla yellow or purplish with dark red or purple centre, 3–6 cm long, stellate-pubescent on the outside, glabrous inside; staminal tube 15–20 mm long; filaments 0.3–0.5 mm long; exserted part of style 6–7.5 mm long, glabrous or very sparsely pilose. Capsule ovoid-acute, 2 cm long, 1.5 cm diameter, densely setose; seeds brown, subreniform, 4 × 2–3 mm, glabrous.

subsp. **diversifolius**; Exell in F.Z. 1: 444 (1961); Edmonds, Distr. *Hibiscus* Sect. *Furcaria* Trop. E. Afr.: 17 (1991); Wilson in Bull. Nat. Hist. Mus. Lond. (Bot.) 29(1): 63 (1999)

Stems usually with one or more longitudinal lines of pubescence, sometimes nearly glabrous and aculeolate with short stout conical prickles. Flowers yellow with maroon centre.

UGANDA. Kigezi District: Lake Mutanda, Oct. 1947, *Purseglove* 2501!; Toro District: 48 km SW of Fort Portal on Kasese Road, 11 March 1969, *Jones* 6998!; Mengo District: Manve Swamp, Mar. 1932, *Eggeling* 484!

KENYA. Naivasha District: Olchoro Oirooiuwa Gorge, 21 Oct. 1952, *Glover & Samuel* 3348!; Nairobi District: Kirichwa Ndogo Valley, 10 June 1944, *Bally* 3407!; South Kavirondo District: 17 km NW of Kisii on Kisumu Road, 1 Mar. 1969, *Jones* 6981!

TANZANIA. Mwanza District: near Ukiriguru, not dated, *Smith* 580!; Kigoma District: near Mugombasi R., N of Mahali Peninsula, 18 July 1959, *Newbould & Harley* 9011!; Songea District: Ndengo, 29 Sep. 1956, *Semsei* 2490!
DISTR. **U** 2, 4; **K** 3–6; **T** 1, 2, 4, 7, 8; tropical and South Africa; Madagascar, India, Australia, Pacific islands
HAB. Swamps, dranage lines in forest and grassland; 1000–2200 m

SYN. *H. scaber* Lam., Encycl. 3: 350 (1792). Type: Mauritius, *Commerson* s.n. (P-LA, holo.)
H. biflorus A. Spreng., Tent. Suppl.: 19 (1828). Type: South Africa, *Zeyher* 241 (BM, drawing)
H. decaisneanus Hochr. in Ann. Conserv. Jard. Bot. Geneve 4: 119 (1900). Type: Ethiopia, Tigre v. Begemder, *Schimper* 1479 (BM, iso.)
H. diversifolius Jacq. var. *angustilobus* Hauman in B.J.B.B. 31: 86 (1961) excl. specim. *Laurent* 11 and *de Witte* 2435. Type: Congo-Kinshasa, Upper Katanga: Sakala Marungu, *Dubois* 1157 (BR, holo.)

subsp. **rivularis** (*Bremek. & Oberm.*) *Exell* in F.Z. 1: 444 (1961); Edmonds, Distr. *Hibiscus* Sect. *Furcaria* Trop. E. Afr.: 17 (1991); Wilson in Bull. Nat. Hist. Mus. Lond. (Bot.) 29(1): 63 (1999). Type: Botswana, Chobe R., Kabulabula, July 1930, *van Son* in Herb. Transv. Mus. 28936 (BM, holo; PRE, iso.)

Stems more densely hairy, more or less uniformly stellate-pubescent to stellate-tomentose. Flowers reddish to purple with dark purple centre.

UGANDA. Ankole District: Igara, Mar. 1939, *Purseglove* 612!; Masaka District: Buddu, 1903, *Dawe* 38!
KENYA. Naivasha District: Korongo Farm, 26 Aug. 1978, *Kerslake & Barrow* 1!; Meru Distict: between Chuka & Embu, 24 July 1936, *Wilkinson* 3266; South Nyeri District: Thiba R., 4 Nov. 1939, *Copley* 377!
TANZANIA. Mwanza District: Mbarika, 13 Sep. 1952, *Tanner* 3750!; Arusha District: Bounday, Senato, Arusha National Park, 16 Sep. 1971, *Arasululu in Richards* 27230!
DISTR. **U** 2–4; **K** 3–4; **T** 1, 2, 4, 7; Angola, Zambia, Malawi, Mozambique, Botswana
HAB. Forest edges, papyrus swamp and other lakeside vegetation; 1400–1900 m

SYN. *H. rivularis* Bremek. & Oberm. in Ann. Transv. Mus. 16, 3: 424 (1935)

13. **Hibiscus berberidifolius** *A. Rich.*, Tent. Fl. Abyss.1: 56 (1847); Hauman in F.C.B. 10: 112 (1963); Edmonds, Distr. *Hibiscus* Sect. *Furcaria* Trop. E. Afr.: 18 (1991); Vollesen in Fl. Eth. & Eritr. 2(2): 196 (1995); Wilson in Bull. Nat. Hist. Mus. Lond. (Bot.) 29(1): 61 (1999). Type: Ethiopia, Sanka Berr, *Quartin Dillon & Petit* 119 (P, lecto.; BR, isolecto.)

Robust erect perennial shrub, 1–4 m tall; stems aculeate, sparsely to densely and uniformly pubescent. Leaves unlobed or shallowly 3-lobed or ovate, 4–8.5 × 1.3–6 cm, apex acute, base truncate to cuneate, margin coarsely serrate, leaf nectary present on midrib at base of abaxial surface; petiole 0.3–3.5 cm long, sparsely aculeate and densely pubescent; stipules linear, 2–4 mm long, pubescent, caducous. Flowers solitary in the leaf axils; pedicel 5–8 mm long, articulated at base, pubescent-hispid; epicalyx of 6–7 linear bracts, 9–13 × 1–2 mm, pubescent or pubescent-hispid, obscurely several-veined; calyx lobes acuminate, 1.4–2.2 cm long, densely hispid in the fused parts, pubescent toward the apex, nectary present or absent on the median vein. Corolla white to deep purple via pale yellow, yellow, pink or red, 3–5.5 cm long, glabrous to sparsely pubescent; staminal tube 2–4 cm long; filaments 2–3 mm long; exserted part of style 6–10 mm long, glabrous. Capsule broadly ovoid, 1.3–1.7 cm long, 1.1–1.7 cm in diameter, densely setose; seeds subreniform 4 × 3 mm, the surface covered with scales.

UGANDA. Ankole District: Ibanda, Jan. 1926, *Maitland* 900!; Toro District: Mubuku Valley, Mt Ruwenzori National Park, 19 Feb 1997, *Lye* 22606!; Mt Elgon, near Sipi, Oct. 1939, *Dale* U13!
KENYA. Northern Frontier District: Maralal, 3 Oct. 1935, *Leakey* 8583!; Trans-Nzoia District: Kitale, 15 Aug. 1956, *Bogdan* 4214!; Kericho District: Kimugu tea estate, Kericho, 4 Dec. 1967, *Perdue & Kibuwa* 9261!

TANZANIA. Ufipa District: Mbesi forest, Sumbawanga, 19 July 1962, *Richards* 16828!; Iringa District: 16 km S of Sao Hill, 18 Aug. 1949, *Greenway* 8427!; Njombe District: Igeri, 19 Dec. 1967, *Robertson* 758!

DISTR. **U** 2, 3; **K** 1, 3, 5; **T** 4, 7; Congo-Kinshasa, Rwanda, Ethiopia

HAB. Forest and forest edges; 1550–3000 m

SYN. *H. diversifolius* Jacq. var. *witteanus* Hochr. in Robyns, B. J. B. B. 18: 276 (1947). Type: Congo-Kinshasa, Kilisti, route de Kibumba, *de Witte* 1320 (G, holo.; BR, iso.)

H. parvilobus F.D. Wilson in Bull. Nat. Hist. Mus. Lond. (Bot.).: 67 (1999). Type: Kenya, Nakuru District: 9 km NE of Londiani, 8 Feb. 1973, *Spjut & Ensor* 3184 (EA!, holo.; BR, K!, iso.)

14. **Hibiscus micranthus** *L. f.*, Sp. Pl. Suppl. 308 (1781); U.O.P.Z.: 296 (1949); F.P.S. 2: 28 (1952); Keay, F.W.T.A. ed. 2, 1(2): 346 (1958); E.P.A.: 564 (1959); Hauman in F.C.B. 10: 124 (1963); Blundell, Wild Fl. E. Afr.: 77 (1987); U.K.W.F. ed 2: 98 (1994); Vollesen in Fl. Eth. & Eritr. 2(2): 210 (1995); Thulin in Fl. Somalia 2: 52, fig. 30 (1999). Type: India, Coromandel, *Konig* s.n. (LINN holo.; K!, microfiche)

Wiry herb 0.4–3 m tall; stems stellate-hispid, occasionally stellate-pubesent, becoming glabrous and longitudinally ridged with age. Leaves ovate, elliptic or lanceolate, 0.6–3 × 0.3–2.5 cm, apex acute, obtuse or rounded, base cuneate, rounded or truncate, with serrate margins, stellate-hispid above and below; petiole 1–11 mm long, sparsely to densely stellate-pubescent; stipules filiform or subulate, 2–6 × 0.2–1 mm, hispid. Flowers solitary in the leaf axils; pedicel 0.3–2.5 cm long, articulated in the upper half, densely to sparsely pubescent, ± glandular above articulation; epicalyx of 5–8 subulate, linear or lanceolate bracts 1.5–2 × 0.3–0.5 mm, pubescent and/or hispid; calyx lobes lanceolate or triangular, 2.5–4.5 × 1–1.5 mm, fused in lower half, densely hispid. Corolla white, fading to pink, 2–6 mm long, stellate-hispid on the outside, glabrous inside; staminal tube 1.5–5 mm long; filaments 0.1–0.3 mm long; exserted part of styles 0.5–1.5 mm long, glabrous. Capsule spherical, 5–7 mm in diameter, sparsely pubescent; seeds reniform, black, 2–3 × 1.5–2 mm, the hairs up to 6 mm long. Fig. 7: 7–9, p. 48.

UGANDA. Karamoja District: Lodoketeminit, 14 Apr. 1959, *Kerfoot* 922! & NW foothills of Debasien, Mar. 1962, *Tweedie* 2330!; Bunyoro District: Bukumi–Butiaba, 17 Oct. 1970, *Katende* 694!

KENYA. Baringo District: Kinyang, 10 July 1976, *Gitonga* 59!; Machakos District: Athi River, near Kibwezi, 25 May 1959, *Napper* 1260!; Kwale District: S of Kinango on Mombasa–Tanga road, 15 Nov. 1962, *Greenway* 10849!

TANZANIA. Mwanza District: Nyambiti, 9 Mar. 1953, *Tanner* 1263!; Tanga District: Tanga, 4 Nov. 1929, *Greenway* 1835! Bagamoyo District: Kisarawe Forest Reserve, 17 Oct. 1965, *Mgaza* 696!; Zanzibar: Zanzibar, 1927, *Toms* 1!

DISTR. **U** 1–4; **K** 1–7; **T** 1–6, 8; **Z**; widespread in tropical and South Africa; Arabia and India

HAB. Grassland and bushland; 0–1700 m

SYN. *H. micranthus* L. f. var. *micranthus* L. f., E.P.A.: 564 (1959)

H. micranthus L. f. race 1 of U.K.W.F.: 196 (1974)

NOTE. This is a widespread and a very variable species. I am however unable to classify the variation, as it seems largely continuous.

15. **Hibiscus meyeri** *Harv.*, Fl. Cap. 1:173 (1861); Exell in F.Z. 1(2): 454 (1961); U.K.W.F. ed. 2: 99 (1994). Type: South Africa, valley of River Omblas, *Drege* s.n. (TCD, holo.)

Erect herb 0.5–2 m tall; stems densely to sparsely stellate-pubescent and/or stellate-hispid, becoming glabrous with age, the bases of the hairs persisting as tubercules. Leaves ovate, narrowly ovate or lanceolate, 1.5–6.3 × 0.4–3 cm, apex rounded, obtuse or acute, base cuneate, rounded or cordate, margins serrate,

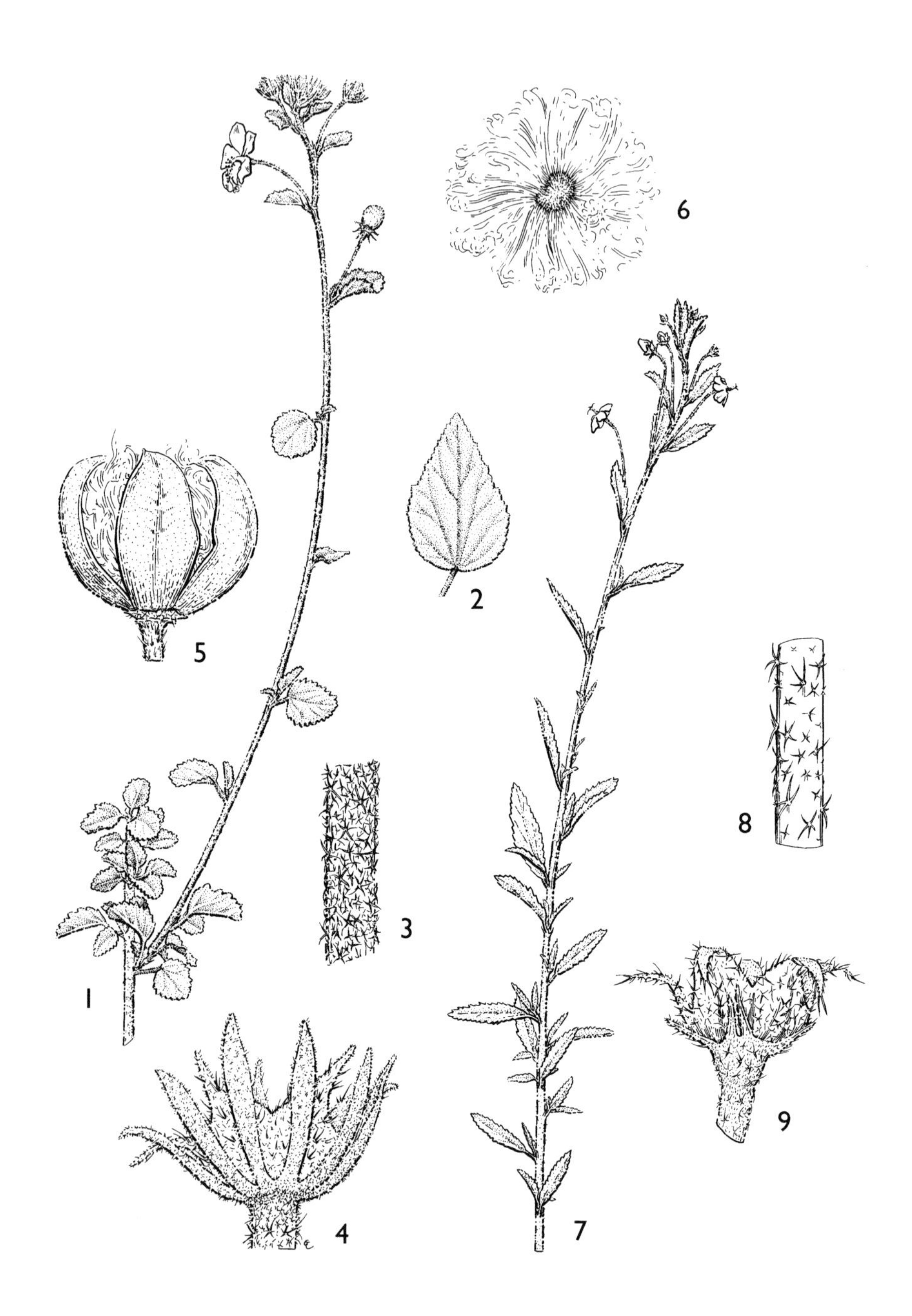

FIG. 7. *HIBISCUS CRASSINERVIUS* — **1**, flowering stem, × $^1/_2$; **2**, leaf, × $^1/_2$; **3**, stem indumentum, × 5; **4**, epicalyx and calyx, × 4; **5**, capsule, × 2; **6**, seed, × 2. *HIBISCUS MICRANTHUS* — **7**, flowering stem, × $^1/_2$; **8**, stem indumentum, × 5; **9**, epicalyx and calyx, × 4. 1, 3, 4 from *De Wilde et al.* 8344; 2, 5, 6 from *Mooney* 5446; 7–9 from *Ash* 1606. Drawn by Eleanor Catherine, and reproduced with permission from Flora of Ethiopia and Eritrea 2, 2.

densely to sparsely stellate-pubescent and stellate-hispid below, stellate-pubescent above, the veins raised on the lower surface, with a gland present on the midrib near the base; petiole 2–13 mm long, sparsely to densely pubescent and/or hispid; stipules linear to subulate or filiform, 4–8 × 0.2–1 mm, ciliate. Flowers solitary in the leaf axils; pedicel 1–2 cm long, articulated in the middle, shortly stellate-hispid, more densely so above the articulation; epicalyx of 7–8 linear bracts 2–3 × 0.2–0.5 mm, pubescent; calyx lobes narrowly triangular, 5–7 × 1–2 mm, acuminate, fused in the lower third, pubescent and hispid. Corolla pink or mauve, 10–16 mm long, stellate-hispid on the outside, glabrous inside; staminal tube 8–14 mm long, lower third ± filament-free; filaments 0.5–1 mm long; exserted part of styles 1–3 mm long, glabrous. Capsule spherical, 7–8 mm in diameter, sparsely pubescent; seeds subreniform, 3 × 2 mm, brown, the hairs 3–6 mm long.

subsp. **meyeri**

Indumentum dense.

UGANDA. Karamoja District: Amudat, April 1961, *Tweedie* 2126! & Lodoketeminit, 9 Nov. 1962, *Kerfoot* 4479!; Toro District: Channel track, near Mweya, Queen Elizabeth National Park, 9 Sep. 1965, *Lock* 698!

KENYA. Northern Frontier District: track leading up valley E of Nguronit Mission Station, 9 Aug. 1979, *Gilbert et al.* 5579! Machakos District: Nguungi Hill, 3.5 km N of Kangonde, 18 May 1969, *Kimani* 187!; Masai District: 16 km N of Magadi on Nairobi road, 9 Oct. 1955, *Greenway* 8846!

TANZANIA. Mwanza District: Ukerewe island or near, 1 Mar. 1928, *Conrads* 169!; Mbulu District: Chemchem river, Lake Manyara National Park, 12 Nov. 1963, *Greenway & Kirrika* 11001!; Iringa District: Kimiramatonge Hill, 15 Dec. 1970, *Greenway & Kanuri* in EA 14811!

DISTR. **U** 1–2; **K** 1–7; **T** 1–7; Congo-Kinshasa, Burundi, Ethiopia, Somalia, Mozambique, South Africa

HAB. Grassland and bushland; 550–2000 m

SYN. *H. pycnostemon* Hochr. in B.J.B.B. 18: 274 (1947), F.C.B. 16: 124 (1963); Thulin in Fl. Somalia 2: 52 (1999). Type: Congo-Kinshasa, Valley of the Rutshuru, Buhembo, *de Witte* 1062 (BR, holo.)

H. pospischilii Cufod. in Ann. Nat. Hist. Mus. Wien 56: 54 (1948). Type: Kenya, Nairobi/Masai District: Athi Plains, *Pospischil* s.n. (?FT, holo.)

H. mendonçae Exell in Bol. Soc. Brot. Ser. 2. 33: 171 (1959). Type: Mozambique, Maputo [Lorenço Marques], Goba, *Mendonça* 3019 (LISC, holo.)

H. micranthus L. var. *grandifolius* Fiori in B.J.B.B. 29: 564 (1959). Type: Ethiopia, Embatcalla, *Fiori* 644 (?FT, holo.)

H. micranthus L. race 2, U.K.W.F.: 196 (1974)

NOTE. Our plants are referable to subsp. *meyeri*, as subsp. *transvaalensis* (Exell) Exell, widespread in South Africa, is less hairy and the bracts of the epicalyx are usually larger.

16. **Hibiscus fuscus** *Garcke* in Bot. Zeit. 7: 854 (1849); Cufod. in B.J.B.B. 59: 562 (1959); Exell in F.Z. 1(2): 449 (1961); Hauman in F.C.B. 10: 127 (1963); Blundell, Wild Fl. E. Afr.: 76 fig.56 (1987); U.K.W.F. ed. 2, 98 (1994); Vollesen in Fl. Eth. & Eritr. 2(2): 206 (1995). Type: South Africa, Natal, *collector unknown* (B†, holo.); South Africa, Natal, Msinsin, Umzinto, *Strey* 4456 (K!, neo., chosen here)

Herb 1–4 m tall; stems densely brown-stellate-setose, bark becoming longitudinally ridged with age. Leaves ovate, 3–9 × 1.4–7.4 cm, unlobed or shallowly 3-lobed, apex rounded or acute, base cuneate or truncate, margins crenate or serrate, simple pubescent above, densely stellate-pubescent below, with or without scattered stellate setae; petiole 0.4–6 cm long, densely brown-stellate-hispid and pubescent; stipules linear-lanceolate, 5–9 × 0.5 mm, hispid. Flowers solitary in the leaf axils; pedicel 2–6 cm long, articulated in the upper third, sparsely to densely brown-stellate-setose, more densely setose above articulation; epicalyx of 8–11 linear or filiform bracts 5–12 × 0.2–0.5 mm, pubescent and hispid; calyx lobes lanceolate, 10–13 × 3–4 mm, sepals fused in the lower third, densely brown-stellate-setose outside, pubescent inside. Corolla white, mauve or pink, 12–25 mm long, sparsely stellate-hispid on the outside, glabrous inside; staminal column 10–18 mm long with up to 12 sterile staminodes at its base; filaments 1–2 mm long; exserted part of styles 1–3 mm long, glabrous or pilose. Capsule ovoid, 9–12 mm long, 9–10 mm in diameter, sparsely tomentose; seeds reniform, brown, 2–3 × 1.5–2 mm, the hairs 5–7 mm long.

a. subsp. **fuscus**

Corolla white, not reflexed when mature; flower concave; staminal column up to 15 mm long.

UGANDA. Karamoja District: Mt Debasien, 1936, *Eggeling* 2728!; Kigezi District: Kachwekano Farm, May 1949, *Purseglove* 2818!; Mbale District: Budadiri, Jan. 1932, *Chandler* 486!
KENYA. Northern Frontier District: Mathews Range, Kichich, 6 Dec. 1960, *Kerfoot* 2437!; Trans Nzoia District: Cherangani foothills, 23 Sep. 1949, *Maas Geesteranus* 6358!; Kiambu District: Kabete, 22 Aug. 1935, *Edwards* 2968!
TANZANIA. Bukoba District: Bunazi, 17 Feb. 1936, *Gillman* 507!; Dodoma District: Salengo Forest, 25 Mar. 1974, *Richards & Arasululu* 29047!; Iringa District: Mufindi, 5 May 1968, *Paget Wilkes* 24!
DISTR. **U** 1–4; **K** 1–7; **T** 1–7; Congo-Kinshasa, Rwanda, Burundi, Ethiopia, Malawi, Mozambique, Zimbabwe, South Africa
HAB. Grassland, bushland, thicket, and forest; 1000–2500 m

SYN. *H. ferrugineus* sensu Hochr. in Ann. Conserv. Jard. Bot. Geneve 4: 84 (1900) *pro parte*, *non* Cav.

NOTE. A neotype is selcted here after efforts to locate an isotype of the material destroyed in Berlin proved unsuccessful.

b. subsp. **naivashense** *Mwachala* **subsp. nov**. a subsp. *fusco* distinguenda petalis majoribus columnisque staminiferis majoribus. Type: Kenya, Naivasha District: escarpment 6.5 km N of Naivasha, *Williams* in EA 12328 (EA!, holo.; K!, iso.)

Corolla mauve or pink, rarely white, petals reflexed when mature so that the flower is convex; staminal column 16–19 mm long.

KENYA. Naivasha District: Mennell Farm, W side of Lake Naivasha, 12 Mar. 1972, *Gillett* 19608! & S side of Lake Naivasha, 27 June 1969, *Polhill* 114/1A!; Machakos District: Kiu, 30 Aug. 1972, *Kennox* 123!
DISTR. **K** 3–4; not known elsewhere
HAB. Grassland; 1800–2200 m

17. **Hibiscus flavifolius** *Ulbr.* in N.B.G.B. 7: 367 (1920); E.P.A.: 561 (1959); Blundell, Wild Fl. E. Afr.: 76 fig. 55 (1987); U.K.W.F. ed. 2, 99 (1994); Vollesen in Fl. Eth. & Eritr. 2 (2): 208 (1995). Type: Tanzania, Masai District: Ngare-Olmotonyi towards Neirascherasch, *Uhlig* 405 (K!, holo.)

Erect herb 0.5–3 m tall; stems densely stellate-pubescent when young, becoming glabrous and longitudinally ridged with age. Leaves ovate to broadly ovate-elliptic, 1.5–6 × 1–4.5 cm, apex acute, obtuse or rounded, base truncate or cuneate, margins crenate or serrate, densely stellate-pubescent above and below; petiole 0.5–2 cm long, stellate-pubescent; stipules subulate, 4–10 × 1 mm, sparsely to densely stellate-tomentose. Flowers solitary in the leaf axils; pedicel 0.5–3.5 cm long, articulated in the upper third, often geniculate at the articulation, rusty-brown pubescent above articulation; epicalyx of 8–12 linear bracts 4–5 × 0.6–1 mm, densely pubescent; calyx lobes lanceolate-acuminate, 8–11 × 2–4 mm, fused in the lower half, densely pubescent. Corolla white, 10–17 mm long, stellate-pubescent on the outside, glabrous on the inside; staminal tube 7–12 mm long; filaments 1–1.5 mm long; exserted part of styles 2 mm long, glabrous. Capsule spherical, 9–10 mm in diameter, pubescent, setose along the sutures; seeds reniform, black, 2–3 × 1.5–2 mm, the hairs white, 5–7 mm long.

UGANDA. Karamoja District: Amudat, 11 June 1959, *Symes* 531! & Kaabong, Apr. 1960, *Wilson* 884! & Timu, Dodoth county, Sep. 1963, *Wilson* 1509!
KENYA. Northern Frontier District: Moyale, 15 km out on Wajir Road, 6 Nov. 1952, *Gillett* 14146!; Laikipia District: 1.6 km S of Kelole dam, 7 Apr. 1969, *Magor* 58!; Masai District: steamjet ridge near Stafford's Camp, Suswa, 1 Jan. 1963, *Glover et al.* 3428!
TANZANIA. Musoma District: Dabaka, about 137 km from Mwanza on Musoma road, 14 Nov. 1962, *Verdcourt* 3303!; Arusha District: Monduli Plateau, 5 Dec. 1944, *Bally* 4125! & Sanya Juu–Engare Nanyuki road, 9 Apr. 1965, *Richards* 20115!
DISTR. **U** 1, 3; **K** 1–4, 6; **T** 1, 2; Ethiopia, Somalia
HAB. Woodland and grassland; 750–2200 m

18. **Hibiscus rhodanthus** *Gürke* in Bull. Herb. Boiss. III: 405 (1895); Exell in F.Z. 1(2): 457 (1961); Hauman in F.C.B. 10: 128 (1963). Type: Angola, Pungo Andongo, near Gintage, *Welwitsch* 4901 (?B†, holo; BM!, K!, LISU, iso.)

Suffrutex 15–60 cm tall, from a tuberous rootstock; stems densely to sparsely stellate-pubescent and/or setose, becoming glabrous and longitudinally ridged with age. Leaves narrowly elliptic, 2.8–7 × 1.8–3.2 cm, apex acute, obtuse or rounded, base cuneate or rounded, margins serrate, stellate-setose above and below; petiole 4–7 mm long, stellate-setose; stipules linear 3–4 × 0.5 mm, setose. Flowers solitary in the leaf axils; pedicel 1.2–5 cm long, articulated in the upper half, stellate-setose; epicalyx of 8–10 linear or narrowly triangular bracts 2–3 × 0.5–0.8 mm, setose; calyx lobes triangular 5–7 × 2–3 mm, fused in the lower third, setose. Corolla red, 15–20 mm long, stellate-setose outside, glabrous inside; staminal tube 7–12 mm long; filaments 1–3 mm long; exserted part of styles 3–4 mm long, glabrous. Capsule spherical, 5–10 mm in diameter, sparsely simple setose; seeds reniform, black, 3–4 × 2 mm, the hairs brown, 5–8 mm long.

UGANDA. Ankole District: Mbarara, Dec. 1925, *Maitland* 42!
TANZANIA. Kigoma District: Mt Livandabe [Lubalisi], 31 May 1997, *Bidgood et al.* 4229!; Chunya District: Lupa Forest Reserve, 153 km N of Mbeya–Itigi road, 8 Sep. 1962, *Boaler* 651!; Songea District: Kiteza Hill, Mpapa, 16 Oct. 1956, *Mgaza* 106!
DISTR. **U** 2–3; **T** 4, 7–8; Congo-Kinshasa, Angola, Zambia, Malawi, Mozambique, Zimbabwe, South Africa
HAB. Plateau grassland and woodland; 1250–2200 m

SYN. *H. welwitschii* Hiern in Cat. Afr. It. Welw. 1: 75 (1896). Type: Angola, Pungo Andongo, Gintage, *Welwitsch* 4901. Type as for *Hibiscus rhodanthus* Gürke
H. carsonii Baker in K.B. 1897: 244 (1897). Types: Malawi, Fort Hill, *Whyte* s.n. (K!, syn.); Zambia, Fwambo, *Carson* s.n. (K!, lecto.)
H. cornetii De Wild. & T. Durand in Bull. Soc. Roy. Belg. 38, 2: 18 (1899). Type: Congo-Kinshasa, upper Lualaba, 1897, *Cornet* s.n. (BR, holo.)

19. **Hibiscus somalensis** *Franch.*, Revoil Çomali: 17 (1882); E.P.A.: 568 (1959); Vollesen in Fl. Eth. & Eritr. 2(2): 208 (1995). Type: Somalia, Warsangeli Mts [Oursangelicus], *Revoil* s.n. (P, holo.)

Shrub 10–60 cm tall, from a tuberous rootstock; stems stellate-hispid, becoming glabrous with age. Leaves elliptic or ovate, 19–38 × 6–20 mm, with 2 or 3 rounded lobes on each side, apex rounded or truncate, base cuneate or rounded, margin crenate or serrate, sparsely stellate-pubescent above and below, stellate-hispid below; petiole 3–12 mm long, stellate-hispid; stipules linear, 2–5 × 0.5 mm, hispid. Flowers solitary in the leaf axils, pedicel 16–33 mm long, hispid and/or pubescent; epicalyx of 8–10 linear bracts 4–5 × 0.5 mm, pubescent and/or hispid, reflexed at anthesis;

calyx lobes lanceolate, 7–9 × 1–1.5 mm, fused in the lower third, hispid. Corolla 10–13 mm long, bright scarlet, stellate-hispid on the outside, glabrous on the inside; staminal tube 5–7 mm long, anthers located on the apical half only; filaments 0.2–0.5 mm long; exserted part of styles 3–5 mm long, glabrous. Capsule spherical, 6–8 mm in diameter, glabrous or sparsely pubescent; seeds reniform, brown, 3 × 3 mm, the hairs 4–6 mm long.

KENYA. Northern Frontier District: Yabichu, near Ramu, 23 May 1952, *Gillett* 13292! & Ramu–Banissa road, 10 km from the turning to Banissa, 4 May 1978, *Gilbert & Thulin* 1428!
DISTR. **K** 1; Ethiopia, Somalia; Arabia
HAB. Dry bushland and scrub; 30–1250 m

20. **Hibiscus aponeurus** *Sprague & Hutch.* in K.B. 1908: 54 (1908); F.P.S. 2: 31 (1952); E.P.A.: 556 (1959); Exell in F.Z. 1(2): 455 (1961); Hauman in F.C.B. 10: 123 (1963); U.K.W.F. ed. 2, 98 (1994); Vollesen in Fl. Eth. & Eritr. 2(2): 209 (1995). Lectotype: Tanzania, Bukoba District: Karagwe, *Grant* 215 (K!, lecto., chosen by Exell in F.Z.)

Erect herb 0.5–3 m tall; stems densely and shortly stellate-hispid, with a longitudinal line of pubescence. Leaves ovate to orbicular in outline, 3–4.5 × 3.5–5.7 cm, unlobed or shallowly 3-lobed, apex obtuse or rounded, base rounded or truncate, rarely cuneate, margin dentate or serrate, densely stellate-hispid above and below, the setae 6–10-armed, simple-pubescent above (rarely glabrous), veins inconspicuous above, prominently raised beneath; petiole 2–25 mm long, densely rufous-stellate-hispid; stipules linear to linear-lanceolate, 3–6 × 0.3–0.4 mm, sparsely hispid. Flowers solitary in the upper leaf axils, or in terminal racemes; pedicel 1.6–6 cm long, articulated in the upper half, rusty brown above articulation, densely stellate-hispid; epicalyx of 8–12 linear bracts 4–7 × 0.5–1 mm, hispid, each with a prominent median vein; calyx lobes triangular to lanceolate to ovate-acuminate, 7–8 × 1.2–2 mm, fused in the lower third, 3-veined, densely stellate-hispid. Corolla bright red, 10–15 mm long, stellate-hispid on the outside, glabrous on the inside; staminal tube 6–10 mm long; filaments 1–2 mm long; exserted part of styles 3–6 mm long, glabrous. Capsule obovoid, 8–10 mm long, 9 mm diameter, sparsely pubescent; seeds reniform, black, 2–3 × 1.5–2 mm, the hairs white, to 6 mm long.

UGANDA. Ankole District: Ruizi river, 3 Nov. 1950, *Jarrett* 352!; Toro District: Mweya, Queen Elizabeth National Park, 27 Nov. 1967, *Lock* 67/149!; Busoga District: Jinga, banks of the Nile, 29 Feb. 1936, *Michelmore* 1181!
KENYA. Nairobi District: Railway Training Centre, 8 May 1983, *Muasya* 148!; Masai District: S slopes of Ngong Hills, 28 July 1956, *Greenway et al.* 9018!; Tana River District: Congolani Central, 4 Apr. 1988, *Medley* 314!
TANZANIA. Musoma District: 24 km Seronera to Soitayai, 27 Mar. 1961, *Greenway* 9911!; Arusha District: Moshi–Arusha Road, 0.8 km past road to Machame, 15 Mar. 1955, *Huxley* 145! & Greater Momela lake, Ngurdoto Crater National Park, 2 Mar. 1966, *Greenway & Kanuri* 12419!
DISTR. **U** 1–4; **K** 1–7; **T** 1–3, 7; Congo-Kinshasa, Rwanda, Burundi, Sudan, Ethiopia, Somalia, Mozambique
HAB. Bushland, thicket, forest edges; 900–2700 m

NOTE. *H.* sp A. of U.K.W.F. is included here until sufficient material is seen to make a decision.

21. **Hibiscus crassinervius** *Hochst.*, Tent. Fl. Abyss. 1: 61 (1847); Sprague in K.B. 1908: 53 (1908); F.P.S. 2: 28 (1952); Vollesen in Fl. Eth. & Eritr. 2(2): 209 (1995). Type: Ethiopia, Mt Scholoda, near Adua, *Schimper* 646 (K!, holo.)

Erect herb to 80 cm; leaves ovate to lanceolate, 1–5 × 0.5–3 cm ide, unlobed, apex rounded, base cordate, truncate or rounded, margin crenate-serrate, densely stellate-tomentose to stellate-hispid above and below; petiole 3–12 mm long, stellate-tomentose to stellate-hispid; stipules linear to filiform, 3–6 × 0.3–0.5 mm, tomentose to hispid. Flowers solitary in leaf axils or in a terminal raceme; pedicel 0.5–4 cm long, articulated above the middle, stellate-tomentose to stellate-hispid; epicalyx of 8–11 linear bracts 3–5 × 0.5–1 mm, stellate-pubescent or stellate-hispid; calyx lobes 4–7 × 1–2 mm, fused in the lower half, acuminate, densely stellate-hispid. Corolla red, 7–12 mm long, stellate-hispid on the outside, glabrous inside; staminal tube ± 5 mm long; filaments 0.5 mm long; exserted part of styles 1–2 mm long, glabrous. Capsule spherical, 5–10 mm in diameter, glabrescent; seeds reniform, black, 2–3 × 1.5–2 mm, lanate, the hairs 5–8 mm long. Fig. 7: 1–6, p. 48.

UGANDA. Acholi District: Chua, 1936, *Eggeling* 2379!
KENYA. Marsabit District: Marsabit, 5 June 1960, *Oteke* 46! & Mt Kulal, Gatab, 17 Nov. 1978, *Hepper & Jaeger* 6872!; Turkana District: Naitamaiong, June 1934, *Champion* T 327!
DISTR. **U** 1; **K** 1, 2; Ethiopia, Somalia
HAB. Forest margin grassland, grassland, rocky bushland; 1300–1800 m

SYN. *H. erianthus* R. Br. in Salt, Voy. Abyss. App. 65 (1814), *nom. nud.*
H. chiovendae Cuf., Miss. Biol. Borana, Racc. Bot., Angiosp.-Gymnosp. 129 (1939).Type: Ethiopia, Arero, *Cufodontis* 332 (W!, holo.; FT, iso.)
H. crassinervius Hochst. var. *flammeus* (Schweinf.) Schweinf. in K.B. 1894, App.: 41 (1894), *nom. nud.*
H. wellbyi Sprague in K.B. 1908: 55 (1908). Type: Ethiopia, Harar to Addis Ababa, *Wellby* s.n. (K!, holo.)
H. crassinervius Hochst. var. *minor* Sprague in K.B. 1908: 54 (1908). Type: Eritrea, Mt Bizen, *Schweinfurth & Riva* 2053 (K, holo.)

22. **Hibiscus shirensis** *Sprague & Hutch.* in K.B. 1907: 47 (1907); Exell in F.Z. 1(2): 451 (1961); Hauman in F.C.B. 10: 125 (1963). Type: Malawi, Banks of Likangola river, *Buchanan* 385 (K!, holo.)

Perennial erect herb 1–2 m tall; stems stellate-pubescent and sparsely to densely brown-stellate-hispid, especially toward the apex, becoming glabrous brown and longitudinally ridged with age. Leaves simple-ovate to shallowly 3-lobed, 4–9 × 2–6 cm, apex acuminate, obtuse or acute, base rounded or truncate, margin crenate or serrate, stellate-pubescent above and below; petiole 1–4 cm long, stellate-pubescent; stipules linear, 4–8 × 0.1–0.5 mm, pubescent. Flowers in axillary cymes, rarely solitary, toward the stem apex; pedicel 1–6.5 cm long, articulated in the top quarter, stellate-pubescent and sparsely to densely brown-stellate-setose; epicalyx of 8 linear bracts 4–8 × 0.6–1 mm, stellate-pubescent on both surfaces with a prominent median vein; calyx lobes lanceolate to ovate-acuminate, 6–10 × 2–5 mm, stellate-pubescent and brown-stellate-hispid on the outside, simple pubescent on the inside, fused in the lower half. Corolla red, 12–16 mm long, stellate-pubescent outside, glabrous inside; staminal tube 5–6 mm long; filaments 2.5–3 mm long; exserted part of styles 3.5 mm long, glabrous. Capsule spherical, 8–11 mm in diameter, sparsely pubescent; seeds reniform, black, 2–3 × 1–2 mm, the hairs 4–6 mm long.

TANZANIA. Kilosa District: Mikumi National Park, 30 Apr. 1968, *Renvoize & Abdallah* 1821!, and June 1968 *Procter* 3903!, and Mugira Track 14 Apr. 1973, *Greenway & Kanuri* 15166!
DISTR. **T** 6, 8; Congo-Kinshasa, Zambia, Malawi, Mozambique, Zimbabwe
HAB. Woodland, forest and thicket; 400–1500 m

SYN. *H. heterochlamys* Ulbr. in N.B.G.B. 7: 367 (1920). Type: Tanzania, Lindi District: Rondo [Mueri] Plateau, *Busse* 2614 (B†, holo.)

23. **Hibiscus debeerstii** *De Wild. & T. Durand* in Compt. Rend. Soc. Bot. Belg. 38: 21(1899); Exell in F.Z. 1 (2): 452 (1961); Hauman in F.C.B. 10: 126 (1963); Brummitt in K.B. 31: 165 (1976). Type: Congo-Kinshasa, Katanga, Pala, *Debeerst* 11 (BR, holo.)

Erect wiry herb to 2 m tall; stems appressed stellate-hispid, becoming longitudinally ridged with age. Leaves ovate to elliptic, 1.4–6.2 × 0.5–3.2 cm, apex acute, obtuse or rounded, base rounded or truncate, margin serrate, shortly stellate-hispid below, stellate- and simple-hispid above; petiole 2–16 mm long, spreading stellate-hispid; stipules subulate, 2–3.5 × ± 0.2 mm, hispid. Flowers solitary in the upper leaf axils; pedicel 1.6–3 cm long, articulated in the upper half, appressed stellate-hispid, the hairs above articulation rusty-brown; epicalyx of 10 linear bracts 4–6.5 × 0.2–0.3 mm, sparsely hispid, exceeding the sepals; calyx lobes ovate-acuminate 4–4.5 × 2 mm, fused in the lower third, densely rusty-brown setose. Corolla 1.4–2.5 cm long, bright red, densely stellate-hispid on the outside, glabrous on the inside; staminal tube 6–10 mm long; filaments 2–3 mm long; exserted part of styles 3–7 mm long, glabrous. Capsule globose, 6–8 mm in diameter, sparsely pubescent; seeds reniform, dark brown, 2 × 1.5 mm, the hairs up to 8 mm long, dark brown.

TANZANIA. Mpanda District: Kasogi, Kungwe-Mahali peninsula, 29 Aug. 1958, *Harley* 9434! & Nkala, Mahali Mts, 15 Apr. 1978, *Uehara* 539!; Mbeya District: 16 km N of Utengule on track to Saza and Lake Rukwa, 26 May 1990, *Carter et al.* 2465!

DISTR. **T** 4, 7; Congo-Kinshasa, Zambia, Malawi, Mozambique

HAB. Open woodland and forest margins; 800–1100 m

SYN. *H. nyikensis* Sprague in K.B. 1908: 56 (1908); Exell in F.Z. 1 (2): 452 (1961). Type: Malawi, Nyika Plateau, *Whyte* 226 (K!, lecto., chosen by Exell)

24. **Hibiscus hildebrandtii** *Sprague & Hutch.* in K.B. 1907: 46 (1907); Vollesen in Fl. Eth. & Eritr. 2(2): 208 (1995); Thulin in Fl. Somalia 2: 51 (1999). Type: Ethiopia, Adda-Gallah, *James & Thrupp* s.n. (K!, iso.)

Herb 0.5–2 m tall; stems densely and shortly stellate-tomentose, becoming glabrous with age. Leaves ovate to broadly elliptic, 2–4.6 × 1–5 cm, apex rounded or obtuse, base cuneate, rounded or cordate, margins subentire, crenate or serrate, densely short-stellate-hispid above and below, veins prominently raised below; petiole 0.3–4 cm long, densely and shortly stellate-hispid; stipules linear or filiform, 2–6 × 0.2–0.5 mm, pilose. Flowers solitary in the leaf axils; pedicel 1.7–4.2 cm, articulated in the upper half, densely short-stellate-hispid; epicalyx of 8–10 linear bracts, 3–5 × 0.3–1 mm, stellate-tomentose; calyx lobes lanceolate, 7–10 × 2–3 mm, fused in the lower half, densely stellate-tomentose, 3-veined. Corolla red, 1.2–2.3 cm long, sparsely stellate-hispid on the outside, glabrous inside; staminal column 1–1.4 cm long; filaments 1 mm long; exserted part of styles 3 mm long, glabrous. Capsule spherical, 8–11 mm in diameter, sparsely pubescent; seeds subreniform, dark brown, 2–3 × 1.5–2 mm, lanate, the hairs 6–10 mm long.

KENYA. Kwale District: Mwachi, 5 km S of Mazeras, 10 Sep. 1953, *Drummond & Hemsley* 4244!; Kilifi District: Lali Hills, 7 Aug. 1967, *Adamson* 52!; Tana River District: Makere bend, Tana River National Primate Reserve, 15 Mar. 1990, *Kabuye et al.* in TPR 442!

TANZANIA. Tanga District: Langoni, Pangani, 19 Jun 1956, *Tanner* 2915!; Rufiji District: Mtemere, 1 May 1976, *Vollesen in MRC* 3505!; Pemba: 17 Mar 1902, *Kaessner* 350!

DISTR. **K** 7; **T** 3, 6; **P**; Ethiopia, Somalia

HAB. Grassland and bushland; 0–500 m

25. **Hibiscus allenii** *Sprague & Hutch.* in K.B. 1907: 45 (1907); Exell in F.Z. 1(2): 458 (1961). Lectotype: Uganda, Victoria Falls, *Allen* 103 (K!, selected by Exell in F.Z.)

Erect or prostrate perennial herb 0.3–1 m tall; stems spreading, simple hispid, with or without a longitudinal line of white pubescence. Leaves ovate in outline, 3-5 × 3.5–6 cm, simple to shallowly 3-5-lobed to deeply 3–5-fid, rounded, obtuse or acute, base rounded or truncate, margin serrate, hispid above, densely stellate-setose below, veins prominently raised below; petiole 1.3–2.5 cm long, spreading-stellate-hispid, with or without a line of pubescence adaxially; stipules linear, 3–6 × 0.3–0.4 mm, hispid. Flowers solitary in upper leaf axils; pedicel 2–3 cm long, articulated in the upper third, rusty brown above articulation, densely hispid; epicalyx of 8? linear bracts, 3 × 0.5 mm, setose; calyx lobes lanceolate, 5–6 × 2 mm, fused in lower third, densely setose. Corolla red, 10–12 mm long, stellate-setose on the outside, glabrous inside; staminal column 9–10 mm long; filaments 1.5 mm long; exserted part of styles 4–5 mm long, glabrous or sparsely pilose. Capsule spherical, 9–10 mm, puberulous; seeds wedge-shaped, dark brown, 2–3 × 1.5 mm, glabrous or sparsely brown-pubescent.

UGANDA. Bunyoro District: Victoria Falls, *Allen* 103!
TANZANIA. Shinyanga District: Shinyanga Block 4A, Jan. 1935, *Burtt* 5116!; Ufipa District: Milepa, Rukwa Valley, 18 Jan. 1947, *Pielou* 57!; Mbeya District: above Chamala mission, 23 Mar. 1988, *Bidgood et al.* 648!
DISTR. **U** 2; **T** 1, 3–4, 7; Zambia, Malawi, Mozambique, Zimbabwe, Botswana, Namibia (Caprivi)
HAB. Grassland, woodland and bushland; 1000–1300 m

26. **Hibiscus zanzibaricus** *Exell* in J.B. 66: 327 (1928); U.O.P.Z.: 298 (1949). Type: Tanzania: Zanzibar, Muyuni, *Vaughan* 39, (BM!, holo; EA, iso.)

Prostrate or erect herb 0.3–1.5 m tall; stems shortly stellate-hispid, pubescent, becoming glabrous and longitudinally ridged with age. Leaves ovate, linear or shallowly to deeply 3-lobed, 0.9–7 × 0.4–5 cm, apex acute, obtuse or rounded, base cuneate, truncate or subcordate, margin crenate or serrate, hispid above and below; petiole 1–5 cm long, stellate-hispid, pubescent or not; stipules linear-subulate, 3–6 × 0.5 mm, sparsely hispid. Flowers solitary in upper leaf axils or in terminal paniculate inflorescences; pedicel 1.2–3 cm long, hispid, articulated in the upper half; epicalyx of 6–8 linear bracts 2 × 0.5 mm, hispid; calyx lobes narrowly lanceolate or lanceolate, 4–6 × 1.5–2 mm, fused in the basal half, with a prominent median vein, hispid. Corolla 8–12 mm long, red, stellate-hispid and pubescent on the outside, glabrous inside; staminal tube 6–10 mm long; filaments 1–2 mm long; exserted part of styles 3–4 mm long, glabrous. Capsule spherical, 6–7 mm in diameter, sparsely pubescent; seeds wedge-shaped, black, 2 × 1.5 mm, the hairs 4–6 mm long.

TANZANIA. Mpanda District: Mango forest, 2 Oct. 1970, *Faulkner* 4455!; Uzaramo District: Dar es Salaam, Oyster Bay, S of beach, 11 Aug. 1974, *Frazier* 1060!; Rufiji District: Kibambawe Camp area, Selous Game Reserve, 3 Aug. 1993, *Luke* 3617!; Zanzibar: Fumba, 12 June 1950, *Oxtoby* 11!
DISTR. **T** 4, 6, 8; **Z**; **P**; Malawi, Mozambique
HAB. Grassland and woodland; 0–500 m

SYN. *H. micranthus* sensu Baker f. in J.L.S. 40: 28 (1911), *non* L. f.
H. migeodii Exell in J.B. 68: 83 (1930) & in F.Z. 1 (2): 453 (1961). Type: Tanzania, Lindi District: Tendaguru, *Migeod* 528 (MB, holo.)

27. **Hibiscus richardsiae** *Exell* in Bol. Soc. Brot. Ser 2, 33: 176 (1959) & in F.Z. 1(2): 458 (1961). Type: Zambia, Lake Tanganyika, Sumba, *Richards* 9033 (K!, holo; BM, iso.)

Perennial erect herb to 60 cm tall, patent stellate-pubescent. Leaves ovate, 1–2.5 × 0.8–1.5 cm, apex acute, margin serrate, base obtuse or rounded, densely pubescent above, sparsely setose below, hairs 2–3-ramous; petiole to 1.5 cm long, stipules 3–5 mm long, brown, subulate. Flowers solitary in leaf axils and forming corymbose clusters at the stem tops, ± 2.5 cm in diameter; pedicel 3–4 cm long, articulated near the top; epicalyx of 8–9 bracts 3–5 mm long, ciliate; calyx 10–11 mm long, lobes lanceolate, to 10 × 3 mm, acute, pilose outside, glabrous inside. Corolla red, petals obovate, 1.5–2 × 1–1.3 cm, stellate-pilose outside, glabrous inside; staminal tube 9–10 mm long, free parts of filaments 1–1.5 mm long; style branches 6.5 mm long. Capsule subglobose, 9–10 mm in diameter, very sparsely pubescent; seeds not seen.

TANZANIA. Ufipa District: Kasanga, 31 Mar. 1959, *Richards* 11026!
DISTR. **T** 4; Zambia, Malawi, Mozambique
HAB. Thick bush; ± 780 m

NOTE. Known only a single collection in the Flora area.

28. **Hibiscus palmatus** *Forssk.*, Fl. Aegypt.-Arab.: 126 (1755); Mast. in F.T.A. 1: 198 (1868); Exell in F.Z. 1(2): 469 (1961); Blundell, Wild Fl. E. Afr. 78 (1992); U.K.W.F. ed. 2: 98 (1994); Vollesen in Fl. Eth. & Eritr. 2(2): 205 (1995); Thulin in Fl. Somalia 2: 50, fig. 28f–i (1999). Type: Yemen, Al Mukham, *Forsskål* 104 (C, holo.)

Perennial prostrate herb; stems hispid to setose. Leaves broadly ovate and 5–11 × 8–15 cm in outline, 3-5-palmatilobed or -palmatipartite, apex (apices) acute, base hastate, margin entire, undulate, crenate, serrate or irregularly pinnatifid; petiole 3–3.5 cm long, sparsely hispid-pilose; stipules linear, 2–5 mm long, sparsely hispid. Flowers solitary, axillary; pedicel 5–9 mm long, articulated at base, hispid; epicalyx of 8 linear or lorate bracts, 8–10 mm long, sparsely hispid; calyx lobes linear-lanceolate, 10–12 mm long, fused in the lower quarter, long hispid on the veins, glabrous elsewhere. Corolla 2–3 cm long, cream to pale yellow, glabrous; staminal tube 7 mm long; filaments to 1–2 mm long; exserted part of style 4–6.5 mm long, style branches glabrous. Capsule ellipsoid, 8 mm long, 4 mm diameter, pubescent-hispid, with an awn 2.5–3.5 mm long; seeds subreniform, brown, 2.5–3 mm long, pubescent. Fig. 8: 6–9, p. 57.

UGANDA. Karamoja District: Lodoketeminit, 7 Apr. 1962, *Kerfoot* 3636! & Amudat, Nov. 1964, *Tweedie* 2965!; Bunyoro District: Bulisa, Jan. 1941, *Purseglove* 1095!
KENYA. Northern Frontier District: Dandu, 2 Feb. 1952, *Gillett* 13013!; Meru District: Meru National Park, Mughwango Plains No. 69, 17 May 1979, *Hamilton* 530!; Teita District: Tsavo National Park East, Voi H.Q., 20 km on Sobo road, 31 Jan. 1967, *Greenway & Kanuri* 12944!
TANZANIA. Mbulu District: Tarangire National Park, Tarangire R., 11 June 1969, *Vesey-Fitzgerald* 6313!; Dodoma District: Kondoa, 24 Jan.1973, *Richards* 28385!; Mbeya District: Igawa, 250 km S of Iringa, 13 Apr. 1962, *Polhill & Paulo* 2000!
DISTR. **U** 1–2; **K** 1–7; **T** 1, 2, 5–7; widespread in tropical Africa, Arabia and India
HAB. Alluvial soils in savanna and woodland; 0–1400 m

SYN. *H. intermedius* A. Rich. in Tent. Fl. Abyss. 1: 58 (1847). Type: Ethiopia, Choho province, *Petit* s.n. (P, holo.)
H. aristivalvis Garcke in Bot. Zeit. 7: 849 (1849). Type: Mozambique, Sena, *Peters* s.n. (B†, holo.)

29. **Hibiscus vitifolius** *L.*, Sp. Pl. 2: 694 (1753); Mast. in F.T.A. 1: 197 (1868); F.P.S. 2: 28 fig. 8 (1952); Keay, F.W.T.A. ed. 2, 1(2): 346 (1958); Exell in F.Z 1(2): 470 (1961); Hauman in F.C.B. 10: 103 (1963); Vollesen in Fl. Eth. & Eritr. 2(2): 205 (1995); Thulin in Fl. Somalia 2: 50, fig. 28a–e (1999). Lectotype "Habitat in India.", Herb. Hermann 4: 39, No. 265, designated by Brenan & Exell in Bol. Soc. Brot., sér. 2, 32 : 70 (1958)

Herb to 2 m tall, stems pilose, with or without tubercules, sometimes red-tinged. Leaves ovate, 3–9 × 2.5–10 cm, unlobed or shallowly to deeply 3–5(–7)-lobed, apex acute or acuminate, base cordate or truncate, margin crenate, serrate or undulate, glabrous to pubescent above and below, stellate-hispid or not; petiole 1–12.5 cm long, glabrescent to pilose, tuberculate or not; stipules linear or filiform, 3–8 × 0.3 mm, sparsely pilose. Flowers solitary in leaf axils and in terminal cymes; pedicel 0.6–3.3 cm, articulated at the middle, pubescent-pilose, rarely tuberculate; epicalyx of 8–12 filiform bracts, 7–12 × 0.1–0.5 mm, pilose; calyx lobes ovate or ovate-lanceolate, 10–18 × 4–10 mm, fused in the lower half, pubescent-hispid. Corolla yellow or lilac with a dark red or maroon blotch at base of each petal, 2.5–6 cm long, glabrous; staminal tube 10–16 mm; filaments 2–3 mm long; exserted part of style 2–5 mm long, glabrous.

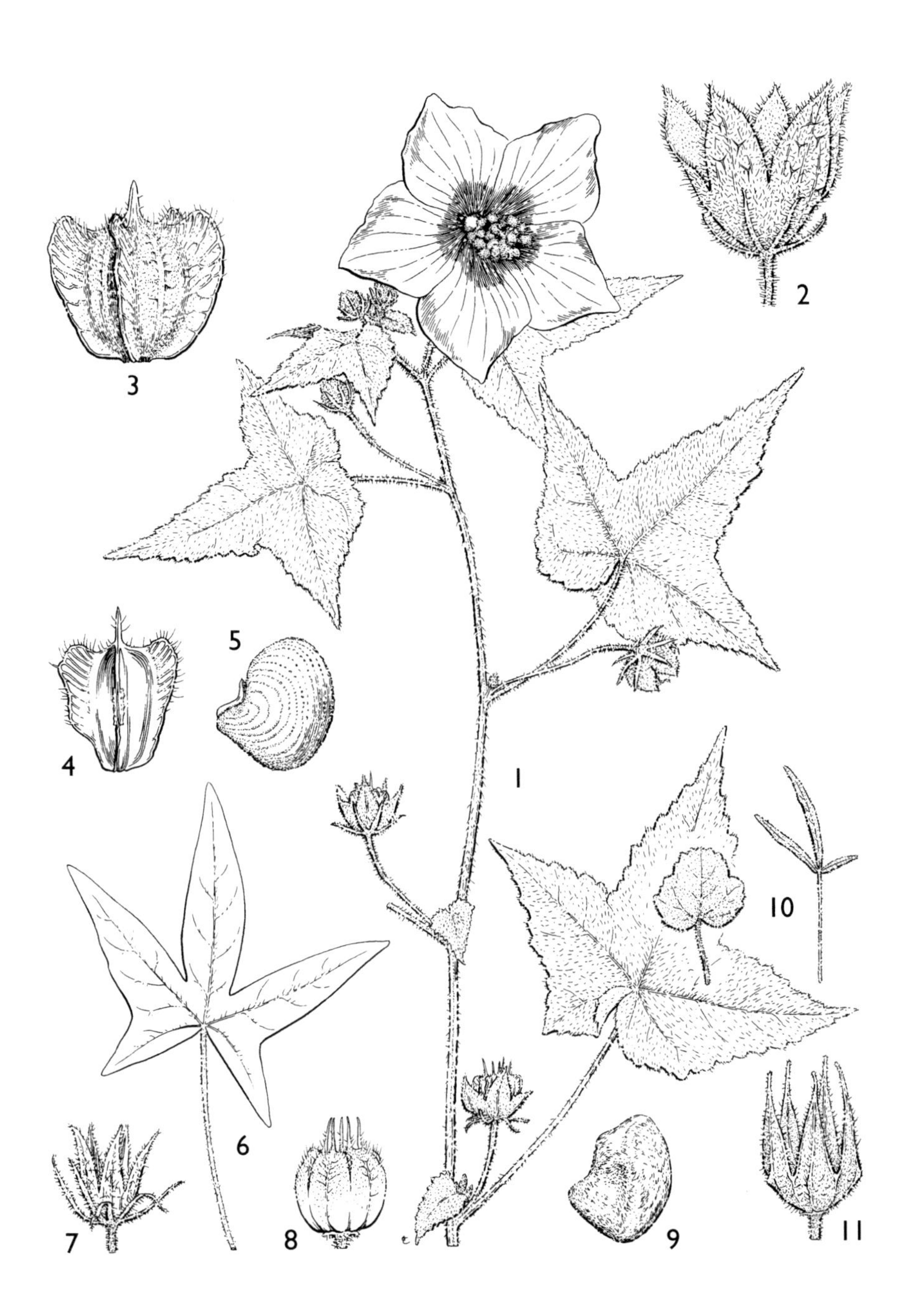

FIG. 8. *HIBISCUS VITIFOLIUS* — **1**, flowering and fruiting stem, × $^1/_2$; **2**, epicalyx and calyx, × 1.3; **3**, capsule, × 1.3; **4**, segment of dehisced capsule, × 1.3; **5**, seed, × 5. *HIBISCUS PALMATUS* — **6**, leaf, × $^1/_2$; **7**, epicalyx and calyx, × 1.3; **8**, capsule, × 1.3; **9**, seed, × 5. *HIBISCUS SIDIFORMIS* — **10**, basal (left) and apical (right) leaf, × $^1/_2$; **11**, calyx, × 1.3. 1–5 from *Getachew* 787; 6–9 from *Brown* 1445; 10–11 from *Gilbert & Thulin* 311. Drawn by Eleanor Catherine, and reproduced with permission from Flora of Ethiopia and Eritrea 2, 2.

Capsule straw-coloured, ovoid-rotund, 10–17 mm long, 9–15 mm diameter, wings 2–3 mm wide, sparsely pubescent, setose along the sutures; seeds black or dark brown, wedge-shaped, 3 × 2 mm, with evenly-spaced longitudinal rows of tubercules all over. Fig. 8: 1–5, p. 57.

1. Leaves and stems tomentose, tomentellous, densely pilose, densely pubescent or hispid abaxially, leaves drying brownish green, not lobed or usually shallowly 3–5(–7)-lobed; plants of grassland and woodland habitats ..... b. subsp. *vulgaris*
   Leaves and stems nearly glabrous to rather sparsely hairy, abaxial surface only hairy on the veins or reticulation, drying dark green, unlobed to deeply 3–5(–7)-lobed; plant of forest habitats 2
2. Pedicel articulated at the middle; corolla yellow with a dark blotch at base; capsule 1–1.7 cm long, 0.9–1.5 cm diameter .... a. subsp. *vitifolius*
   Pedicel articulated above the middle; corolla lilac with a dark maroon blotch at base; capsule 0.8 cm long, 0.8 cm diameter .... c. subsp. *lukei*

a. subsp. **vitifolius**; *Brenan & Exell* in Bol. Soc. Brot. Ser. 2. 32: 73 (1958)

Pedicel articulated at the middle. Corolla yellow with a dark blotch at base. Capsule 1–1.7 cm long, 0.9–1.5 cm diameter.

UGANDA. Kigezi District: Mitano Gorge, Oct. 1947, *Purseglove* 2550!; Toro District: Bundibugyo, 8 Dec. 1925, *Maitland* s.n.!; Mengo District: Mabira, Oct. 1916, *Dummer* 2989!
KENYA. Northern Frontier District: Dawa river, Murri, 27 June 1951, *Kirrika* 93!; Mt Elgon, Dec. 1930, *Lugard* 454!; Kwale District: Jadini Beach, 29 km S of Mombasa, 26 Aug. 1953, *Drummond & Hemsley* 3989!
TANZANIA. Lushoto District: Lutindi Forest Reserve, 16 Nov. 1986, *Iversen et al.* 86804!; Mpanda District: Mahale Mountains, Apr. 1981, *Hasegawa* 15!; Zanzibar island, 1857, *Bouton* s.n.!
HAB. *Acacia* bushland, grassland, woodland, forest edges, mountain forest; 0–2200 m
DIST. **U** 1–4; **K** 1, 3–4, 6–7; **T** 3–4, 7–8; **Z**; Cameroon, Congo-Kinshasa, Ethiopia, Angola, Malawi, Mozambique, Zimbabwe, South Africa; Sri Lanka

SYN. *H. jatrophaefolius* A. Rich., Fl. Abyss. 1: 58 (1847). Type: Ethiopia, Aderbati, Tigre province, *Petit* s.n. (P, holo.)

b. subsp. **vulgaris** Brenan & Exell in Bol. Soc. Brot. Ser. 2, 32: 73, t. 89, fig. 2 (1958). Type: Angola, *Welwitsch* 5236 (BM, holo.)

Leaves and stems tomentose, tomentellous, densely pilose, densely pubescent or hispid abaxially, drying brownish green, not lobed or usually shallowly 3–5(–7)-lobed.

UGANDA. Karamoja District: Amaler, base of Mt Debasien, 1936, *Eggeling* 2547!; Toro District: 18 km Congo border on main Rutshuru road, 12 May 1961, *Symes* 691!; Teso District: Serere, 9 Dec. 1964, *Smith* 67!
KENYA. Baringo District: 10 km Marigat to Kabarnet, 17 Sep. 1986, *Robertson* 4297!; Fort Hall District: N of Thika river and W of Nairobi–Murang'a [Fort Hall] road, 31 Mar. 1968, *Faden* 68/029!; Masai District: road from Keekorok gate to Narok, 16 Aug. 1971, *Kokwaro* 2720!
TANZANIA. Masai District: Engaruka road, 21 Feb. 1970, *Richards* 25487!; Lushoto District: Shagayu–Mtae road, 20 Oct. 1964, *Mgaza* 624!; Kondoa District: Kikori Tsetse Research Station, 14 Oct. 1932, *Burtt* 4427!
HAB. Dry bushland and grassland; 0–2000 m
DISTR. **U** 1–4; **K** 1–4, 6–7; **T** 1–7; tropical Africa and South Africa; Asia

SYN. *H. vitifolius* L. var. *vitifolius sensu* Keay, F.W.T.A. ed. 2, 1(2): 346 (1958), *non* Brenan & Exell in Bol. Soc. Brot. Ser. 2, 32: 73 (1958)

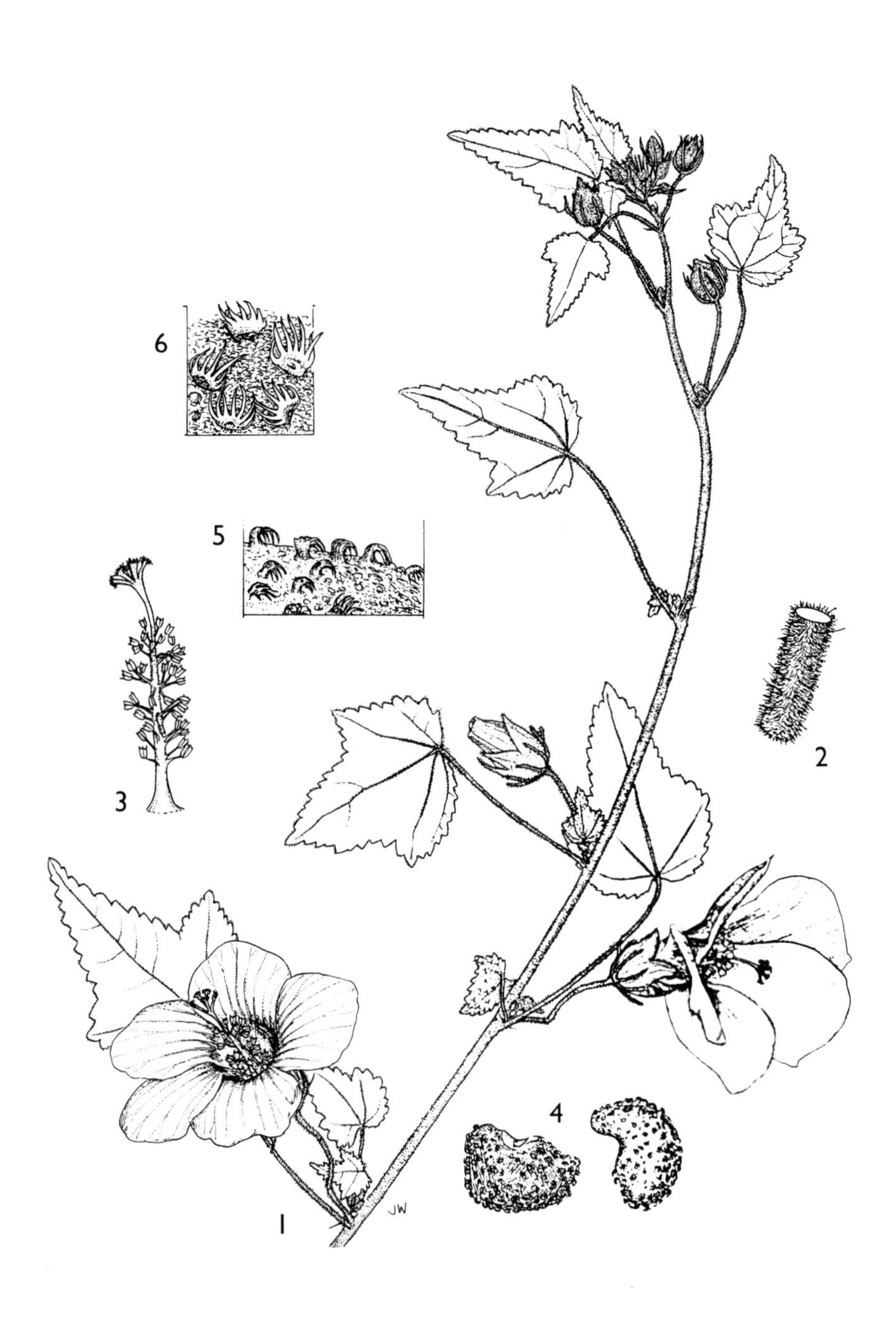

Fig. 9. *HIBISCUS KABUYEANA* — **1**, habit, × $^{2}/_{3}$; **2**, stem hairs × 16; **3**, staminal tube & style, × 16; **4**, seeds, × 6; **5–6**, seed surface, × 20 and × 30. 1–3 from *Haylett* 24; 4–6 from *Bally* 7691. Drawn by Juliet Williamson.

c. subsp. **lukei** *Mwachala & Cheek* in K.B. **58**: 499 (2003). Type: Kenya, Meru District: Ngaia forest, Camp 142, 5 May 2001, *Luke & Luke* 7393 (EA! holo.; K! iso.)

Pedicel articulated above the middle. Corolla lilac with a dark maroon blotch at base. Capsule 0.8 cm long, 0.8 cm diameter.

KENYA. Meru District: Ngaia Forest Reserve, Camp 142, 5 May 2001, *Luke & Luke* 7393!
HAB. Understorey of dry upland forest, with *Croton megalocarpus, Baphia keniensis* and *Uvariodendron anisatum*; ± 1080 m
DISTR. **K** 4; only known from the type

30. **Hibiscus kabuyeana** *Mwachala* **sp. nov.** *H. vitifolius* similis ab eo tubo staminali longiore (16–33, non 10–15 mm), corolla longiore (5–8 non 2.5–5 cm), seminibus angulatis-reniformibus pectinatis-squamatis (non levibus-reniformibus papillatis), floribus pendulis (non erectis) differt. Typus: Kenya, Taita District: Tsavo National Park, Voi Gate campsite, *Greenway & Kanuri* 12683 (EA!, holo., K!, iso.)

Erect herb 0.5–2.5 m tall; stems terete, glabrescent to pubescent-pilose with or without tubercules, sometimes red-tinged. Leaves ovate or narrowly ovate, 2.3–7.2 × 1.6–6.5 cm, unlobed to 3- to 5-lobed, apex acute or acuminate, base cordate or truncate, margins crenate, serrate or undulate, glabrous to pubescent, stellate-hispid or not; stipules linear-filiform, 2–5 × 0.3 mm, pilose; petiole 1.5–9 cm long, pubescent to pilose, tuberculate or not. Flowers solitary in leaf axils and in terminal cymes; pedicel 1–1.5 cm long, articulated in the lower half, pubescent-pilose; epicalyx of 8–12 linear-filiform bracts, 4–10 × 0.1–0.5 mm, pubescent, glandular or not; sepals triangular, 1–1.5 cm long, fused in the lower half, sparsely pubescent with scattered stellate setae. Corolla 5–8 cm long, pink with maroon base to each petal, glabrous; staminal tube 1.5–3.5 cm long; filaments 3–5 mm long; exserted part of style 5–12 mm long, pilose. Capsule broadly ovoid-rotund, 10–13 mm long and 10–12 mm diameter, wings 2–3 mm wide, sparsely pubescent, setose along the sutures straw-coloured; seeds reniform 3 × 2 mm, brown, covered with pectinate scales. Fig. 9, p. 59.

KENYA. Northern Frontier District: Mathews Range foothills, 25 Jan. 1969, *Curry & Glen* 18!; Kitui District: Ndui–Voo location, 13 Aug. 1968, *Kimani* 46!; Teita District: Voi gate camp site, Tsavo National Park East, 7 Dec. 1966, *Greenway & Kanuri* 12683!
TANZANIA. Pare District: Mkomazi Game Reserve, Kisima area, 30 Apr. 1995, *Abdallah & Vollesen* 95/61!; Lushoto District: near Kerenge, 29 June 1953, *Drummond & Hemsley* 3088!; Tanga District: Kibuguni, 25 Nov. 1936, *Greenway* 4771!
DISTR. **K** 1, 4, 7; **T** 3; Somalia
HAB. Open woodland, *Acacia-Commiphora* bushland; 100–1750 m

NOTE. Similar to *H. vitifolius* but differs in the longer staminal tube, longer corolla, and in its seeds that are angular-reniform (not smooth-reniform), purple-brown (not chestnut-brown) and with scattered pectinate scales (not evenly-spaced longitudinal rows of papillae) all over. The flowers in this species are pendent unlike those of *Hibiscus vitifolius* that are erect.

31. **Hibiscus schizopetalus** (*Boulger*) *Hook. f.* in Bot. Mag. 106: t. 6524 (1880); T.T.C.L.: 302 (1949); Cheek in Taxon 38: 261 (1989). Lectotype: Kenya, Mombasa District: Nyika hinterland behind Mombasa, *Kirk* s.n. (K!, chosen by Cheek in Taxon (1989))

Shrub 2–4.5 m tall; stems glabrous, green when young, becoming longitudinally ridged and grey with age. Leaves ovate or lanceolate, 4–10 × 1–5 cm, apex acute or acuminate, base cuneate to obtuse, rarely rounded, margin undulate, serrate or crenate, glabrous; petiole glabrous with a line of white pubescence adaxially, 4–37 mm long; stipules lanceolate, 0.5–3 × 0.1–0.5 mm, with or without scattered stellate hairs. Flowers solitary in the leaf axils; pedicel articulated in the lower half, sparsely pubescent; epicalyx of 8 triangular bracts, 1–3 × 0.5 mm, glabrous or sparsely pubescent; calyx lobes fused into a 12–15 mm long, 2-lipped tube, the lips toothed or not, glabrous or sparsely pubescent. Corolla 3–4 cm long, red, fimbriate, reflexed, glabrous; staminal column 4–5 cm long, the lower half filament-free; filaments 3–10 mm long; exserted parts of styles 6–23 mm long, glabrous. Capsule obovoid or cylindrical, 19–36 mm long, 15 mm diameter, puberulous; seeds triangular, brown, 2–4 × 2–3 mm, pubescent.

KENYA. Kwale District: Kwale Coast, 1936, *Jex-Blake* s.n.! & Mwachi, 5 km S of Mazeras, 10 Sep. 1953, *Drummond & Hemsley* 4256!; Kilifi District: Kombeni R. valley, edge of Kaya Fimboni, 21 Aug. 1989, *Robertson & Luke* 5811!

TANZANIA. Bagamoyo District: Kikoka Forest reserve, 27 Mar. 1964, *Semsei* 3721!; Kilwa District: 59 km from Kilwa Kivinje on Lindi Road, 29 July 1954, *Wilkinson* 75!; Lindi District: 6.5 km N of Lindi, 9 Dec. 1955, *Milne-Redhead* & *Taylor* 7481!

DISTR. **K** 7; **T** 3, 6, 8; widely cultivated

HAB. Coastal bushland and forest; 0–200 m

SYN. *H. rosa-sinensis* L. var. *schizopetalus* Boulger, Gard. Chron. 12: 372 (1879)

32. **Hibiscus calyphyllus** *Cav.*, Diss. Bot. 5: 283, t. 140 (1788); Hochr. in Ann. Conserv. Jard. Bot. Geneve, 4: 99 (1900); Ulbr. in Engl. Pflanzenw. Afr. 3, 2: 397 fig.186 U-W (1921); F.P.S. 2: 27 (1952); E.P.A.: 558 (1959); Exell in F.Z. 1(2): 459 tab. 89 (1961); Hauman in F.C.B. 10: 99 (1963); Blundell, Wild Fl. E Afr: 76, fig 446 (1992); K.T.S.L.: 173 (1994); U.K.W.F. ed. 2: 99 (1994); Vollesen in Fl. Eth. & Eritr. 2(2): 192 (1995). Type: Ethiopia, Tigray, Djeladjeranne, *Schimper* 510 (MAD lecto.; K!, P, isolecto.)

Shrub up to 3 m tall, unarmed; branches tomentose or pubescent when young, becoming glabrous, bark striate. Leaves concolorous, 3-lobed or simple, ovate to narrowly ovate, 3.8–21 × 1.8–22 cm, apex acuminate, base cordate to truncate, margins crenate-serrate, sparsely pubescent above, pubescence not obscuring the leaf surface above, stellate-pubescent to stellate-tomentose below, the hairs sparse or obscuring the leaf surface below; petiole terete, 1.5–15.5 cm long; stipules filiform, 5–20 × 1–2 mm, glabrous. Flowers solitary in leaf axils; pedicel 0.7–1.7 cm long, pubescent, articulated at base; epicalyx of five bracts, widest at or near the middle and narrowing to the base and toward the apex, 10–25 × 3–8 mm, stellate-pubescent, always longer than calyx; calyx lobes ovate or narrowly ovate, 10–25 × 5–9 mm, joined to half their length, stellate-pubescent, each with one main rib-like vein and a minor one on either side of it. Corolla 3–5.7 cm long, yellow with dark red spot at base, stellate-pubescent on the outside, glabrous inside; staminal tube 1–1.8 cm long; filaments 0.1–0.3 cm long; exserted length of style 0.5–0.1 cm long, style branches ± glabrous. Capsule ovoid, 1.7–2.6 cm long, 1–2 cm diameter, stellate-setose, valves awned; seeds reniform, 3 × 2.5 mm, appressed tomentose. Fig. 10: 6–8, p. 62.

UGANDA. Karamoja District: Kidepo National Park, Dodoth County, 29 Aug. 1972, *Synnott* 1215!; West Nile District: 1.2 km SE of Metu Rest Camp, 15 Sep. 1953, *Chancellor* 271!; Mengo District: edge of Mabira forest, 12 Nov. 1938, *Loveridge* 79!

KENYA. Nakuru District: Nakuru National Park, 9 June 1972, *Gillett* 19778!; North Kavirondo District: near Forester's house, Kakamega–Kaimosi road, 15 Oct. 1953, *Drummond & Hemsley* 4795!; Masai District: on Lenkutoto river, Nguruman North, 15 July 1976, *Fayad* 119!

TANZANIA. Mwanza District: Ukerewe Island, 16 Apr. 1929, *Conrads* in EA 13352!; Mbulu District: near Main Gate, Lake Manyara National Park, 2 Dec. 1963, *Greenway & Kirrika* 11111!; Lushoto District: Soni Valley between Soni and Miombo, 1 Sep. 1960, *Mgaza* 365!

DISTR. **U** 1–4; **K** 1–7; **T** 1–4, 6; tropical and Southern Africa; Indian Ocean Islands

HAB. Forest edges and glades, riverine forest, bushland; 0–2600 m

SYN. *H. calycinus* Willd., Sp. Pl. ed. 4, 3: 817 (1800). Type as for the species

H. grandifolius Hochst. in A. Rich., Tent. Fl. Abys. 1: 61 (1874). Type: Ethiopia, Djeladjeranne, *Schimper II*, 510 (P, syn.; FT, K, isosyn.)

H. wildii Suesseng. in Proc. Trans. Rhod. Sci Ass. 43: 103 (1951). Type: Zimbabwe, Marandellas, 12 Aug. 1941, *Dehn* 202 (M, holo.)

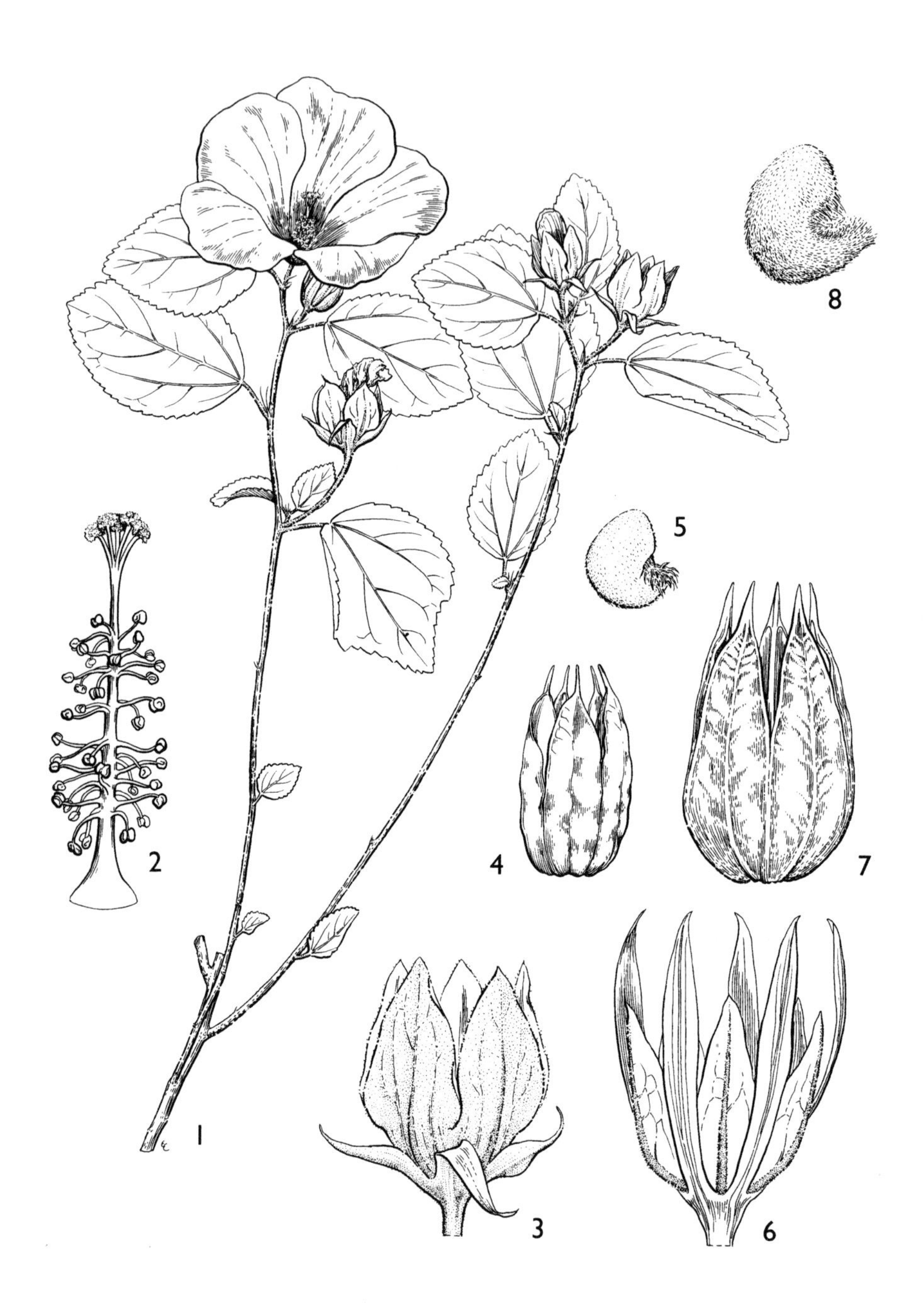

FIG. 10. *HIBISCUS OVALIFOLIUS* — **1**, flowering stem, × $^1/_2$; **2**, staminal column, × 2; **3**, epicalyx and calyx, × 1.3; **4**, capsule, × 1.3; **5**, seed, × 4. *HIBISCUS CALYPHYLLUS* — **6**, epicalyx and calyx, × 1.3; **7**, capsule, × 1.3; ; **8**, seed, × 4. 1, 3–5 from *Gilbert* 1069; 2 from *Turton* 18; 6–8 from *Ash* 2764. Drawn by Eleanor Catherine, and reproduced with permission from Flora of Ethiopia and Eritrea 2, 2.

33. **Hibiscus ovalifolius** (*Forssk.*) *Vahl* in Symb. Bot. I. 50 (1790); Vollesen in Fl. Eth. & Eritr. 2(2): 192 (1995); Thulin in Fl. Somalia 2: 42, fig. 23 (1999). Type: Ethiopia, Tigray, *Petit* 320 (P, holo.; K! iso.)

Suffrutex 0.5–3 m tall; stems sparsely to densely stellate-pubescent. Leaves ovate in outline or shallowly 3-lobed, 3–13 × 3–13 cm, apex rounded or obtuse, if lobed, the apices of the lobes also rounded or obtuse, base cuneate, truncate or shallowly cordate, margin serrate, sparsely pubescent-hispid above, densely so below; petiole 2.5–8 cm long, sparsely to densely pubescent; stipules linear or filiform, 8–16 × 1–2 mm, tapering to a fine tip, glabrous, pubescent or hispid. Flowers solitary in leaf axils; pedicel 1.2–2.5 cm long, articulated at base, sparsely to densely pubescent, hispid or not; epicalyx of 5 ovate to linear to linear-lanceolate bracts, widest above the base, 10–27 × 2–6 mm, sparsely to densely stellate-pubescent; calyx lobes triangular to ovate, 1.5–2.5 × 1–1.5 cm, stellate-pubescent, hispid or not, chartaceous or foliose, enlarging in fruit and completely enclosing the capsule. Corolla yellow with maroon patch at base, 4.5–7 cm long, glabrous or sparsely pubescent; staminal column 1.0–1.8 cm long; filaments 3–4 mm long; exserted part of style 4–6 mm long, glabrous. Capsule ovoid, 17 mm long, 12 mm diameter, glabrous or sparsely pubescent, enclosed in the calyx; seeds pale brown, reniform, 3 × 2 mm, tomentose, with hyaline setae around the hilum. Fig. 10: 1–5, p. 62.

UGANDA. Karamoja District: base of Mt Moroto, 6 Oct. 1952, *Verdcourt* 792!; Mbale District: near Apoli, 3 km N of Malaba river, 14 July 1958, *Kerfoot* 301!

KENYA. Northern Frontier District: Barecha at the turning 4 km S of Banissa, 5 May 1978, *Gilbert & Thulin* 1479!; Meru District: Golo Murera River, Meru National Park, 17 May 1978, *Hamilton* 260!; Teita District: E side of Voi river, S of Mwandongo forest, 16 June 1998, *Mwachala et al.* 936!

TANZANIA. Kigoma District: Bulimba, 25 May 1975, *Kahurananga et al.* 2795!; Ufipa District: Sumbawanga, 15 Mar. 1957, *Richards* 8755!; Iringa District: Magangwe, 10 Apr. 1970, *Greenway & Kanuri* 14314!

DISTR. **U** 1, 3; **K** 1–7; **T** 1–5, 7; Ethiopia

HAB. Forest, woodland, bushland, scrub, grassland; 500–2300 m

SYN. *Urena ovalifolia* Forssk., Fl. Aegypt.-Arab.: 124 (1775). Type: Yemen, Taizz [Taaes], *Forrskahl* 1763 (C, lecto.)

Hibiscus calycosus A. Rich., Tent. Fl. Abyss. I: 62 (1847). Type: Ethiopia, Chire Province, *Petit* 320 (K, iso.)

H. lanzae Cufod. in Miss. Biol. Borana, Racc. Bot., Angiosp-Gymnosp.: 131 (1939). Type: Ethiopia, Arero, *Cufodontis* 301 (FT, holo.)

34. **Hibiscus platycalyx** *Mast.* in F.T.A. 1: 202 (1868); Exell in F.Z. 1(2): 460 (1961); Vollesen in K.B. 35: 375 (1980). Lectotype: Tanzania, Mikindani District: Rovuma river, *Kirk* s.n. (K!), selected by Exell in F.Z. (1961)

Shrub to 3 m tall, stems grey-white, shortly stellate-pubescent when young, becoming glabrous. Leaves discolorous, narrowly ovate, 2.8–14 × 4.7–5.5 cm, apex acuminate, base shallowly cordate, margins dentate or crenulate, white-stellate-pubescent below, glabrous above; petiole 5–9 cm long, sparsely stellate-pubescent; stipules subulate, 8 mm long, 1 mm wide at base, tapering to a filiform tip. Flowers solitary in leaf axils; pedicel 1–1.3 cm long, articulated at base, shortly stellate-pubescent; epicalyx of 5 ovate bracts, 2–3 × 1.5–1.8 cm, sparsely stellate-pubescent; calyx lobes triangular, 3–3.5 × 1.5–2.5 cm, sparsely stellate-pubescent outside, glabrous within, fused in the lower half. Corolla yellow with a maroon patch at base, to 6 cm, pubescent outside, glabrous inside; staminal tube 12–18 mm long; filaments 1–3 mm long; exserted part of styles 2–4 mm long, glabrous. Capsule conical, 3.6–4 cm long 2.3–2.5 cm diameter, densely stellate-setose; seeds reniform, brown, 5 × 3 mm, tomentose.

TANZANIA. Ulanga District: Selous Game Reserve, Kibambawe upriver, S bank, 4 Aug. 1993, *Luke* 3655!; Lindi District: Lake Lutamba, 5 Apr. 1935, *Schlieben* 6230!; Kilwa District: Kingupira, 8 May 1975, *Vollesen in MRC* 2274!
DISTR. **T** 6, 8; Mozambique, Zimbabwe, Botswana, South Africa
HAB. Thicket; 0–150 m

SYN. *H. mossambicensis* M.L. Gonçalves in Garcia de Orta 4(1): 9 (1979). Type: Mozambique, Shamo, Mar. 1859, *Kirk* s.n. (K!, syn.)

35. **Hibiscus seineri** *Engl.*, Pflanzenw. Afr. 3, 2. 397 (1921); Exell in F.Z. 1(2): 460 (1961); Vollesen in K.B. 35: 375 (1980)& in Fl. Eth. & Eritr. 2(2): 192 (1995). Type: Botswana, Mabele a pudi, *Seiner* II. 310 (B†, holo.); Neotype: Botswana, Kwebe Hills, *Lugard* 95 (K!), selected by Vollesen in K.B. (1980)

Shrub 1–1.5 m tall; stems white stellate-tomentose when young, becoming glabrous with age. Leaves ovate-acuminate, 6–8 × 3.3–4.5 cm, apex acute or rounded, base rounded or shallowly cordate, margin serrate or crenate, stellate-pubescent below, sparsely so above, discolorous; petiole 2.4–10 cm long, stellate-pubescent; stipules filiform, 0.5–0.9 × 0.1 cm, tapering to a filiform tip. Flowers solitary in leaf axils; pedicel 0.8–1.5 cm long, articulated at base, stellate-pubescent; epicalyx of 5 lanceolate bracts, 1.5–2 × 0.2–0.5 cm, stellate-pubescent, always longer than the sepals; calyx lobes 5, ovate to triangular, 1–1.3 × 0.4–0.6 cm, stellate-pubescent with a prominent rib-like midvein and two weaker lateral ones. Corolla yellow with a maroon blotch at base, 4.5–6 cm long, sparsely stellate-hairy; staminal tube 1.4–2 cm long; filaments 3 mm long; exserted part of style 3.5–7 mm long, glabrous. Capsule conical, 1.5–2.5 cm long, 1.5 cm diameter, densely stellate-setose; seeds reniform, pale brown, 5 × 3–4 mm, densely tomentose.

TANZANIA. Shinyanga District: Kizambi, 10 May 1945, *Greenway* 7434!; Mpwapwa District: 11 km S of Gulwe on Kibakwe Track, 9 Apr. 1988, *Bidgood et al.* 977!; Iringa District: track between Mbagi and Mdonya, Ruaha National Park, 15 Feb. 1966, *Richards* 21329!
DISTR. **T** 1, 5, 7; Ethiopia (one collection), Zambia, Mozambique, Botswana, Namibia, South Africa
HAB. Woodland and bushland; 700–1400 m

36. **Hibiscus faulknerae** *Vollesen* in K.B. 35(2): 374 (1980). Type: Tanzania, Tanga District: Pangwe to Kange, *Faulkner* 1765 (K!, holo.)

Shrub up to 2.5 m tall; stems stellate white-pubescent-tomentose, becoming glabrous with age. Leaves distinctly discolorous, ovate to broadly ovate-acuminate, 7.8–16 × 3.2–8.2 cm, never lobed, apex acute or rounded and mucronate, base shallowly cordate, margin sharply dentate, rarely almost entire, with scattered hairs above, densely greyish tomentose beneath with the hairs concealing the surface; petiole 4–9.5 cm long, sparsely to densely stellate-pubescent; stipules linear, widened at base, 5–15 mm long, deciduous, but base sometimes persisting, sparsely pubescent or glabrous. Flowers solitary in the upper leaf axils, sometimes nodding; pedicel 5–15 mm long, articulated at base, stellate-pubescent, angular in the upper part, with two minute bracts at base; epicalyx of five bracts, linear-oblanceolate with long acuminate apex, 15–24 mm long, adnate to calyx from base to bottom of sinuses between lobes, stellate-tomentose, rarely sparsely so; calyx lobes triangular, 12–18 mm long, acute, joined for about a third to halfway up, pubescent except for the tomentose veins; each lobe with one very prominent median vein and two weaker lateral ones. Corolla yellow with the basal 1.5–2 cm part dark red-purplish, 3.5–5 cm long, pubescent outside and ciliate at base, glabrous inside; staminal tube ± 2 cm long; filaments 3 mm long; exserted part of styles 5–7 mm long, sparsely villose. Capsule ovoid, 1.5–2.4 cm long, densely greyish-yellowish tomentose, valves acuminate; seeds reniform, whitish brown, 3–4 × 3 mm, densely sericeous.

KENYA. Kwale District: Lungalunga–Msambweni Road, Marenge Forest, 18 Aug. 1953, *Drummond & Hemsley* 3855!; Kwale District: Dzombo Mountain, 8 Apr. 1968, *Magogo & Glover* 797!; Lamu District: Pangani, N of road, 5 Mar. 1977, *Hooper & Townsend* 1209!

TANZANIA. Tanga District: Mkaramo, Pangani, 10 Jan. 1957, *Tanner* 3380!; Morogoro District: Mtibwa Forest Reserve, Nov. 1953, *Semsei* 1437!; Uzaramo District: University of Dar es Salaam, 28 May 1986, *Kisena* 278!

DISTR. **K** 7; **T** 3, 6; not known elsewhere

HAB. Lowland evergreen forest and riverine forest; 0–300 m

37. **Hibiscus ludwigii** *Eckl. & Zeyh.*, Enum. Pl. Afr. Austr. Extratrop. 1: 39 (1834); Harv., F.C. 1: 171 (1860); Mast. in F.T.A. 1: 203 (1868); Hochr. in Ann. Conserv. Jard. Bot. Geneve 4: 161.(1900); F.P.S. 2: 27 (1952); Exell in F.Z. 1: 461 (1961); Hauman in F.C.B. 10: 101 (1963); Blundell, Wild Fl. E. Afr.: 77 (1992); Vollesen in Fl. Eth. & Eritr. 2 (2): 194 (1995). Type: South Africa, Uitenhage, *Cooper* 1456 (K!, holo.)

Shrub or robust herb 1–3.2 m high; young stems stellate-hispid, older stems glabrescent. Leaves 3-5-lobed or unlobed, 6–12 × 7–12.5 cm, acuminate, apex acute, base cordate, margin serrate, pubescent to pubescent-hispid above and below, with scattered stellate setae; petiole 1.5–12 cm long, pubescent-hispid, terete; stipules linear or filiform, 3–8 mm long, hispid. Flowers solitary in leaf axils and in terminal racemes; pedicel 5–12 mm long, articulated at base, densely pilose; epicalyx of 5 or 6 ovate to ovate-lanceolate bracts, 12–18 × 4–8 mm, densely to sparsely pubescent, many-veined, shorter than sepals in fruit; calyx lobes acuminate, 18–22 × 6–8 mm, densely to sparsely pilose, 3–7 veined. Corolla yellow, with a maroon spot at base, 4–6.5 cm long, puberulous and sparsely stellate-hispid; staminal tube 3 cm long; filaments 4 mm long; exserted part of styles 5 mm long, sparsely to densely villous. Capsule ovoid or broadly ovoid, beaked, 2.3 cm long, 1.8 cm diameter, densely setose and pilose, straw-coloured; seeds reniform, black, 2.5–3 × 2 mm, with parallel lines of minute stellate tubercules.

UGANDA. Karamoja District: Mt Morongole, 11 Nov. 1939, *A.S. Thomas* 3277!; Mt Elgon, 28 Oct. 1923, *Snowden* 818!; Mengo District: Lake Nabugabo, Aug. 1935, *Chandler* 1322!

KENYA. Elgeyo District: Cherangani Hills, Kabwibich, 3 Aug. 1969, *Mabberley & McCall* 117!; South Kavirondo District: SW Mau forest, 14 Jan. 1961, *Kerfoot* 758!; Masai District: Orengitok, about 20 km from Narok on road to Olokurto, 17 May 1961, *Glover et al.* 1212!

TANZANIA. Arusha District: Seneto, Arusha National Park, 22 July 1969, *Vesey-Fitzgerald* 6368!; Lushoto District: Mazumbai forest, 7 July 1966, *Semsei* 4059!; Rungwe District: Poroto mountains, by road from Tukuyu, 23 May 1957, *Richards* 9875!

DISTR. **U** 1–3; **K** 1, 3, 5, 6; **T** 2, 3, 6, 7; Ethiopia, Malawi, Zimbabwe, South Africa

HAB. Open areas in *Juniperus* forest, forest edges and grassland; 1300-2700 m

SYN. *H. sp. C* of U.K.W.F. ed. 2: 99 (1994), *pro parte*

38. **Hibiscus macranthus** *Hochst.* in Flora 24 (1841); A. Rich., Tent. Fl. Abyss I: 55 (1847); Keay, F.W.T.A. ed. 2, 1(2): 347 (1958); Exell in F.Z. 1(2): 461 (1961); Hauman in F.C.B. 10: 102 (1963); Blundell, Wild Fl. E. Afr.: 77 (1987); U.K.W.F. ed. 2: 99 (1994); Vollesen in Fl. Eth. & Eritr. 2(2): 194 (1995). Syntypes: Ethiopia, Scholoda, *Schimper* III: 1883 (B†, P, K!, syn.) and *Schimper* I, 362 (P, K!, syn.)

Shrub 0.5–3.1 m tall; stems sparsely to densely stellate-hispid. Leaves ovate, 4–7.5 × 3–7.5 cm, unlobed or shallowly 3-lobed, apex acute, base cordate, margin serrate, pubescent and sparsely stellate-hispid above, densely so below, the hairs mostly 3-armed; petiole 2.2–8.5 cm long, densely to sparsely pilose, hispid or not, grooved in upper half; stipules lanceolate, 3–6 × 0.4–1 mm, hispid. Flowers solitary in the lower leaf axils and in terminal racemes; pedicel 0.6–1.2 cm long, articulated at base, densely stellate-setose; epicalyx of 5 (rarely 6) ovate or lanceolate bracts, 1.5–1.8 × 0.5–0.8 cm, stellate-pubescent, longer than the sepals; calyx lobes ovate to narrowly ovate-acuminate, 1–1.3 × 0.5 cm, glabrescent outside, pubescent within, with a strong rib-like mid-vein and two weaker laterals. Corolla yellow with maroon patch at base, 6–10 cm long, pubescent or glabrous, rarely sparsely setose; staminal column 5–6.5 cm long; filaments 2–5 mm long; exserted part of style 3-7 mm long, villous. Capsule broadly ellipsoid, 10–13 mm long, 9–12 mm in diameter, pubescent and densely stellate-setose, ruminate, beaked, the beak up to 8 mm long; seeds reniform, dark brown, 3 × 2 mm, covered in stellate tubercules.

UGANDA. Kigezi District: Kachwekano farm, Kigezi, May 1949, *Purseglove* 2867!; Mbale District: Bufumbo, Bugisu, Nov. 1932, *Chandler* 1034!; Mengo District: Kampala, 17 Sep. 1923, *Maitland* 713!

KENYA. SW Elgon, 14 June 1958, *Symes* 387!; Kiambu District: Muguga, 21 km W of Nairobi, 8 Feb. 1956, *Verdcourt* 1444!; Kakamega District: Irembe, Isukha Location, 12 July 1960, *Paulo* 543!

TANZANIA. Mbeya District: N Poroto Mts, Mporoto sawmill, 18 Mar. 1932, *St. Clair Thompson* 957!; Njombe District: 26 km S of Njombe, 10 July 1956, *Milne-Redhead & Taylor* 11109! & Nyumbanitu Forest, 21 July 1982, *Magogo* 2216!

DISTR. **U** 2–4; **K** 1–7; **T** 2, 7; Congo-Kinshasa, Burundi, Sudan, Eritrea, Ethiopia

HAB. Forest edges, cleared forest, bushland; 500–2500 m

39. **Hibiscus dongolensis** *Delile* in Voy. Meroe Bot.: 59 (1826); Garcke in Reise Mossamb. Bot. 1: 1261 (1861); F.P.S. 2: 26 (1952); E.P.A.: 558 (1959); Exell in F.Z. 1(2): 458 (1961); Vollesen in Fl. Eth. & Eritr. 2 (2): 192 (1995); Thulin in Fl. Somalia 2: 42, fig. 24 (1999). Type: Ethiopia, Sidamo region, Lake Awasa, *Ellenbeck* 1723 (B†, holo.)

Shrub up to 2 m tall; stems very sparsely stellate-pubescent or glabrous. Leaves ovate to ovate-lanceolate, 2.5–10 × 1.5–7 cm, rarely 3-lobed, apex acute, base rounded or truncate, margin coarsely serrate or crenate, glabrous or very sparsely stellate-pubescent; petiole 3–7 cm long, with an adaxial line of pubescence; stipules linear or filiform, 5–10 × 0.1–0.3 mm, glabrous. Flowers solitary in the leaf axils, occasionally also in a terminal raceme; pedicel 5–10 mm long, finely pubescent, articulated at base; epicalyx of 5 strap-shaped bracts, 10–20 × 1–2 mm, glabrous; calyx lobes narrowly triangular, 18–20 mm long, fused in the lower quarter, glabrous, each with a rib-like median vein and a weaker one on either side, glabrous. Corolla yellow with maroon spot at base, 4–6 cm long, stellate-pubescent in the lower half, sparsely so above; staminal tube 20–25 mm long; filaments 2 mm long; exserted part of style 4–6 mm long, pubescent. Capsule subglobose, 0.9–1.5 cm long and 1.5–2 cm diameter, stellate-setose, valves awned or not; seeds angular-reniform, 2.5–3 × 2.5–3 mm, stellate-setulose, rarely glabrous.

UGANDA. Karamoja District: Lotome–Moroto km 4, 6 Oct. 1952, *Verdcourt* 775! & Madang Hill, Jie, 30 June 1955, *J. Wilson* 91!; Teso District: Agu, 5 May 1965, *Smith* 555!

KENYA. Northern Frontier District: Marsabit–Isiolo road, 14 May 1970, *Magogo* 1366!; Machakos District: Athi River, 18 July 1952, *Webster* in EA 10177!; Masai District: Lenyamu, 40 km from Nairobi on Magadi road, 16 June 1962, *Glover & Samuel* 2853!

TANZANIA. Mbulu District: Ndara River, Lake Manyara National Park, 3 Dec. 1963, *Greenway & Kirrika* 11124!; Korogwe District: Makuyuni, 20 Sep. 1961, *Semsei* 3311!; Mwanza District: Nyambiti, 26 Mar. 1953, *Tanner* 1319!

DISTR. **U** 1, 3; **K** 1, 4–6; **T** 1–3, widespread in tropical and South Africa

HAB. Woodland and streamsides in grassland and bushland; 550–1700 m

SYN. *H. lunarifolius* Willd. var. *dongolensis* (Caill.) Hochr. in Ann. Conserv. Jard. Bot. Geneve 4: 161 (1900). Type: Sudan, Kordofan, Milbes *Kotschy* 279 (K, iso.)

40. **Hibiscus lunarifolius** *Willd.* in L., Sp. Pl. ed. 4, 3: 811 (1800); F.T.A. 1:202 (1868); Exell in F.Z. 1(2): 459 (1961); U.K.W.F. ed. 2: 99 (1994); Vollesen in Fl. Eth. & Eritr. 2 (2): 194 (1995). Type: India, Tranquebar, *Klein* s.n. (B-W, holo.)

Herb 0.5–1.5 m tall; stems sparsely to densely stellate-hispid. Leaves ovate to broadly ovate, 3–8 × 4–10 cm, unlobed or shallowly 3-lobed, apex acute, base rounded or shallowly cordate, margin serrate or crenate, sparsely stellate-hispid above and below; petiole 3–12 cm long, glabrous or sparsely pubescent, with a line of pubescence adaxially; stipules linear, 4–6 × 0.2–0.3 mm, sparsely hispid. Flowers solitary in leaf axils; pedicel 23–60 mm long, articulated at base, sparsely or densely pubescent; epicalyx of 5–6 linear bracts 10–15 × 1–1.5 mm, sparsely pubescent; calyx lobes 8–17 × 5–10 mm, acuminate, fused for ± half their length, with three prominent rib-like veins, sparsely pubescent, stellate-hispid or not. Corolla yellow with a maroon spot at base, 3.5–6 cm long, stellate-pubescent outside, glabrous inside; staminal tube 1.5–2 cm long; filaments 2–3 mm long; exserted part of style 2–3 mm long, villous. Capsule ellipsoid, 1.5–2 cm long, 1 cm in diameter, densely stellate-setose, valves awned; seeds black, reniform, 2 × 1 mm, with scattered pectinate scales on a background of concentric rings.

UGANDA. Busoga District: Lolwi Island, Lake Victoria, 16 May 1964, *Jackson* U92!; Teso District: Serere, Kyere rock, 1926, *Maitland* 1314!

KENYA. Masai District: Old Narok road, behind Ngong Hills, 21 June 1959, *Napper* 1285! & Chyulu plains, Lenyekati Swamp, 29 Mar. 1993, *Luke* 3552!; Teita District: Mwandongo forest, 16 June 1998, *Mwachala et al.* in EW 933!

TANZANIA. Musoma District: Mugango, 20 Apr. 1959, *Tanner* 4169!; Masai District: Olgarien, 20 Dec. 1962, *Newbould* 6397!; Pangani District: Jassini, Madanga, 27 May 1957, *Tanner* 3514!

DISTR. **U** 3, 4; **K** 1, 3–7; **T** 1–3, 6; Zimbabwe; India

HAB. Seasonally flooded woodland and disturbed habitats; 600–1500 m

SYN. *H. sp. C* of U.K.W.F. ed. 2: 99 (1994), *pro parte*

NOTE. This species is probably introduced from India. The material studied matches the Indian material well. Furthermore, this plant is only known from disturbed/ruderal habitats in our area. I do not find *Hibiscus* sp. C. of U.K.W.F. ed. 2 to be distinct from this species.

41. **Hibiscus panduriformis** *Burm. f.* in Fl. Ind. 151, t. 47, f. 2, (1768); Mast. in F.T.A. 1: 203 (1868); F.P.S. 2: 28 (1952); Keay, F.W.T.A. ed. 2, 1(2): 346 (1958); Exell in F.Z. 1(2): 463 (1961); Hauman in F.C.B. 10: 104 (1963); Thulin in Fl. Somalia 2: 48 (1999). Type: Burm. f., Fl. Ind. t. 47, f. 2 (icono.)

Erect herb 1–3 m tall; stems tomentose and sparsely to densely long-setose, becoming glabrous, brown and reticulately ridged with age. Leaves discolorous, linear to ovate, 4–18 × 3–14 cm, simple or shallowly 3–7 lobed, apex obtuse to acute, base shallowly to deeply cordate, margins serrate, sparsely to densely pubescent above and below; petiole 5–14 cm long, densely pubescent and setose; stipules filiform, 8–12 × 0.1–0.2 mm, ciliate. Flowers 1–4 in leaf axils; pedicel 3–25 mm long, articulation obscure, densely pubescent with scattered setae; epicalyx of 8–9 linear bracts with expanded tips, 6–13 × 1–2 mm, pubescent on the inside and on the outside; calyx lobes broadly linear, 11–13 × 2–4 mm, rounded at the apex, fused in the lower $^{1}/_{2}$, pubescent, each prominently 3-veined. Corolla yellow with maroon blotch at base, 3–4 cm long, densely stellate-pubescent on the outside, glabrous inside; staminal tube 9–10 mm long, the filaments 1–1.5 mm long; exserted part of styles 1–1.5 mm long, glabrous or sparsely pilose. Capsule ovoid, 15–18 mm long, 10–14 mm diameter, densely setose, ± equal to the calyx; seeds wedge-shaped, brown, 3 × 2 mm, tomentose. Fig. 11: 8–10, p. 72.

UGANDA. Karamoja District: Moroto township, Apr. 1951, *J. Wilson* 341!; Ankole District: 5 km E of Kazinga village, 16 May 1969, *Lock* 69/184!; Mbale District: Nabbongo, Bugisu, 22 July 1965, *Smith* 561!

KENYA. Baringo District: Lake Baringo, S shore near mouth of Tiggeri R., 19 June 1953, *Bally* 9024!; Kisumu District: Maboko Island, 27 Oct. 1939, *Glasgow* 24!; Tana River District: near old course of Tana, N of Garsen, 6 Mar. 1977, *Hooper & Townsend* 1235!

TANZANIA. Mwanza District: Ukiriguru, *Smith* 563; Mbulu District: Lake Manyara, 5 May 1982, *Prins* 105!; Kilosa District: Mwanemboga, 17 km from HQ, 2 July 1973, *Greenway & Kanuri* 15308!; Pemba: Ole, 26 Oct 1929, *Vaughan* 879!

DISTR. **U** 1–3; **K** 3–7; **T** 1–4, 6–7; **P**; widespread in tropical Africa, Asia and Australia

HAB. Seasonally flooded areas in grassland and woodland; 700–1400 m

SYN. *H. multistipulatus* Garcke in Bot. Zeit. 7: 849 (1849). Type: 'East Africa', no collector mentioned (probably B†)

NOTE. No authentic specimen of Burman has been found: therefore the plate of Burman must serve as the type.

42. **Hibiscus trionum** *L.*, Sp. Pl. ed. 2: 697 (1753); Harv. in Fl. Cap. 1: 176 (1860); Mast. in F.T.A. 1: 196 (1868); F.P.S. 2: 22 (1952); Keay, F.W.T.A. ed. 2, 1(2): 346 (1958); Exell in F.Z.: 1(2): 446 (1961); Vollesen in Fl. Eth. & Eritr. 2(2): 200 (1995); Thulin in Fl. Somalia 2: 48, fig. 25h–i (1999). Type: "Habitat in Italia, Africa", not designated

Spreading herb to 1 m tall, erect or prostrate; stems terete, glabrous to sparsely setose with a line or two of pubescence. Leaves simple, palmatifid or deeply 3–5-palmatipartite, the segments irregularly pinnatifid, ovate in outline, 2–9 × 3–12 cm, apex acute or rounded, base cordate or truncate, margin dentate, rarely entire, hispid, pubescent or above and below; petiole 2–7.5 cm long, sparsely to densely pubescent with scattered stellate setae; stipules linear, 5–8 × 1–1.5 mm, hispid; epicalyx of 10–15 linear bracts 10–15 × 1–1.5 mm, glabrous to pubescent-hispid. Flowers solitary in the leaf axils; pedicel 15–60 mm long, articulated in the upper quarter, stellate-hispid; calyx lobes narrowly to broadly triangular, 10–15 mm long, united in the lower half, papery, pilose-hispid on the outside, pilose within with pubescent margins. Corolla yellow with a maroon spot at base, 25–30 mm long; glabrous to stellate-hairy outside, glabrous within; staminal tube 4–10 mm long; filaments 3 mm long; exserted part of style 3 mm, silky. Capsule dark brown to black when dry, ellipsoid, 10–12 mm long, 10 mm diameter, sparsely setose, enclosed in the accrescent calyx; seeds grey-brown, wedge-shaped, 2–3 × 1–2 mm, tuberculate.

UGANDA. Karamoja District: Kangole, July 1954, *J. Wilson* 143! & July 1957, *J. Wilson* 372!; Teso District: Agu, 27 Feb. 1966, *Smith* 577!

KENYA. Baringo District: Lake Baringo S shore, 19 July 1953, *Bally* 9027!; Nairobi District: Nairobi National Park area 2, 25 May 1961, *Verdcourt & Polhill* 3168!; South Nyanza District: Mbita Point, March/May 1985, *Dissemond* 109!

TANZANIA. Mwanza District: Mwanza, 16 Apr. 1926, *Davis* 259!; Masai District: Ngorongoro crater, 6 Apr. 1966, *Goddard* 167!; Dodoma District: Rehaya area, Kondoa, 18 January 1973, *Richards* 28288!

DISTR. **U** 1, 3; **K** 1–6; **T** 1–2, 5; tropical and South Africa; North Africa, S Europe, Asia and Australia

HAB. Grassland associated with water courses and areas of poor drainage in black cotton soil; 100–2200 m

NOTE. Although Fawcett & Rendle in Fl. Jamaica 5: 140 (1926) indicated material in LINN as type, they did not distinguish between sheets 875.39 and 875.40, neither of which carries the relevant Species Plantarum number (i.e. "20"). Both appear to be post-1753 additions to the collection, and are not original material for the name. Later authors have often incorrectly treated one or other collection as the type.

43. **Hibiscus masasiana** *Mwachala* **sp. nov.** *H. rhabdotospermus* similis sed foliis 3–5-lobatis cum concretionibus calcareis adaxialibus ab ea differt. Typus: Tanzania, Tunduru District: by road 96 km from Masasi, *Richards* 17903 (K!, holo.)

Erect herb 30–50 cm tall, stems stellate-hispid. Leaves ovate in outline, 0.5–5 × 1–1.5 cm, 3–5-lobed, base cordate to truncate, apices of the lobes rounded, margins coarsely dentate, sparsely stellate and/or simple-hispid above and below, with flattened calcareous concretions below; petiole 1–9 cm long, sparsely hispid, with a line of pubescence adaxially; stipules filiform, 3–6 × 0.2–0.3 mm, hispid. Flowers solitary in leaf axils or clustered at the tips of branches; pedicel 10–12 mm long, articulated at base, densely stellate-hispid; epicalyx of 8–10 filiform bracts 3.5–5 × 0.3–0.4 mm, pubescent; calyx lobes triangular, 7–10 × 3 mm, united for the basal third of their length, sparsely stellate-hispid. Corolla yellow with a maroon blotch at base, 3.5–4 cm long, sparsely stellate-pubescent on the outside, glabrous inside. Capsule conical, 7 mm long, 7 mm diameter, strongly beaked, glabrous proximally, hispid distally; seeds black, subreniform, 2 × 1.5 mm, lepidote.

TANZANIA. Tunduru District: by road 96 km from Masasi, 18 Mar. 1963, *Richards* 17903!; Masasi District: NE of Masasi, Pangani Hill, 11 Mar. 1991, *Bidgood et al.* 1899!
DISTR. **T** 8; not known elsewhere
HAB. Rock outcrops in woodland; 500–900 m
NOTE. This species is only known from the above cited collections

44. **Hibiscus rhabdotospermus** *Garcke* in Bot. Zeit. 7: 839 (1849); Mast. in F.T.A. 1: 200 (1868); F.P.S. 2: 24 (1952); Exell in F.Z. 1(2): 466 (1961); Vollesen in Fl. Eth. & Eritr. 2(2): 202 (1995). Type: Sudan, 1839, *Kotschy* 86 (K!, lecto.; BR, E, WA, iso.)

Erect herb 0.5–1.5 m tall; stems pubescent or tomentose, sparsely simple or stellate-hispid. Leaves ovate, 7–15 × 4–12 cm, acuminate, base cordate to truncate, margins crenate or serrate, glabrous or sparsely pubescent above and below, with or without scattered stellate setae; petiole 7–10 cm long, pubescent; stipules filiform, 4–8 × 0.2–0.5 mm, pubescent. Flowers solitary in leaf axils or clustered at the ends of branches; pedicel 18–22 mm long, articulated in the top quarter, pubescent; epicalyx of 11 filiform-linear bracts, 7–9 × 0.3–0.5 mm, pubescent; calyx lobes ovate-acuminate, 15 × 2–6 mm, united for $^1/_3$ of their length, stellate-pubescent. Corolla white, cream or pink, 3.5 cm long, stellate-pubescent outside on the lower half, glabrous inside; staminal tube 17 mm long; filaments 0.5 mm long; exserted part of styles 1.5 mm long. Capsule elliptic or rotund, 7–10 mm long, 6–8 mm diameter, pubescent; seeds black, subreniform, 3 × 2 mm, lepidote.

KENYA. Turkana District: 22 km from Lokori on Sigor road, 28 May 1970, *Mathew* 6425!; Tana River District: Kora National Reserve, 29 Dec. 1982, *van Someren* 887! & Kora National Reserve, Poacher's Hill, 17 Dec. 1984, *Mungai & Rucina* 490/84!;
DISTR. **K** 2, 4, 7; Sudan, Angola, Malawi, Mozambique, Zimbabwe, Botswana and Namibia
HAB. In sandy or stony ground and among rocks on inselbergs; 300–700 m

SYN. *H. cordatus sensu* Hochst. in Fragm., Fl. Aethiop.-Aegypt.: 45 (1854), *non Hibiscus cordatus* D. Dietr. (1847); Hochr. Ann. Conserv. Jard. Bot. Geneve 4: 164 (1900). Type as for *H. rhabdotospermus* Garcke

45. **Hibiscus caesius** *Garcke* in Bot. Zeit. 7: 850 (1849); Peters in Reise Mossamb. Bot. 1: 125 (1861); Ulbr. & Fr. in R.E. Fr., Wiss. Ergebn. Schwed. Rhod. Kongo Exped. 1: 145 (1914); Garcia in Bol. Soc. Brot. Ser. 2, 20: 40, tab. 89, fig. 9 (1946). Type: Mozambique, Tete, *Peters* s.n. (?B†, holo.)

Perennial scrambling herb to 2 m tall; stems occasionally red-tinged or blotched, glabrous, sparsely hispid. Leaves 3–5-foliolate, 3–8 × 4–7 cm, leaflet apices acute to acuminate, base rounded or cuneate, margins serrate, glabrous above, hispid below, the hairs 3-armed; petiole 3–8 cm long, hispid; stipules subulate, 6–13 × 0.3 mm, sparsely hispid. Flowers solitary in leaf axils; pedicel 6–14 mm long, articulated in the upper half, hispid; epicalyx of 8–10 subulate bracts 2–2.5 × 0.1 cm, hispid; calyx lobes lanceolate, 2 × 0.5 cm, each with 3 prominent veins, fused in the bottom third, glabrous outside, hispid on the veins, pubescent in the upper half inside. Corolla yellow with a maroon spot at base, 3–5.5 cm long, sparsely pubescent and/or stellate-hispid; staminal column 1–1.8 cm long; filaments 1–3 mm long; exserted part of styles 2 mm long, sparsely pilose. Capsule straw-coloured, round, 10–13 mm long, 8–10 mm diameter, beaked, setose; seeds black, triangular, 2 × 2 mm, sparsely pubescent.

TANZANIA. Mbeya District: Songwe valley ± 2 km N of Mbeya–Tunduma road, 25 Mar. 1988, *Bidgood et al.* 687! & Songwe gorge, by hot springs, W of Mbeya, 27 May 1990, *Carter et al.* 2481!
DISTR. **T** 7; Zambia, Malawi, Mozambique, Zimbabwe, Botswana, South Africa; Australia
HAB. *Combretum* bushland on limestone; ± 1200 m

SYN. *H. caesius* Garcke var. *genuinus* Hochr. in Ann. Conserv. Jard. Bot. Geneve 4: 160 (1900). Type: as for the species

46. **Hibiscus obtusilobus** *Garcke* in Bot. Zeit. 7: 837 (1849); Mast. in F.T.A. 1: 197 (1868); F.P.S. 2: 22 (1952); Vollesen in Fl. Eth. & Eritr. 2(2): 204 (1995). Type: Sudan, Kordofan, Arasch Cool, *Kotschy* s.n. (B† holo.), *Kotschy* 183 (K!, lecto., BM!, E, M, P, isolecto.)

Herb 0.5–2.5 m tall, stems stellate-hispid or glabrous. Leaves shallowly to deeply 3-lobed, 5–8 × 6–8 cm, the lobes rounded, apex rounded, base truncate or cordate, margin crenate, very sparsely stellate-hairy above and below or glabrous; petiole 5–7 cm long, sparsely stellate-hispid; stipules linear, 3–6 × 0.2 mm, hispid. Flowers solitary in leaf axils (rarely in axillary cymes); pedicel 2–3 cm long, articulated in the top quarter, shortly stellate-hispid; epicalyx of 7 linear bracts, 2–3 × 0.5–1 mm, pubescent, shorter than the calyx; calyx lobes lanceolate or triangular, 5–8 × 2–3 mm, fused in the lower half, stellate-hispid, each 3-veined. Corolla white or cream-coloured, fading pink, 6–10 mm long, glabrous. Capsule straw-coloured, ellipsoid, 8–11 mm long, 6–7 mm diameter, hispid; seeds black, reniform, 2 × 1.5 mm, papillose.

KENYA. Turkana District: Ayangyangi swamp, 17 km S of Kangetet, 23 May 1970, *Mathew* 6325! & 12 June 1970, *Mathew & Gwynne* 6770!
DISTR. **K** 2; Sudan, Ethiopia, Zambia, Mozambique, South and West tropical Africa
HAB. Seasonally flooded *Acacia* woodland; 550 m

SYN. *H. amblycarpus* Hochst., Fragm. Fl. Aethiop.: 45 (1854). Type as for *Hibiscus obtusilobus*

47. **Hibiscus physaloides** *Guill. & Perr.* in Fl. Seneg. tent. 1: 52 (1831); Harv. in Fl. Cap. 1: 172 (1860); Mast. in F.T.A. 1: 199 (1868); C.F.A. 1, 1: 171 (1937); F.P.S. 2: 24 (1952); Keay, F.W.T.A. ed. 2, 1(2): 346 (1958); Exell in F.Z. 1 (2): 464 (1961); Vollesen in Fl. Eth. & Eritr. 2(2): 204 (1995). Type: "Senegambia", *Leprieur* 75 (P, holo.)

Herb to 2 m tall; stems pubescent and hispid. Leaves deltoid, ovate, lanceolate or linear, 6–18 × 5–18 cm, simple to shallowly 3–5-lobed, the lobes triangular, apex acute or acuminate, base cordate, bearing calcareous, off-white concretions abaxially, margins serrate, crenate or dentate, stellate-pubescent-hispid above and below, the hairs above various, those below 3-armed; petiole 3–10 cm, pubescent-hispid; stipules filiform, 4–6 × 0.1 mm, pubescent and hispid. Flowers solitary in leaf axils, or in terminal cymose clusters; pedicel 4–7 cm long, articulated in the upper half, pubescent-hispid, the hairs simple and spreading; epicalyx of 7–10(–12) filiform bracts 7–11 × 0.3 mm, pubescent and hispid; calyx lobes ovate-lanceolate, 16 × 5–8 mm, tip cuneate to acuminate, fused in the lower half, stellate-pubescent and hispid outside, the hairs both stellate and simple, pubescent-hispid inside. Corolla yellow with a maroon spot at base, 5–7 cm long, pubescent-hispid outside, glabrous inside; staminal column 1–1.5 cm long; filaments 1 mm long; exserted part of styles 1 mm long, glabrous. Capsule ovoid, beaked, 0.5–1.6 cm long, 0.7–1.5 cm diameter, densely hispid, pilose along the sutures, straw-coloured; seeds dark brown, triangular-reniform, 3 × 2 mm, tuberculate.

UGANDA. Toro District: Bwamba hot springs, 21 Dec. 1969, *Lye* 4735!; Teso District: Serere, Dec. 1931, *Chandler* 103!; Mengo District: Entebbe, May 1937, *Chandler* 1634!
KENYA. Kwale District: Shimba Hills, Tanga road near Madabara river, 8 Nov. 1968, *Magogo & Estes* 1202!; Mombasa District: Nyali Estate, 5 June 1934, *Napier* 6354!; Teita District: Kiruwuko on Wanganga–Msau road, 10 June 1998, *Mwachala et al.* EW 750!
TANZANIA. Mbulu District: Lake Manyara National Park, between the southern boundary and Majimoto, 31 May 1965, *Greenway & Kanuri* 11803!; Mpanda District: Kungwe-Mahali peninsula, 25 Aug. 1958, *Harley* 9427!; Iringa District: Msembi–Mbagi track, 3.5 km, 25 Mar. 1970, *Greenway & Kanuri* 14200!; Zanzibar: Kisim-Kazi, 27 May 1960, *Faulkner* 2567!
DISTR. **U** 2–4; **K** 7; **T** 2, 4, 6, 7; **Z**; widespread in tropical and South Africa; Indian Ocean islands
HAB. Woodland, abandoned cultivation; 0–1100 m

SYN. *H. variabilis* Garcke in Peters, Reise Mossamb. Bot. 1: 126 (1861). Type: Mozambique, Querimba Is., *Peters* s.n. (B†, holo.)
H. physaloides Guill. & Perr. var. *genuinus* Hochr. in Ann. Conserv. Jard. Bot. Geneve 4: 162 (1900). Type as for *Hibiscus physaloides* Guill. & Perr.

48. **Hibiscus corymbosus** *A. Rich.*, Tent. Fl. Abyss. I: 57 (1847); F.P.S. 2: 24 (1952); U.K.W.F. ed. 2: 99 (1994); Vollesen in Fl. Eth. & Eritr. 2(2): 204 (1995). Type: Ethiopia, Sana to Ferferra, *Schimper* II: 787 (P, holo.; BR, FT, K!, MO, iso.)

Erect suffrutex herb 0.5–1.5 m tall; stems pilose, pubescent or rarely glabrous. Leaves discolorous, ovate in outline, 10–16 × 2–12 cm, apex obtuse, acute or rounded, base rounded, truncate or cordate, margins serrate, shallowly 3–5-lobed to deeply (2)3–5-fid, shortly stellate-hispid above and below, the leaf surface scabrid; petiole 0.5–3 cm long, pubescent; stipules linear, 3–6 × 0.2–0.3 mm, pubescent or hispid. Flowers solitary in upper leaf axils and/or in terminal corymbs; pedicel 2–4 cm long, articulated in the upper half, densely stellate-pubescent; epicalyx of 8 linear bracts, 2 × 0.5 mm, hispid and/or stellate-pubescent; calyx lobes narrowly triangular-acuminate, 8–10 × 2 mm, stellate-pubescent inside and outside. Corolla yellow, 2–5.5 cm long, sparsely stellate-pubescent outside, glabrous inside; staminal column 1.5 cm long; filaments 3 mm long; exserted part of styles 6–13 mm long, branches densely pilose. Capsule globose, 1–1.3 cm long, 0.7–1 cm diameter, stellate-tomentose, the hairs along the sutures simple and/or spreading; seeds brown, triangular, 2 × 2 mm, tuberculate.

UGANDA. Karamoja District: foot of Sipi escarpment, 3 July 1958, *Irwin* 420!; Teso District: Serere, 1930, *Liebenberg* 1574!; Mengo District: Busana, Bugerere, Apr. 1932, *Eggeling* 420!
KENYA. Meru District: a few miles from start of Meru Game Reserve on Kinna–Maua road, 11 Sep.1963, *Verdcourt* 3747!; South Kavirondo District: Kisii, July 1934, *Napier* 6608! & Migori, hill above Migori Police Station, 30 May 1967, *Hanid & Kaniaruh* 807!
TANZANIA. Shinyanga District: Shinyanga, Jan 1933, *Bax* 91!; Ufipa District: 33 km on Namanyere–Karonga road, 5 Mar. 1994, *Bidgood et al.* 2653!; Songea District: Lilenga Hill, 19 Apr. 1956, *Milne-Redhead & Taylor* 9055B!
DISTR. **U** 1–4; **K** 4–6; **T** 1, 4, 7, 8; Congo-Kinshasa, Sudan, Ethiopia
HAB. Grassland and woodland; 1000–1500 m

SYN. *H. corymbosus* A. Rich. var. *integrifolia* Chiov. in Ann. Di Bot. 9: 52 (1911). Type: Ethiopia, Gondar, Asoso, *Chiovenda* 2691 (FT, holo.)
H. corymbosus A. Rich. var. *palmatilobata* Chiov. in Ann. Di Bot. 9: 52 (1911). Type: Ethiopia, Mt Inceduba, near Gondar, 26 Aug 1909, *Chiovenda* 1604 (?FT, holo.)

49. **Hibiscus articulatus** *A. Rich.*, Tent. Fl. Abyss. 1: 60 (1847); Mast. in F.T.A. 1: 200 (1868); Oliv. in Trans. Linn. Soc. 29: 36, t.13 (1872); Hochr. in Ann. Conserv. Jard. Bot. Genève, 4: 159 (1900); F.P.S. 2: 26 (1952); Keay, F.W.T.A. ed. 2, 1(2): 347 (1958), Exell in F.Z. 1(2): 463 (1961); U.K.W.F. ed 2.: 99 (1994); Vollesen in Fl. Eth. & Eritr. 2(2): 204 (1995). Type: Ethiopia, Gafta, *Schimper* II: 1201 (P, holo., BM, FT, K! iso.)

FIG. 11. *HIBISCUS ARTICULATUS* — **1**, flowering stem, × $^1/_2$; **2**, leaf, × $^1/_2$; **3**, epicalyx and calyx, × 1.3; **4**, capsule, × 1.3; **5**, seed, × 5. *HIBISCUS AETHIOPICUS* — **6**, habit, × $^1/_2$; **7**, epicalyx and calyx, × 1.3. *HIBISCUS PANDURIFORMIS* — **8,** epicalyx and calyx, × 1.3; **9**, capsule, × 1.3; **10**, seed, × 5. 1 from *Gilbert & Getachew* 2894; 2–3 from *Thulin & Asfaw* 4035; 4–5 from *Andrews* 860; 6 from *Chojnacki in Mooney* 8884; 7 from *Gilbert* 4339; 8–10 from *Gilbert* 1637. Drawn by Eleanor Catherine, and reproduced with permission from Flora of Ethiopia and Eritrea 2, 2.

Perennial procumbent herb; stems densely to sparsely hispid, becoming glabrous with age. Leaves ovate, 4–8 × 1–9 cm, simple, 3-lobed or deeply 3-fid, the divisions linear or linear-lanceolate, apex obtuse, truncate or rounded, base rounded or cuneate, margins serrate, hispid above and below, occasionally sparsely so, the hairs various; petiole 0.5–2 cm long, glabrous or stellate-hispid; stipules linear, 2–7 × 1 mm, glabrous, hispid along the margins. Flowers solitary in the leaf axils; pedicel 1.7–5.8 cm long, articulated in the upper half, pubescent, hispid or not; epicalyx of 7–10 linear bracts 3–5 × 1–1.5 mm, glabrous, pubescent or hispid; calyx lobes lanceolate, 10–20 × 3–3.5 mm, fused in the bottom $^1/_3$, sparsely to densely pubescent, hispid or not, each with a prominent median vein and a weaker one on either side. Corolla white, cream or yellow, 2.5–3.5 cm long, sparsely stellate-pubescent; staminal tube 0.8–1.2 cm long; filaments 2–3 mm long; exserted part of styles 2–8 mm long, glabrous or sparsely pilose. Capsule ovoid, beaked, 1–1.3 cm long, 0.7–1 cm diameter, sparsely to densely stellate-pubescent, setose along the sutures, brown; seeds triangular, dark brown or black, 1.5–2 × 2 mm, very shortly tomentose. Fig. 11: 1–5, p. 72.

UGANDA. Acholi District: Madi Opei, Apr. 1943, *Purseglove* 1502!; Ankole District: Queen Elizabeth National Park, Rutanda area, W of Kibona swamp, 26 Mar. 1968, *Lock* 68/54!; Mbale District: between NE foothills of Mt Elgon and Kadam, 12 Mar. 1958, *Symes* 324!

KENYA. Trans Nzoia District: NE Elgon, June 1955, *Tweedie* 1314!; Meru District: Meru National Park, Kiolu plain, 10 Nov. 1978, *Hamilton* 341!; Kwale District: Shimba Hills, Kwale–Mombasa road, 4 May 1968, *Magogo & Glover* 986!

TANZANIA. Tanga Dictrict: Kishidani, 13 Apr. 1965, *Faulkner* 3481!; Kilosa District: Msada, 7 km from HQ, *Greenway & Kanuri* 15285!; Kilosa District: Mwega river, 13 km from HQ, 29 June 1973, *Greenway & Kanuri* 15293!; Zanzibar: Mbweni, 22 June 1930, *Vaughan* 1129!

DISTR. **U** 1–3; **K** 3–4, 7; **T** 1–2, 4, 6, 8; **Z**; Ethiopia, Zambia, Malawi, Mozambique, Zimbabwe, Botswana, Namibia (Caprivi Strip), South Africa

HAB. Seasonally water-logged grassland and flood-plains; 0–1300 m

SYN. *H. articulatus* A. Rich. var. *stenolobus* Hochst. in F.T.A. 1: 20 (1868). Type: Ethiopia, Mt Walcha, *Schimper* III: 1620 (K!, syn.; FT, P, isosyn.)

H. rhodesicus Baker f. in J.B. 37: 424 (1899). Type: Zimbabwe, Bulawayo, *Rand* 69 (BM, holo.)

H. eburneopetalus Baker f. in J.B. 75: 100 (1937). Type: Mozambique, Mucombeze, *Le Testu* 911 (P, holo.)

50. **Hibiscus aethiopicus** *L.* in Mant. Pl. Alt.: 258 (1771); Harv. in Fl. Cap. 1: 174 (1860); Hochr. in Ann. Conserv. Jard. Bot. Geneve, 4: 98 1900); Engl., Pflanzenw. Afr. 3, 2: 396 (1921); E.P.A.: 555 (1959); Exell in F.Z. 1(2): 468 (1961); U.K.W.F. ed 2: 99 (1994); Vollesen in Fl. Eth. & Eritr. 2(2): 202 (1995). Type: South Africa, Cape of Good Hope, "Habitat ad Cap. b. spei.", not designated

Suffrutex from a woody rootstock 6–60 cm tall; stems appressed stellate-hispid, becoming glabrous with age. Leaves ovate to elliptic, rarely lanceolate, 2.2–5.5 × 1–3 cm, apex acute, obtuse or truncate, base cuneate or rounded, margin entire or coarsely dentate, stellate-hispid above and below, the hairs mostly 4-armed; petiole 4–7 mm long, stellate-pubescent and/or -hispid; stipules linear, 4–10 × 2–3 mm, sparsely hispid. Flowers solitary in leaf axils; pedicel 4.5–7 cm long, articulated in the top quarter, stellate-hispid, densely so above articulation; epicalyx of 10 linear bracts 10–15 × 1.5–2 mm wide, stellate-hispid; calyx lobes ovate-acuminate to lanceolate, 13–17 × 3–5 mm, fused in the bottom third, with a prominent median vein and a weaker one along each edge, stellate-hispid along the veins, pubescent on the inside. Corolla yellow, rarely pink, 1.7–3.5 cm long, sparsely hairy or glabrous; staminal tube 0.6–1 cm long; filaments 2–2.5 mm long; exserted part of styles 0.3–0.7 cm long, glabrous. Capsule round, 0.9–1 cm long, 0.9 cm diameter, setose in the upper half, shorter than calyx; seeds dark brown, reniform, 3–3.5 × 2–2.5 mm, tomentose. Fig. 11: 6, 7, p. 72.

UGANDA. Karamoja District: Kokumongole, 28 May 1939, *Thomas* 2871!; Kigezi District: Ishasha river Camp, 12 May 1961, *Symes* 686!; Busoga District: Bukoli county, by the Bugiri–Buswale road, N of Luvunya valley, about 21 km SSE of Bugiri, 17 Dec. 1952, *Wood* 553!

KENYA. Trans Nzoia District: Seboti Hill, SE Elgon, 11 May 1955, *Symes* 360!; Machakos District: Western Division, 53 km from Machakos township, 1 Jan. 1968, *Mwangangi* 529!; Masai District: Chyulu Hills, 19 Oct 1969, *Gillett & Kariuki* 18798!

TANZANIA. Musoma District: Mara River guard post, 24 Dec. 1964, *Greenway & Turner* 11778!; Mbulu District: Chajafa Hill, 24 Apr. 1929, *Burtt* 2111!; Iringa District: Mt Image, 60 km E of Iringa, 9 Dec. 1994, *Goyder et al.* 3925!

DISTR. **U** 1–4; **K** 2–7; **T** 1–2, 4, 6–7; from Eritrea and Ethiopia to South Africa; Yemen

HAB. Grassland and bushland; 900–2200 m

SYN. *H. asperifolius* Eckl. & Zeyh., Enum. I: 38 (1834). Type: South Africa, near Silo on Klipplaat river, Tambukiland, Dec. 1834, *Ecklon & Zeyher* s.n. (?B, holo.)

H. ambelacensis Schweinf. & Ulbr. in N.B.G.B. 8: 160 (1922). Type: Ethiopia, near Maldi, Ambelaco, *Schweinfurth* 496 (B†, holo.)

51. **Hibiscus lobatus** (*Murr.*) *Kuntze* in Rev. Gen. 3. 2: 19 (1898); C.F.A. 1, 1: 176 (1937); F.P.S. 2: 31 (1952); Hochr., Fl. Madag., Malvac.: 42, t. 12 fig. 1–3 (1955); Keay, F.W.T.A. ed. 2, 1(2): 346 (1958); Exell in F.Z. 1 (2): 445 (1961); Vollesen in Fl. Eth. & Eritr. 2(2): 205 (1995). Type: Murray, Comm. Soc. Reg. Sc. Goetting. 6, t. I (icono.)

Herb up to 1 m high; stems tomentose to pilose. Leaves ovate in outline, 5–9.7 × 4–9 cm, unlobed to obscurely 3-lobed to distinctly 3-lobed, apex acute to acuminate, base truncate or cordate, margin undulate, crenate or serrate, sparsely tomentose-hispid above, sparsely stellate-hispid below; petiole 4–9 mm long, pubescent; stipules linear-lanceolate, 4–12 mm long, hispid; pedicel 0.5–1.6 cm long, articulated in the top third, glandular-pubescent; calyx lobes triangular, 4–7 × 1.5 mm, fused in the lower $^{1}/_{3}$, each with a prominent median vein and two lateral veins, sparsely pubescent. Corolla 6–15 mm long, white, glabrous; staminal tube ± 6 mm long; filaments 6 mm, long; exserted part of style 2 mm long, glabrous. Capsule ovoid-ellipsoid, 8–14 mm long, valves lanceolate, sparsely tomentose, setose along the sutures; seeds black, wedge shaped, 1.5 × 1 mm, tuberculate.

UGANDA. Karamoja District: Namalu, July 1956, *J. Wilson* 1800!; Busoga District: Lubolo Hill, 5 km N of Namwiwa, 17 July 1953, *Wood* 835!

TANZANIA. Kigoma District: Tubira Forest, 26 Apr. 1994, *Bidgood & Vollesen* 3189!; Mpwapwa, 19 Feb. 1931, *Hornby* 364!; Iringa District: Mpululu Hill, 11 Mar. 1970, *Greenway et al.* 14069!

DISTR. **U** 1–3; **T** 1, 3–7; Ethiopia, Zambia, Mozambique, Zimbabwe

HAB. Woodland and thicket and forest, rock outcrops and riverine vegetation; 200–1250 m

SYN. *Solandra lobata* Murr. in Comment. Soc. Regiae Sci. Gott. 6: 20 (1783–84)

Hibiscus solandra L'Hérit., Stirp. Nov. 1: 103, t. 49 (1788); Hochr. in Ann. Conserv. Jard. Bot. Genève 4: 128 (1900). Type: Ethiopia, Djeladjeranne, *Schimper* III 1676 (P, lecto.)

Laguna abyssinica Hochst., Tent. Fl. Abyss. 1: 71 (1847). Type as for *Hibiscus solandra*

NOTE. The basionym for this species is *Solandra lobata* Murr. In the absence of a recognized aunthetic specimen, the plate of Murray is the type of the species.

52. **Hibiscus sidiformis** *Baill.* in Bull. Soc. Linn. Par. I: 518 (1855); Hochr., Fl. Madag., Malvac.: 44, t. 12 fig. 4–5 (1955); Exell in F.Z. 1 (2): 445 (1961). U.K.W.F. ed 2: 98 (1994); Vollesen in Fl. Eth. & Eritr. 2(2): 205 (1995); Thulin in Fl. Somalia 2: 50, fig. 28j–k (1999). Type: Madagascar, *Humblot* 645 (P, holo.)

Herb up to 1 m tall; stem simple-pubescent and hispid. Leaves ovate to ovate-lanceolate, simple to digitately trifoliolate, 1–5 × 1–1.5 cm, the leaflets linear to linear-lanceolate, ± equal, apex acute or acuminate, base cordate or truncate, margin entire, sub-entire or dentate, glabrous to sparsely pubescent-hispid above and below;

petiole (1.5–)2–5 cm long, pubescent and hispid; stipules linear, 2–4 × 0.5 mm, hispid, caducuous. Flowers solitary in the leaf axils; pedicel 1.4–4.5 cm long, articulated in the top third, pubescent-hispid; epicalyx lacking or rudimentary, to 0.1 mm long; calyx lobes linear-lanceolate, 7–12 mm long, united in the lower third, sparsely pubescent-hispid, each with a prominent median vein and two less prominent lateral ones. Corolla yellow, 10–15 mm long; glabrous to very sparsely stellate-hispid outside, glabrous within; staminal tube 8 mm long; filaments 2 mm long; exserted part of style 1–1.5 mm long, glabrous. Capsule straw-coloured, ellipsoid, 6–8 mm long, 4–6 mm diameter, setose, half to two-thirds as long as the enclosing persistent calyx, desintegrating at maturity, valves acuminate, papery; seeds black, wedge-shaped, (1–)2 × 2 mm, tuberculate. Fig. 8: 10, 11, p. 57.

KENYA. Northern Frontier District: Dandu, May 1952, *Gillett* 13112!; Machakos District: 3 km WNW of Kyulu Station, 11 Apr. 1966, *Gillett* 17270!; Tana River District: 92 km on the Garissa–Nairobi road, 14 May 1978, *Gilbert & Thulin* 1738!

TANZANIA. Mbulu District: Endabash, Manyara National Park, 20 Jan. 1971, *Richards & Arasululu* 26476!; Kilosa District: Ilonga, 18 Mar. 1950, *Disney* 17!; Iringa District: Msembi Rest House grounds, 30 Mar. 1970, *Greenway & Kanuri* 14222!

DISTR. **K** 1, 3–4, 7; **T** 2, 6, 7; tropical and southern Africa; Madagascar

HAB. Rocky areas in dry woodland; 350–1100 m

SYN. *Solandra ternata* Cav., Diss. 5: 279, t. 136/2 (1788). Type: Senegal, *Adanson* 143 (P-JU, holo.)
Hibiscus ternatus (Cav.) Mast. in Oliv., F.T.A. 1: 206 (1868), *non Hibiscus ternatus* Cav. (1787)
H. ternifolius F.W. Andrews, F.P.S. 2: 31 (1952), *nom. nov.* for *Solandra ternata* and *Hibiscus ternatus* (Cav.) Mast.

UNCERTAIN SPECIES

Hibiscus torrei *Baker f.* in J. Bot. 75: 101 (1937). Type: Mozambique: Niassa, Vila Cabral, July 1934, Torre 435 (COI, holo.; BM, iso.)

Wilson in Bull. Nat. Hist. Mus. Lond. (Bot.) 29(1): 59 (1999) mentions this species as occurring in southern Tanzania, and cites two specimens: TANZANIA. Iringa District: Uhafiwa, 3 August 1989, *Kayombo* 795 A & Luhega forest near Uhafiwa, 10 June 1989, *Lovett et al.* 3287. I have seen neither of these, but according to the Tropicos website the second was identified as *H. noldeae* by Fryxell: which is also puzzling, as that taxon is not known from **T** 7 either. As I have not seen the specimens *H. torrei* is restricted to northern Mozambique, and is only known from the type.

4. ABELMOSCHUS

Medick., Malven-Fam.: 45 (1787); Borss. Waalk. in Blumea 14: 89–105 (1966)

Annual or perennial herbs or subshrubs. Leaves usually palmately 3–5-lobed; stipules subulate or linear, deciduous. Flowers solitary or in terminal pseudoracemes due to reduction of the leaves. Epicalyx lobes 4–16, free or basally connate, persistent or deciduous. Calyx adnate to corolla, spathaceous, irregularly splitting into 2–5 lobes or teeth during anthesis, circumscissile at base, falling together with the corolla. Corolla white, pinkish or yellow with purple centre. Staminal column with anthers all over. Ovary pubescent, 5-locular, each locule with many ovules. Capsule oblong-ellipsoid or fusiform-cylindrical, loculicidally dehiscent, remaining attached to receptacle. Seeds reniform, glabrous or hairy.

About 6 species (estimates of 10 or even 15 are exaggerated) in temperate and tropical parts of Asia, East Africa and Australia but several now widely cultivated in the rest of the world, especially *A. esculentus*, grown for its edible young fruits (lady's fingers, okra, gumbo, bhindi).

1. Epicalyx lobes falling after dehiscence of the fruit, linear to linear-lanceolate 2
   Epicalyx lobes falling before corolla expands, often more broadly lanceolate; capsule 3–3.5 cm long with membranous valves .... 1. *A. ficulneus*
2. Mature capsule fusiform, with ± woody valves, usually sulcate, (7–)10–25 cm long; pedicel 0.5–1.5 cm (2.5 cm in fruit) long .... 2. *A. esculentus*
   Mature capsule ovoid to oblong, with thinner valves, terete or slightly angular, up to 8 cm long; pedicel 1.5–7.5(–19) cm long .... 3. *A. moschatus*

1. **Abelmoschus ficulneus** (*L.*) *Wight*, Cat. Ind. Pl.: 14 (1833); Wight & Arn., Prodr.: 53 (1834); Hochr. in Candollea 2: 86 (1924) & Fl. Madag. 129: 6, t. 1. figs. 5, 6 (1955); Borss. Waalk. in Blumea 14: 101 (1966); Abedin in Fl. W. Pak. 130: 26 (1979); Vollesen in Op. Bot. 59: 35 (1980) & in Fl. Eth. & Eritr. 2(2): 212, fig. 82.9.1–5 (1995); Philcox in Rev. Fl. Ceylon 11: 309 (1997). Type: Dillenius, Hort. Elth., t. 157, fig. 190 (lecto.)

Erect annual or subshrubby herb 0.5–1.5 m tall with tap-root. Stem usually densely pubescent with simple hairs at apex, rarely with small bulbous-based prickles. Leaves ± round in outline, 2–16 cm long and wide, cordate at base, palmately 3–5-lobed; lobes obovate to spatulate or sometimes ± round, 1.5–10 × 1–7 cm, rounded at apex, with simple hairs and some 3-armed hairs, mainly beneath; petiole 2–21 cm long; stipules linear, 5–12 mm long. Flowers solitary in leaf axils or upper in pseudoracemes; pedicel 1–1.5 cm long, becoming up to 3.5 cm in fruit; epicalyx lobes 5–6, usually falling before the flowers expand, linear to lanceolate, 4–12 × 0.5–1.5 mm; calyx 1.5 cm long, with simple hairs, 5-lobed, the lobes linear, 3 mm long. Corolla white turning pink with dark purple centre; petals obovate, 2–3 × 2–3 cm. Capsule oblong to ellipsoid, 3–4 cm long, 1.5–2 cm diameter, 5-angular, essentially rounded at the apex but sometimes with a short acumen, with short stiff hairs, the valves ± membranous; seeds globose, 3 mm, glabrous, with some stellate hairs or sometimes (in Ethiopia) with long crisped hairs, the surface with concentric lines.

UGANDA. Karamoja District: Bokora County, Morulinga, July 1957, *J. Wilson* 365!
KENYA. Tana River District: near Lango ya Simba, 6 Aug. 1988, *Robertson & Luke* 5372!
TANZANIA. Kilwa District: Kingupira Forest, 17 May 1976, *Vollesen* MRC 3613!
DISTR. **U** 1; **K** 7; **T** 8; Nigeria, Chad, Sudan, Ethiopia, Zambia, Mozambique; Madagascar, India, Pakistan, Sri Lanka, Malesia from Java to New Guinea and North Australia
HAB. Grassland on clay, flood plains, river banks, roadsides; 20–1350 m

SYN. *Hibiscus ficulneus* L., Sp. Pl.: 695 (1753); Keay, F.W.T.A. ed. 2, 1(2): 348 (1958)

2. **Abelmoschus esculentus** (*L.*) *Moench*, Meth. Pl.: 617 (1794); Hochr. in Candollea 2: 86 (1924) & in Fl. Madag. 129: 7, t. 2, figs. 1–4 (1955); Exell in F.Z. 1: 423, t. 84 (1961); Borss. Waalk. in Blumea 14: 100 (1966); Abedin in Fl. W. Pak. 130: 25 (1979); Vollesen in Fl. Eth. & Eritr. 2 (2): 212 (1995); Philcox in Rev. Fl. Ceylon 11: 306 (1997). Type: 'India', Linnean Herb 873.31 (LINN, lecto.)

Stout erect annual herb, 0.5–2.7 m tall; stems sometimes tinged red, often fistular, setulose with stiff simple hairs, becoming woody at base. Leaves elliptic to round in outline, 5–25 × 5–30 cm, cordate at base, angular or 5–7-lobed, the lobes triangular, ovate, oblong or lanceolate, coarsely serrate to crenate, with scattered stiff simple hairs on both sides; petiole 5–35 cm long; stipules 5–15 mm long. Flowers solitary in the leaf axils; pedicel 0.5–1.5 cm, accrescent to 2.5(–5) cm long; epicalyx lobes

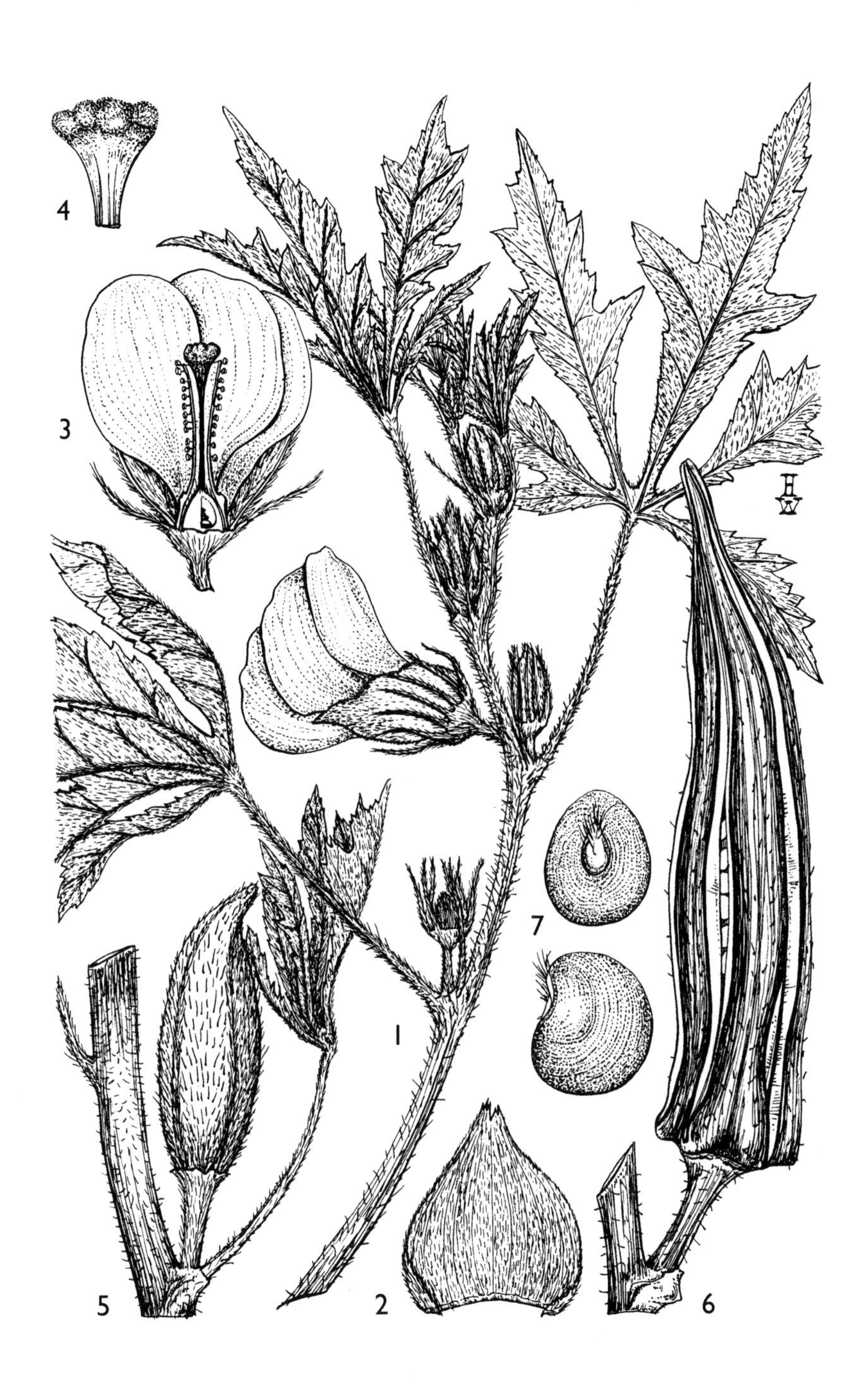

FIG. 12. *ABELMOSCHUS ESCULENTUS* — **1**, flowering stem, × $^2/_3$; **2**, calyx opened out, × 1; **3**, longitudinal section of flower, × 1; **4**, stigma, × 3; **5**, young fruit, × $^2/_3$; **6**, mature fruit, × $^2/_3$; **7**, seed, side and front view, × 4. 1–4 from *Pielou* 112; 5, 7 from *Bullock* 3900; 6 from *Davis* 249. Drawn by Heather Wood.

7–10(–12), linear to lanceolate, 0.5–18(–25) × 1–2.5 mm, falling when capsule dehisces; calyx 2–3(–4) cm long, acuminate in bud, rough with stiff simple hairs, with 5 short linear teeth. Corolla white or mostly yellow with dark purple centre; petals obovate, 3.5–4.5 (?–8) × 3–4 cm; ovary conical to ovoid, 1.2 cm long, 5(–9)-locular. Capsule fusiform, 7–25 (?–30) cm long, 1.3–3 cm diameter, rounded or ± angular, sulcate, with scattered simple hairs or glabrous; seeds dark brown or grey, 5–15 per cell, depressed globose to reniform, 3–6 mm long, striped, with concentric lines of minute stellate hairs, minutely warty, glabrous or pilose. Fig. 12, p. 77.

UGANDA. 'Nile district', *Dawe* 841!
TANZANIA. Mwanza District: Mwanza, 1929, *Davis* 249!; Ufipa District: Rukwa Valley, 25 Feb. 1947, *Pielou* 112! & Milepa, 26 May 1951, *Bullock* 3900!
DISTR. **U** 1; **T** 1, 4; very widely cultivated throughout the tropics and not truly wild anywhere for certain; but almost certainly of tropical Asian origin and *not* from Africa as stated in several crop books; naturalized, especially in Ufipa District
HAB. Open grassland, seasonally flooded plains with *Panicum* etc.; also in flood plains within *Brachystegia* woodland; cultivated in Uganda and West Tanzania; near sea-level–900 m

SYN. *Hibiscus esculentus* L., Sp. Pl.: 696 (1753); Mast. in F.T.A. 1: 207 (1868); Stuhlmann in Deutsch Ost-Afr. 10: Beitr. Kulturgesch. Ostafr.: 142 (1909); U.O.P.Z.: 296 (1949); Exell & Mendonça, C.F.A. 1: 178 (1951); Keay, F.W.T.A. ed. 2, 1(2): 348 (1958); Purseglove, Trop. Crops, Dicot.: 368, fig. 58 (1968)

3. **Abelmoschus moschatus** *Medik.*, Malven-Fam. 46 (1787); Hochr. in Candollea 2: 86 (1924); Borss. Waalk. in Blumea 14: 90 (1966); Abedin in Fl. W. Pak. 130: 26 (1979); Philcox in Rev. Fl. Ceylon 11: 307 (1997). Type: Herb. Clifford: 349, Hibiscus 4 (BM-000646497) lecto., designated by Borssum Waalkes in Blumea 14 : 92 (1966)

Herb or subshrubby herb, 0.2–1.5 m tall with tap-root or tuber; stems usually not fistular, plant mostly hispid with simple hairs and also with minute stellate hairs, rarely glabrous. Leaves very variable, angular to palmately 3–7-lobed, often (and in most of the East African material seen) with upper leaves essentially 3-lobed, 6–30 cm long and wide, hastate or sagittate, the lateral lobes with a small triangular lobe on lower side at base, mostly coarsely serrate; stipules linear, 6–12 mm long; petiole 2–30 cm long. Flowers solitary in the leaf axils; pedicel 1.5–8 cm long, reaching 19 cm in fruit but usually much shorter; epicalyx lobes 6–10, linear to lanceolate, 8–25 × 1–2.5(–5) mm; calyx 2–3.5 cm long. Corolla yellow with deep purple throat, less often white or pink; petals (3–)6–9 cm long, 2–5.5 cm wide. Capsule ovoid-ellipsoid to fusiform, 5–8 cm long, 2–3.5 cm diameter, the valves coriaceous or thinner, not sulcate, hispid with simple hairs; seeds brown or black with paler densely minutely postulate concentric ridges, ovoid-reniform, 3–4 mm long, glabrous or stellate-pubescent.

subsp. **moschatus**; Borss. Waalk. in Blumea 14: 91 (1966)

Herb with slender tap root; stems mostly retrorsely hispid or glabrous. Epicalyx segments usually appressed in fruit. Corolla yellow with purple centre. Capsule 5–8 cm long; seeds glabrous or nearly so.

var. **moschatus**; Borss. Waalk. in Blumea 14: 91 (1966)

Stems always hispid, mostly evenly tinged with red; epicalyx segments 7–10, linear, 8–15 × 1–2 mm.

TANZANIA. Pemba: Chake Chake, Sep. 1929, *Vaughan* 868! & SW of Pemba, Panza I., 13 Feb. 1929, *Greenway* 1392! & Jendeli, 19 Feb. 1929, *Greenway* 1493!
DISTR. **Z**; **P**; Congo-Kinshasa, Cabinda, Angola; India, Indochina, S China, Malesia and Pacific Is., widely cultivated elsewhere in the tropics and occasionally naturalised

HAB. Cultivated in gardens and found near villages; near sea-level

SYN. *Hibiscus abelmoschus* L., Sp. Pl.: 696 (1753); Hiern, Cat. Afr. Pl. Welw. 1: 75 (1896); U.O.P.Z.: 295 (1949); Exell & Mendonça, C.F.A 1: 177 (1951)

NOTE. Borssum Waalkes divides *A. moschatus* into three subspecies, one with two varieties. East African material is the widespread typical variety. Williams (U.O.P.Z.): 295 (1949)) states 'An annual West Indian herb, wild in Zanzibar and Pemba'. It is certainly not native in any of these three territories but presumably naturalised in Zanzibar and Pemba. The seeds produce an oil having a musk-like odour used in perfumery and the pods are used unripe as a vegetable in the same way as *A. esculentus*.

5. ROIFIA

Verdc., **gen. nov.** a *Hibisco* dignoscenda indumento simplici pro parte maxima, fructu schizocarpio loculicide in 3 findens, loculis binis non-aperientibus seminibus inclusis (non capsulis 5-findentibus)

Fioria sensu Vollesen in Fl. Eth. & Eritr. 2 (2): 214 (1995)

Perennial herb or subshrub with mainly simple indumentum. Leaves shallowly to deeply 3–5-lobed, coarsely dentate; stipules subulate. Flowers solitary in the leaf axils; epicalyx of 8–10 free lobes. Calyx 5-lobed to below the middle. Ovary 5-locular; locules with 2 ovules; style 5-branched. Fruit a loculicidal schizocarp with 5 longitudinal wings along the sutures dehiscing into three parts, two 3-winged parts consisting of a whole indehiscent carpel and two halves and one 2-winged part consisting of two halves which dehisce; seeds 1–2 per locule, reniform.

A single species in NE Africa and Arabia. Mattei proposed a new genus *Fioria* (Boll. R. Ort. Bot. Palermo N.S. 2: 71 (1917)) for three species of *Hibiscus* (*Hibiscus vitifolius* L., *Hibiscus dictyocarpus* Webb and *Hibiscus pavonioides* Fiori) but although I suspect he would have considered the first-mentioned to be the type he did not formally choose a type species for his genus. Kearney in Leafl. Western Bot. 7 (11): 272 (1955) discussed the genus and maintained it. Vollesen (Fl. Eth. & Eritr. 2 (2): 214 (1995)) restricted the genus to *Hibiscus dictyocarpus* without a strict mention of a type species but there is an earlier formal selection of *Hibiscus vitifolius* as the 'type' of *Fioria* by Fryxell in Howard, Fl. Lesser Antilles 5 (2): 213 (1989) (see also Sivarajan and Pradeep, Malvaceae of S. Peninsula India: 24 (1996)) and Fryxell in Brittonia 49: 230 (1997). The lectotype is clearly *Hibiscus vitifolius*. Vollesen's argument that *Fioria* in his sense is clearly not congeneric with *Hibiscus vitifolius* indicate that a new genus is needed for *Hibiscus dictyocarpus*. If *Senra* and *Kosteletzkya* are to be kept separate from *Hibiscus* then *Fioria* sensu Vollesen must be also.

Roifia dictyocarpa (*Webb*) *Verdc.* **comb. nov.** Type: Sudan, Kordofan, Arasch-Cool, *Kotschy* 124 (FT, holo.; K!, iso.)

Erect herb or shubshrub, 0.45–1(–1.5) m tall, pubescent or pilose. Lamina broadly ovate or broadly elliptic to ± round in outline, 1–8.5 × 1–9 cm, shallowly to deeply 3–5-lobed, lobes triangular, rounded to acute at the apex, coarsely dentate; petiole 0.5–8 cm long. Pedicels 1.5–6.5 cm long, articulated above or below the middle; epicalyx lobes linear to slightly spatulate, (3–)7–15(–20) mm long; calyx 1–1.8 cm long, covered with stellate hairs with conspicuous yellow incrassated bases and also densely pubescent at apex and margins of the triangular lobes. Corolla white to yellow with red or purple centre (some field notes do not mention dark centre); petals 1–2.5 cm long; staminal column 5–8 mm long. Capsule oblong-ellipsoid, 7–12 mm long, 8–9 mm diameter, glabrous with distinct raised reticulate venation; seeds 2.5–3.5 mm long, shortly pubescent. Fig. 13, p. 80.

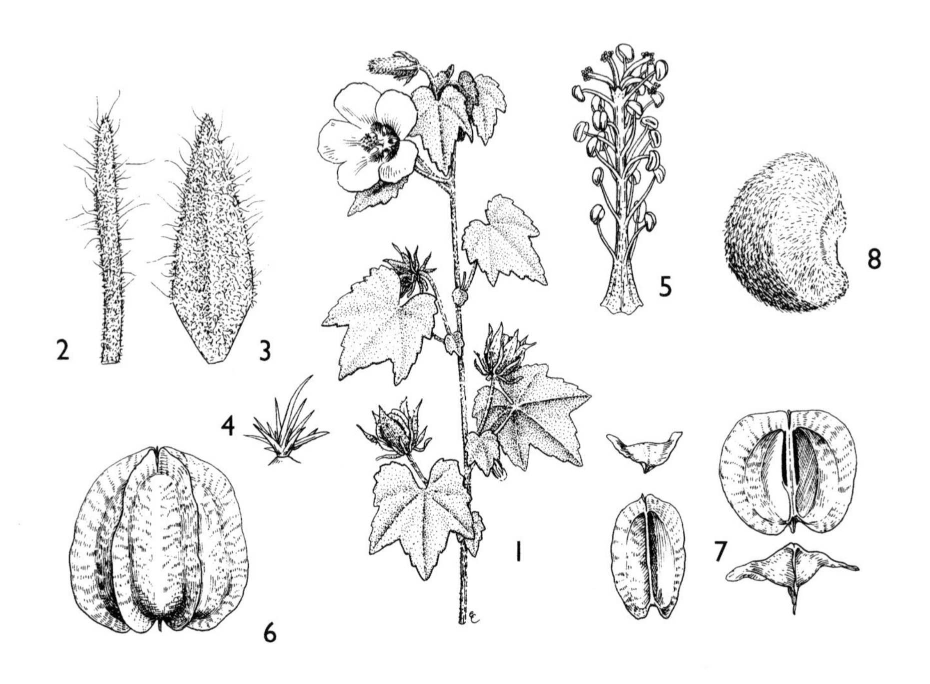

FIG. 13. *ROIFIA DICTYOCARPA* — **1**, flowering stem, × $^1/_2$; **2**, epicalyx bract, × 3; **3**, calyx lobe, × 3; **4**, calyx indument, × 20; **5**, staminal column, × 3; **6**, fruit, × 2; **7**, dehisced fruit, × 3; **8**, seed, × 7. 1 from *Adamson* 617; 2–5 from *Ash* 1210; 6–8 from *Gillett* 25208. Drawn by Eleanor Catherine, and reproduced with permission from Flora of Ethiopia and Eritrea 2, 2.

KENYA. Northern Frontier District: Garissa District: 46 km from Garissa on Hagadera road, 29 May 1977, *Gillett* 21216!; Kitui District: below Yatta Gap, Athi Camp, 28 Jan. 1942, *Bally* 1712!; Tana River District: Nairobi–Garissa road, 25 km E of Hatama Corner, 9 May 1974, *Gillett & Gachathi* 20543!

DISTR. **K** 1, 2, 4, 7; Sudan, Ethiopia and Somalia; Saudi Arabia

HAB. Open *Acacia bussei–A. mellifera* bushland with grassland, floodplain bushland and scattered tree grassland with *Acacia, Commiphora, Cordia, Terminalia* etc. usually on grey alluvial soil; 200–750 m

SYN. *Hibiscus dictyocarpus* Webb, Fragm. Fl. Aethiop. Aegypt.: 46 (1854); Thulin, Fl. Somalia 2: 50, fig. 29 (1999)

H. pavonioides Fiori in Bull. Soc. Bot. Ital. 1913: 46, fig. 1 (1913). Type: Somalia, Benadir, *Fiori* 3, 19 (FT, syn.)

Fioria dictyocarpa (Webb) Mattei in Boll. R. Ort. Bot. Palermo N.S. 2: 74 (1917); Vollesen in Fl. Eth. & Eritr. 2 (2): 214, fig. 82.10 (1995)

F. pavonioides (Fiori) Mattei in Boll. R. Ort. Bot. Palermo N.S. 2: 74 (1917)

6. **KOSTELETZKYA***

C.Presl, Reliq. Haenk. 2:130, t. 70 (1835); *nom. conserv.*

Subshrubs or perennial or annual herbs, erect, ascending, climbing or procumbent; indument stellate, glandular or simple, but always with a line of minute curved hairs decurrent from one side of the leaf base and extending ± to the next node, this often somewhat obscured by other pubescence. Leaves petiolate, unlobed or shallowly to deeply 3–5-lobed, the margin variously toothed or crenate or serrate or subentire; stipules linear or linear-subulate. Flowers single in the axils of the upper leaves or, by reduction of the upper leaves, forming open panicles or racemes or spiciform racemes, or the flowers coalesced into pedunculate glomerules. Epicalyx of 5–13 segments, these ± free. Calyx enlarging somewhat in fruit, lobed for about $^2/_5$–$^5/_6$ its length. Corolla pink or white or yellow and sometimes with a red spot at base, coarsely pubescent dorsally where exposed in bud. Staminal column 5-toothed at the apex, filamentiferous nearly throughout or mostly in the apical $^1/_3$. Style single, 5-fid at the orifice of the staminal column, stigmas capitate. Fruit a depressed, 5-locular, 5-angulate, often 5-crested capsule, the capsule valves often setose on the margins, separating from one another and from the fruiting axis at maturity, rarely tardily so. Seeds 1 per locule, reniform-obovate to reniform-globose, glabrous or pubescent and often with curved, concentric lines. Chromosome number X= 19.

Seventeen species from tropical Africa, the northern neotropics and the Philippines, with one species extending into temperate coastal areas in the United States, Europe and western Asia; five occur in the Flora area; usually in (seasonally) moist areas.

Kosteletzkya has come to embrace a number of discordant elements, and some of these must be excluded. Among them is the taxon treated here as *Hibiscus vitifolius* L. [= *Kosteletzkya vitifolia* (L.) M. R. Almeida & N. Patil].

This genus has never been treated as a whole, although a number of regional floras have each included several species, e.g. F.W.T.A., ed. 2 (1958); Exell in F.Z. (1961); Hauman in F.C.B. (1963); Fryxell in Syst. Bot. Monogr. 25 (1988). The five East African species, though variable, offer few problems in circumscription or identification. Experimental hybrids among some of them have been produced in plants under cultivation (Blanchard, Proc. Indiana Acad. Sci. 86: 407 (1977)), and there is evidence of allopolyploid speciation in the genus as well (Blanchard, Amer. J. Bot. (Suppl.) 2: 941–942 (1985)), but no specimens among many hundreds that were examined could be interpreted as naturally occurring interspecific hybrids.

Plants in most *Kosteletzkya* species, and in all Flora species, are self-compatible and their flowers are each open for only a single day. Over the course of the day the stigmas, which are at first more-or-less approximate, gradually recurve into the anther mass, ensuring self-pollination in the absence of outcrossing. Exogenous pollen halts the recurving process in at least one species (Ruan *et al.*, S. Afr. J. Bot. 70: 640–645 (2004)). *K. borkouana* takes this recurving to extremes by usually selfing before the corolla has expanded in the morning.

During dehiscence of the fruits, two or three adjacent valves often cohere and separate from the fruiting axis as a unit. Presumably they adhere to animal fur and feathers, affording a means of dispersal in the wild.

Some species of *Kosteletzkya* have been described as stinging or urticating, including the East African *K. begoniifolia* and *K. grantii*, but no structural or chemical evidence has been reported. Most of the species are reported to have indigenous medicinal, ritual or other uses, notably *K. grantii.*

1. Leaves linear to narrowly lanceolate; flowers creamy white, usually drying yellowish; seeds reniform-globose, pubescent ... 1. *K. buettneri*
   Leaves broader, often angulate or lobed; flowers white to pink, with or without a darker center, drying pink or bluish; seeds reniform-obovate, pubescent or minutely papillose or glabrous ... 2

* by Orland J. Blanchard, Jr.

2. Petals dark toward the base; pedicel 0.8–4.5 cm long, flowers and fruits never forming spike-like racemes or peduncled few-flowered glomerules 3
   Petals pale toward the base; flowers and fruits, or some of them, subsessile in spike-like racemes or few-flowered glomerules, these often pedunculate 4
3. Petals 18–28 mm long; staminal column dark; capsule 8–12 mm wide, valve margin strongly angled in outline; seeds 3.2–3.6 mm long and glabrous .... 2. *K. begoniifolia*
   Petals 8–20(–22) mm long; staminal column pale; capsule 5.5–9.5 mm wide, valve margin rounded in outline; seeds 2.2–2.7 mm long and minutely papillose or papillose-pubescent .... 3. *K. adoensis*
4. Petals short, less than twice the length of the calyx; fruit much wider than high, bristly, the fruiting calyx ± rotate; seeds pubescent .... 4. *K. borkouana*
   Petals longer, more than twice the length of the calyx; fruit $^{2}/_{3}$ or more as high as wide, predominantly minutely tomentose, embraced by the calyx; seeds glabrous ...... 5. *K. grantii*

1. **Kosteletzkya buettneri** *Gürke* in Büttner, Verh. Bot. Ver. Province Brandenburg 31: 92 (1890) & in P.O.A.: 268 (1895) & in Kunene-Sambesi-Exped.: 301 (1903); Exell & Mendonça, C.F.A. 1:157 (1937); F.P.S. 2:33 (1952); Keay, F.W.T.A. ed. 2, 1(2): 349 (1958); Exell in F.Z. 1: 472, fig. 91 (1961); Gledhill, Ch. List Fl. Pl. Sierra Leone: 12 (1962); Hauman in F.C.B. 10: 141 (1963); Blanchard in Rhodora 76: 64 (1974); F.P.U.: 55 (1975); Berhaut, Fl. Illustr. Sénégal 6: 225 (1979); Liberato in Fl. Guinée-Bissau: 19 (1983). Type: Angola, Malange, *Teusz* 381 (B†, syn.; K!, lecto., chosen here)

Coarse perennial herb or subshrub 0.5–2.5 m tall, erect or ascending, stems stellate- and often simple-pubescent, glabrate basally; roots fibrous-thickened. Leaf blades linear to lanceolate, (3–)6–22 × (0.2–)0.5–3.5 cm, rounded or broadly cuneate to truncate at base, rarely cordate or subhastate, margin irregularly coarsely but shallowly crenate-serrate or subentire, apex acute or narrowly so, sparingly to moderately pubescent above with stellate hairs of two sizes, more densely pubescent below, weakly three-veined in the basal half; petiole 0.2–3.5(–6) cm long; stipules linear, 2.5–10 mm long, ± setose. Flowers single in the axils of upper cauline leaves or in open ± leafy axillary racemes; pedicel 0.5–4.6 cm long, abruptly more densely pubescent distally, the transition point becoming joint-like in fruit; epicalyx broadly funnelform to broadly campanulate, of 8–12 linear segments, these of somewhat irregular length, to 2–6 mm long, setose; calyx funnel-form to campanulate, 4–6.5 mm long, lobed for × to $^{4}/_{5}$ its length, the lobes triangular, acute. Corolla creamy-white or very rarely with a pink blush, yellowish at the extreme base, drying yellowish, rotate to broadly funnelform, the petals broadly obovate, 12–20 × 10–16 mm, apically rounded; staminal column 7–14 mm long, anthers 28–36, throughout the distal $^{3}/_{4}$ or more of the staminal column; style-branches exserted 0.8–2 mm, white, stigmas white or rarely pale pink. Fruit reddish brown to dark brown at maturity, depressed, pentagonal, 9–13 mm in diameter, about $^{1}/_{2}$ as high, hispid, the stiff, simple hairs mixed with minute, fine stellate hairs, the margin of each capsule valve ± rounded in outline; seeds reniform-globose, (3.1–)3.4–4.4 mm long, minutely and densely stellate-pubescent.

UGANDA. Lango District: Dokolo, Nov. 1931, *Herb. Bot. Uganda* 2363!; Teso District: Soroti, 17 Sept. 1954, *Lind* 390!; Mengo District: Bulamagi, 16 May 1953, *G. & S. Wood* 745!

TANZANIA. Ngara District: ?Kibirizi, 8 March 1961, *Tanner* 5873!; Buha District: 64 km from Kibondo on Kasulu road, 24 Nov. 1962, *Verdcourt* 3444!; Mpanda District: Uruwira–Tabora road, Uruwira Forest, Mtambo R., 30 Sept. 1970, *Richards & Arasululu* 26214!

DISTR. **U** 1, 3, 4; **T** 1, 4; West Africa from Senegal to Sudan and south to Namibia and Botswana
HAB. Marshes in sandy to dark muddy soils, marshy borders of rivers and lakes, and periodically flooded savannas and wooded savannas; 650–1850 m

SYN. *Kosteletzkya flava* Baker in J. L. S. 30:74 (1894); Ulbr. in V. E. 3(2): 405 (1921); F.W.T.A. 1: 270 (1928). Type: Sierra Leone, Niger River, Erimakuna, *Elliot* 5256 (K!, lecto., BM!, US!, iso.)
Hibiscus liebrechtsianus De Wild. & T. Durand in B.S.B.B. 38(2): 22 (1899); De Wild. in Ann. Mus. Congo Belge Bot. Sér. I, 1:141, t. 81 (1901) & in B.J.B.B. 5: 33 (1915) & Contr. Fl. Katanga: 128 (1921). Type: Congo-Kinshasa, Bumba, *Dewèvre* 900 (BR!, lecto.)

NOTES. Hauman identified *Büttner* 72 as holotype, but Gürke cited two other collections as well, and designated none as the type. One of these, *Pogge* s.n., from Lulua, has not been found. An isosyntype of *Teusz* 381 is the only known extant specimen among these syntypes, and it is here chosen as lectotype.

This species, with its long narrow leaves, is unique and distinctive within the genus. The petals, which are creamy-white (rarely with a pinkish blush) when fresh, turn yellowish when spent, or in dried specimens.

2. **Kosteletzkya begoniifolia** (*Ulbr.*) *Ulbr.* in Fries & Fries in N.B.G.B. 8:684 (1924); T.T.C.L. 5(1): 112 (1940); E.P.A.: 571 (1959); Exell in F.Z. 1: 474 (1961); Hauman in F.C.B. 10: 141 (1963); U.K.W.F.: 200, fig. (1974); Vollesen in Fl. Eth. & Eritr. 2(2):215 (1995). Type: Tanzania, Arusha District: Meru above Arusha, *Braun* 3997 (B†, holo., K!, lecto.; EA!, iso.)

Coarse, perennial herb or subshrub 0.4–2.8 m tall, ascending or somewhat scrambling, stems hispid with simple or few-armed stellate yellowish hairs, these sometimes accompanied by minute, fine stellate hairs; roots fibrous-thickened. Leaf blades ovate to orbicular in outline, or the lowest broadly transversely ovate, 3–10.5 × 3–11.5 cm, cordate to truncate at base, margin unevenly serrate to crenate-serrate, lower leaves often three-lobed or three-angulate, the lateral lobes in the distal $^1/_2$ to $^1/_3$ of the blade and broadly acute to short-acuminate at apex, pubescence as in the stems, the stiff hairs often somewhat appressed; petiole 0.4–7.4 cm long, less than half the blade in upper leaves, to $^3/_4$ the blade or more below; stipules linear, 3–8 mm long, setose. Flowers single or sometimes paired in axils of upper leaves or forming few-flowered axillary racemes or panicles by reduction of leaves; pedicel 1–4.5 cm long, often somewhat deflexed distally after anthesis, abruptly more densely pubescent distally, this point becoming joint-like at fruiting; epicalyx spreading, of 9–10 linear segments 5–9 mm long, densely setose; calyx campanulate to broadly funnelform at anthesis, somewhat 5-angled at the level of the sinus, 7–14 mm long, lobed for $^2/_3$ to over $^4/_5$ its length, the lobes triangular, acute. Corolla deep pink to nearly white, pinkish-red to deep red near the base and in lines radiating distally along the veins, rotate or broadly funnelform, the petals broadly and asymmetrically obovate, 18–28 × 12–22 mm, apically rounded; staminal column (7–)10–15 mm long, anthers 30–38, distributed in the distal half of the staminal column or also with a few scattered closer to the base; style-branches deep pink, exserted 2–5 mm, stigmas deeper pink. Fruits yellowish-brown, pentagonal, 8–12 mm wide, $^2/_3$ to more than $^4/_5$ as high as wide, tardily dehiscent, hispid on the valve sutures, elsewhere mostly glandular- and stellate-pubescent or if simple hairs present these much finer than on the sutures, the margin of each capsule valve in outline strongly acutely angulate near apex; seeds dark brown with pale slightly raised concentric lines, reniform-obovate, 3.2–3.6 mm long, glabrous. Fig. 14, p. 84.

UGANDA. Mbale District: Bugisu, July 1926, *Maitland* 1218! & Budadiri, Jan 1932, *Agric. Dept. Th.* 671! & Budadiri, Jan. 1935, *Chandler* 512!
KENYA. Trans-Nzoia District: NE Elgon, Dec. 1949, *Tweedie* 820!; Naivasha District: Lake Naivasha Sep. 1969, *Mrs. E. Polhill* 122A!; Meru District: Meru, June 1951, *Hancock* 75!
TANZANIA. Masai District: Ngorongoro crater floor, April 1941, *Bally* 2333!; Arusha District: Arusha Nat. Park, [Ngurdoto Crater Nat. Park], 15 Oct. 1965, *Greenway & Kanuri* 12162!; Mpanda District: Kasangazi, 24 July 1958, *Mahinde* 128!

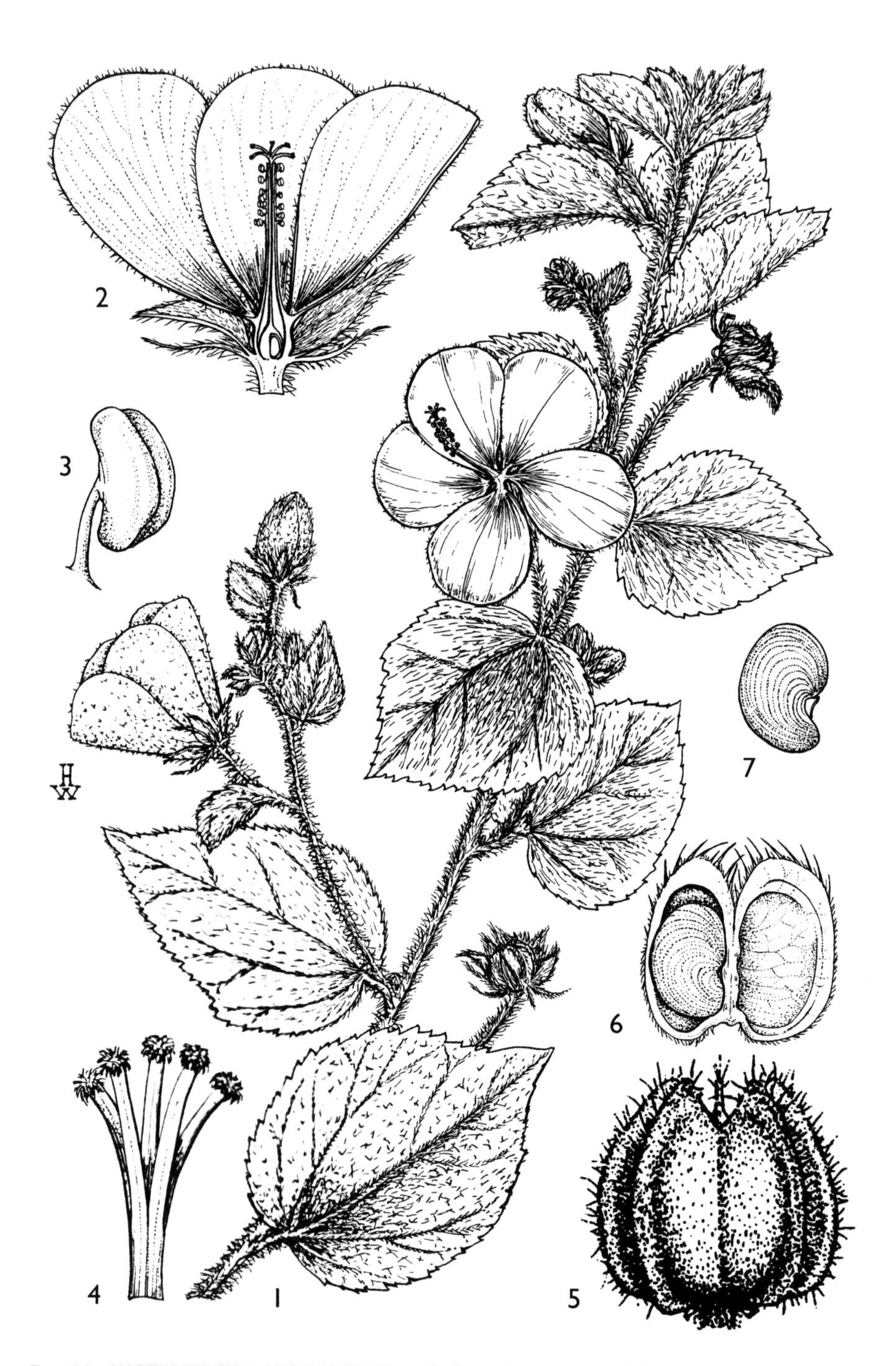

FIG. 14. *KOSTELETZKYA BEGONIIFOLIA* — **1**, flowering stem, × $^2/_3$; **2**, longitudinal section of flower, × $1^1/_2$; **3**, stamen, × 20; **4**, stigma, × 10; **5**, fruit, × 10; **6**, section of fruit, × 3; **7**, seed, × 4. 1–4 from *Fries & Fries* 1870; 5 from *Ash* 1840; 6–7 from *Mahinde* 128. Drawn by Heather Wood, except 5 which is by Eleanor Catherine.

DISTR. **U** 3; **K** 3, 4; **T** 2, 4; Cameroon, Sudan, Ethiopia, Congo-Kinshasa, Zambia
HAB. Clearings and openings in upland and montane forests, swamps, marshes and moist thickets; 1450–2350 m

SYN. [*Kosteletzkya grantii* sensu Garcke in Linnaea 38: 697 (1874) & 43: 53 (1880); Fiori in Nuov. Giorn. Bot. Ital. 47: 33 (1940) as "*grantzii*", *non* Masters]
Hibiscus begoniifolius Ulbr. in N.B.G.B. 7: 182 (1917) & in V. E. 3(2): 403 (1921)

NOTES. Collected first by Schimper in Ethiopia in 1863, this gathering was initially misidentified as *K. grantii* by Garcke (1874). Ulbrich described the taxon as a new species of *Hibiscus*, but he later transferred it to *Kosteletzkya*.
This large-flowered species has distinctive fruits that form five apico-lateral points. The plant sometimes co-occurs with *K. adoensis*.

3. **Kosteletzkya adoensis** (*A. Rich.*) *Mast.* in Oliver, F.T.A. 1: 194 (1868); Gürke in P.O.A.: 268 (1895); Ulbr. in V.E. 3(2): 405, fig. 186 A, B (fr.) (1921) & in Fries & Fries in N.B.G.B. 8:683 (1924); Exell & Mendonça, C.F.A. 1(1): 158 (1937); T.T.C.L. 5(1): 112 (1940); F.P.N.A.: 602 (1948), excl. specim. *Bequaert* 5478; Hochr. in Fl. Madagascar, Fam. 129: 108, fig. 27.1–3 (1955); Keay, F.W.T.A. ed. 2, 1(2): 349 (1958); E.P.A.: 570 (1959); Exell in F.Z. 1:474 (1961); Hauman in F.C.B. 10: 137, fig. 14 (1963); U.K.W.F.: 200, fig. (1974); F.P.U.: 55 (1975); Troupin, Fl. Rwanda 2: 379 (1983); Vollesen in Fl. Eth. & Eritr. 2(2): 215 (1995). Type: Ethiopia, near Adoa, *Schimper* 341 (P!, Herb. Steudel, lecto., chosen here; BM!, BR!, FT!, L!, LE!, LISU!, M, P!, UC!, W!, iso.)

Perennial herb or subshrub, climbing, scrambling or procumbent, sometimes rooting at the nodes, 0.5–3 m long, stems pubescent with spreading simple or few-armed stellate hairs; roots fibrous-thickened. Leaves ovate to orbicular in outline, the lowest usually broadly transversely ovate, 2.5–6(–9) × 2.5–8(–11) cm, cordate or broadly cordate at base, sometimes the upper leaves truncate to broadly cuneate, margin unevenly serrate or crenate-serrate, most leaves 3(–5)-angulate-lobate or -lobed, the principal lateral lobes at about the distal $^1/_2$–$^1/_3$ of the blade, apices acute to short-acuminate, sparingly pubescent above with mostly simple, ± appressed hairs, beneath more densely pubescent, usually with a mixture of simple hairs and coarse 3-armed hairs, sometimes underlain with much finer stellate hairs; petiole 0.3–5 cm long, $^1/_2$ to equalling the blade except the uppermost; stipules linear, 4–8 mm long, setose. Flowers single in axils of upper leaves or in groups of 2–3, or often on lateral branches with reduced leaves; pedicel 0.8–3 cm long, often somewhat deflexed distally after anthesis, abruptly more densely pubescent and often thickened distally, this transition becoming joint-like at fruiting; epicalyx broadly funnelform or rotate or the segments slightly recurved, these 5–9, linear-subulate, 3–7 mm long, setose; calyx funnelform-campanulate, 4–6.5(–8) mm long, lobed for $^2/_3$–$^5/_6$ its length, the lobes triangular, acute to short-acuminate. Corolla creamy white to pink to mauve, dark purplish near the base, this sometimes radiating in lines distally, funnelform to rotate, the petals broadly and asymmetrically obovate, 8–22 × 4–18 mm, apically rounded; staminal column 4–14(–18) mm long, anthers 10–35, distributed in the distal $^1/_2$ to $^1/_3$ of the staminal column; style-branches pale pink to white, exserted 1.3–4 mm. Fruits olivaceous- to yellowish-brown, depressed-pentagonal, 5.5–9.5 mm wide, $^2/_5$ to $^2/_3$ as high as wide, the capsule valves transversely rugose, sometimes with a darker median longitudinal stripe, hispid on the valve margins, otherwise glandular- and finely stellate-pubescent, long simple hairs few if any, the valve margins rounded in outline or somewhat angulate dorso-laterally; seeds olivaceous- to yellowish-brown, reniform-obovate, sometimes slightly impressed laterally, 2.2–2.7 mm long, sometimes with faint concentric lines, the surface papillose, grading to papillose-pubescent, the basally thickened hairs short, minute, curved.

UGANDA. Kigezi District: Mutanda L., 23 April 1970, *Lye* 5233!; Mbale District: Sipi, 30 Aug. 1932, *A.S. Thomas* 402!

KENYA. Turkana District: Murua Nysigar [Moruassigar], 15 Feb. 1965, *Newbould* 7138!; Trans-Nzoia District: S Elgon, Sept. 1939, *Tweedie* 480!; Teita District: Taita Hills, E of Mwandongo Forest, 6 Nov. 1998, *Mwachala* 800!

TANZANIA. Arusha District: Kilinga Forest, 20 Sept. 1971, *Arasululu* in *Richards* 27247!; Mpanda District: Ntakatta Forest, 10 June 2000, *Bidgood, Leliyo & Vollesen* 4625!; Songea District: Matengo Hills, NE of Mpapa by Luhekea R., 25 May 1956, *Milne-Redhead & Taylor* 10451!

DISTR. **U** 2, 3; **K** 2–7; **T** 1–4, 6–8; West Africa from Sierra Leone to Ethiopia and south to Zimbabwe; Madagascar

HAB. Clearings, secondary growth and forest edges in montane gallery forest, grassland thickets and scrub, often near streams, marshes and lake edges, and adventive into pastures, gardens and abandoned cultivated ground, in partial shade or full sun; 1000–2700 m

SYN. *Hibiscus adoensis* A.Rich., Tent. Fl. Abyss. 1: 54 (1847)
H. terniflorus Garcke in Bot. Zeitung (Berlin) 7: 833 (1849). Type: "Africa orientalis" (B†, holo.)
Polychlaena adoensis (A. Rich.) Garcke in Schweinf. Beitr. Fl. Aethiop.: 56 (1867)
Polychlaena adoensis var. *hispidissima* Garcke in Schweinf., Beitr. Fl. Aethiop.: 56 (1867). Type: Sudan, Khartoum in Sennar, *Heuglin* s.n. (B†, holo.)
Kosteletzkya hispida Bak. in J.L.S. 20: 98 (1883), *non* C. Presl (1835), *nom. illegit.* Type: Madagascar, *Lyall* 192 (K!, lecto.)
Pentagonocarpus adoensis (A. Rich.) Kuntze, Rev. Gen. Pl. 1: 73 (1891)
Pentagonocarpus bakerianus Kuntze, Rev. Gen. Pl. 1: 73 (1891). Type as for *Kosteletzkya hispida* Bak.
Hibiscus katangensis De Wild. in B.J.B.B. 5:33 (1915). Type: Congo-Kinshasa, Shinsenda, *Bequaert* 426 (BR!, holo.)
Kosteletzkya adoensis (A. Rich.) Mast. forma *repens* Hauman in B.J.B.B. 31: 89 (1961). Type: Congo-Kinshasa, Nioka, *De Craene* 222 (BR!, holo.)

NOTES. *Hibiscus adoensis* was based on collections of Schimper and Quartin-Dillon. No specimen at Paris shows direct connection with A. Richard, but *Schimper* 341 from Herb. Steudel was undoubtedly available to him. Duplicates of this earliest known gathering of an African *Kosteletzkya* are widely distributed, so it has been chosen here as lectotype. The specimen label was apparently the source of the local name "Sugott" cited by Richard. On the other hand the November date given by Richard for flowering and fruiting cannot be matched among any of the collections at hand, Quaratin- Dillon's or Schimper's. Those with stated dates give September or October.

Garcke's taxa *Hibiscus terniflorus* and *Polychlaena adoensis* var. ß *hispidissima* are included here with some reservation since no authentic specimens of either are known to be extant. The fact that the latter is from Khartoum in Sudan makes its identity further uncertain since *K. adoensis* is not otherwise known from N of the Imatong Mountains in that country. Concerning *Hibiscus terniflorus,* Garcke himself later noted (E.J. 21: 379–401 (1896)) that the name was a synonym of the earlier *K. adoensis.* One might accept his conclusion, trusting that Garcke would have had a good grasp of the species to be willing to make that concession, but a *Schimper* 32? specimen at F that is labeled *Polychlaena adoensis* in Garcke's hand is in fact a *Sida*!

The name *K. adoensis* var. *hirsuta* Oliver has appeared in various places in the literature (Oliver in Trans. Linn. Soc. London ser. 2, 2: 329 (1887), Oliver & Baker in Johnston, Kilima-Njaro Exped. Appendix: 338 (1886), De Wildeman, Ann. Soc. Sci. Bruxelles 40: 113 (1921) & Fl. Katanga: 129 (1921)) but it does not seem to have been published validly.

A record from Zanzibar (*Bouton* s.n.) is doubtful.

Some specimens from the Kigoma, Mpanda and Iringa districts in southwestern Tanzania are notable for their unusually deeply 3–5-lobed leaves or exceptionally large flowers or both.

4. **Kosteletzkya borkouana** *Quézel* in Bull. Soc. Hist. Nat. Afrique N. 48: 87 (1957) & in Mém. Inst. Rech. Sahar. 4: 161, fig. VIIIA (1958). Type: Chad, Borkou, Tigui, *Quézel* s.n. (AIX!, lecto., chosen here; AIX!, iso.)

Coarse perennial herb or subshrub, erect or ascending, 0.7–3 m tall, stems hispid with simple or few-armed stellate pale or yellowish hairs; roots fibrous-thickened. Leaves broadly ovate to orbicular in outline, or the lowest broadly transversely ovate, (1.5–)3–10.5 × (1.5–)3–14.5 cm, cordate or broadly so at base, margin unevenly serrate to crenate-serrate, 3-lobed or 3-angulate, except the uppermost, the lobes in the distal × to $^1/_3$ of the blade and broadly acute at apex, scabridulous above and below, pubescence above of mostly simple hairs, below of mostly triradiate stellate hairs, sometimes sparingly intermixed with shorter, finer hairs; petiole 1–4.5 cm long, less than half the blade in the upper leaves, to $^{3}/_{4}$ the blade in the lower; stipules linear, 2–6 mm long, setose. Flowers single in axils of the upper leaves or more often paired or in threes, the pedicel often coalesced for part or all of their lengths and the flower-cluster thus pedunculate; pedicel or peduncles 0.5–3 cm long, abruptly more densely and coarsely pubescent distally, this point becoming joint-like at fruiting; epicalyx broadly funnelform, of 7–10 linear segments, mostly 2–4(–5) mm long, setose; calyx funnelform-campanulate, 4–5.2 mm long, lobed for mostly $^{2}/_{3}$ to $^{4}/_{5}$ its length, the lobes triangular, acute. Corolla pale pink to nearly white, more pink distally, funnelform, the petals broadly and obliquely obovate, 6–8.5 × 4–6 mm; staminal column 4.5–7 mm long, anthers 8–10, mostly or entirely restricted to the distal $^{1}/_{3}$–$^{1}/_{4}$ of the staminal column, often dehisced before the buds open; style-branches exserted 0.6–1.2 mm, white, glabrous, often recurved into the anther-mass before the buds open, stigmas pale pink to pink. Fruit brown to dark brown at maturity, somewhat depressed, pentagonal, 8–9.5 mm wide, $^{1}/_{2}$ as high as wide or slightly more, valves faintly transversely rugose, hispid throughout, the hairs longest on the valve sutures, otherwise usually underlain with much finer glandular and stellate hairs, the valve margins broadly ovate and acute in outline; seeds dark brown, reniform-obovate, 2.8–3.2 mm long, with pale slightly raised concentric lines, these ± obscured by minute curved simple hairs.

UGANDA. Kigezi District: Lake Edward plains, Nov. 1950, *Purseglove* P.3516!; Masaka District: Bukoto Co., mainland facing Bukakata, 25 May 1972, *Lye* 6989!; Mengo District: Kiagwe, Namanve Swamp, March 1932, *Eggeling* 221!

TANZANIA. Buha District: Ibanda mbuga, 50 km W of Murungu, 26 Aug. 1950, *Bullock* 3214!; Kigoma District: Kitwe Point, 6 km S of Kigoma, 23 April 1994, *Bidgood & Vollesen* 3158!

DISTR. **U** 1, 2, 4; **T** 4; Chad, Congo-Kinshasa

HAB. Marshes and swamps, usually among graminoids, often on or near major water bodies or water courses; 650–1350 m

NOTES. Though first described in 1957 from Northern Chad, a dozen or so separate collections of this taxon had already been made in East Africa and adjacent Congo-Kinshasa, the first in 1910. These have until now generally passed unnoticed in floras and herbaria as *K. adoensis* or *K. grantii.*

Quézel collected two specimens, one in flower and another, apparently on a later date in the same year, in fruit. I have chosen the flowering specimen, which Quézel also used subsequently to illustrate the species, as the lectotype.

Despite its history of obscurity, probably aided by the remoteness of the type collection, this species is quite distinctive. Its erect habit, frequently coalesced pedicel, pale pink to white flowers lacking a darker "eye," and its larger more pubescent capsules and seeds distinguish it from *K. adoensis*. Cytologically, *K. borkouana* is a tetraploid whereas *K. adoensis* is a diploid, based on chromosome counts of multiple collections of both species (Blanchard, unpublished data). The other East African species with which this species has been confused (*K. grantii*) has larger flowers and glabrous seeds.

5. **Kosteletzkya grantii** (*Mast.*) *Garcke* in Linnaea 43:53 (1880), excl. spec. cit., see Note; Gürke in P.O.A.: 268 (1895), excl. Abyssinian distr.; Gürke & Ulbr. in Z.A.E. 2: 502 (1914); Ulbr. in V.E. 3(2): 405 (1921), excl. Abyssinian distr.; Exell & Mendonça, C.F.A. 1(1): 158 (1937); F.P.N.A.: 602 (1948); F.P.S. 2:33 (1952); Keay, F.W.T.A. ed. 2, 1(2): 349 (1958); Gledhill, Ch. List Fl. Pl. Sierra Leone 12 (1952); Hauman in F.C.B. 10: 140 (1963), excl. specim. *Brédo* 1880, *Liben* 421; F.P.U.: 55 (1975); Berhaut, Fl. Illustr. Sénégal 6: 227 (1979); Liberato in Fl. Guiné-Bissau: 18 (1983). Type: Uganda, Ukidi, Ugani and Madi, *Speke & Grant* 162? [sic] (K!, lecto., chosen here)

Coarse perennial herb or subshrub, erect or ascending, (0.5–)1–2.5(–3) m tall, stems hispid with multiple-armed stellate hairs; taprooted or the roots fibrous-thickened. Leaves ovate to orbicular in outline, the lower often broadly transversely ovate, 4–12 × 2.5–11.5 cm, broadly cordate to truncate (rounded to broadly cuneate) at base, margin crenate-serrate or crenate-dentate, usually coarsely and sometimes irregularly or doubly so, lower and middle leaves usually three-lobed or -angulate, lateral lobes at about the distal $^1/_3$ of the blade, apices broadly acute to short-acuminate, ± scabrid above with simple to 3-armed stellate hairs, the hairs beneath more dense and underlain by finer stellate pubescence; petiole 0.2–5.6 cm long, less than $^1/_2$ the blade, often much shorter in upper leaves; stipules linear or linear-subulate, 5–12 mm long, setose. Flowers usually nearly sessile in terminal and axillary glomerules or spike-like racemes, these sometimes appearing pedunculate by reduction or dropping of the subtending leaves; pedicel 0.1–0.7 cm long, densely pubescent; epicalyx broadly funnelform, of 7–13 linear-subulate rather unequal segments 4–13 mm long, conspicuously setose; calyx funnelform-campanulate, 6.5–9 mm long, lobed for about $^2/_5$–$^2/_3$ its length, the lobes triangular, acute to short-acuminate. Corolla pink, rarely somewhat darker basally, very rarely white, rotate or the petals somewhat reflexed, these obliquely obovate or narrowly obovate, 18–26 × 7–16 mm; staminal column 12–21 mm long, anthers 28–40, distributed in the distal $^1/_3$ to $^1/_2$ of the staminal column; style-branches exserted 1.5–4 mm, pink, stigmas deep pink. Fruit brown to yellowish-brown, pentagonal, somewhat embraced by the calyx, 8–9.5 mm in diameter, $^2/_3$ to nearly as high, minutely and densely stellate- and glandular-pubescent, any long simple hairs confined to the sutures, the valve margins hemi-obcordate in outline; seeds brown, narrowly reniform-obovate, 3–4.2 mm long, glabrous, with pale, slightly raised concentric lines.

UGANDA. Lango District: Lango, Sep. 1940, *Purseglove* 1007!; Toro District: Dorwa R., Dec. 1925, *Maitland* 1003; Teso District: Teso, Aug. 1932, *Chandler* 823!

KENYA. District: North Kavirondo District: between Soi R. and Busia, 8 Aug. 1938, *Evans & Erens* 1655!

TANZANIA. District: Kigoma District: Mkuti, 30 May 1979, *Grabner* s.n.!

DISTR. **U** 1–3; **K** 5; **T** 4; Sierra Leone to Sudan and south to Congo-Kinshasa and Angola

HAB. On rocky or sandy soils in grasslands and open savannas, and apparently adventive in clearings, pastures and cultivated ground; 400–2400 m

SYN. *Hibiscus grantii* Mast. in Oliver, F.T.A. 1: 203 (1868); Oliver in Trans. Linn. Soc. London 29: 36, t.12 (1873)

Pentagonocarpus grantii (Mast.) Kuntze, Revis. Gen. Pl. 1: 73 (1891)

Fugosia grantii (Mast.) Hochr. in Annuaire Conserv. Jard. Bot. Genève 4: 173 (1900)

Hibiscus adenosiphon Ulbr. in N.B.G.B. 7: 181 (1917) & in V.E. 3(2): 403, fig. 187 H–K (calyx, fr., seed) (1921) & in N.B.G.B. 8: 169 (1922). Type: Congo-Kinshasa, Lake Tanganyika, *Kassner* 3029 (B†, syn., BM!, lecto.)

H. vilhenae Cavaco in Bull. Mus. Natl. Hist. Nat. sér. 2, 26: 639 (1954). Type: Angola, NE of Lunda, Dundo, *Gossweiler* 13678 (P, holo., seen on website)

Kosteletzkya adenosiphon (Ulbr.) Cufod. in B.J.B.B. 29 (suppl.): 571 (1959)

NOTES. Some confusion arose over the application of *Hibiscus grantii* because in 1874 Garcke made the combination *Kosteletzkya grantii* (Mast.) Garcke but at the same time identified the name with a *Schimper* specimen (no. 1480) of what we now know as *K. begoniifolia*. Ulbrich concluded that the combination by Garcke was somehow improperly made and in 1924 pointed out the correct application of *K. grantii*, but he attributed the combination to himself. More recently, Cufodontis (1959) apparently thinking that Ulbrich's combination was a later homonym rather than a misunderstanding of the consequences of Garcke's publication, concluded that *K. grantii* as applied to the present plant was illegitimate. He therefore resurrected Ulbrich's *Hibiscus adenosiphon* and made a new and unnecessary combination in *Kosteletzkya*.

Ulbrich cited five syntypes for *Hibiscus adenosiphon*, two *Schubotz* collections (5 and 40) and two *Thorbecke* collections (700 and 753), in addition to *Kassner* 3029. All were destroyed at Berlin and a duplicate of the *Kassner* number at BM is the only one to have surfaced.

This plant is especially distinctive for its densely spike-like racemes or axillary glomerules, pink flowers with often recurved petals and decurved staminal column, and a calyx that embraces the fruit.

7. SENRA

Cav., Monad. Class Diss. 2: 83 (as *Serra*)*, 104, t. 35 fig. 3 (1786)

Annual or perennial herbs, sometimes shrubby; indumentum of stellate and simple hairs. Leaves shallowly 3-lobed; stipules filiform. Flowers solitary or in short axillary inflorescences often forming terminal panicles. Epicalyx of 3 bracts, becoming scarious in fruit. Calyx 5-lobed to below the middle. Ovary 5-locular, each locule with 2 ovules; style 5-branched. Fruit a loculicidal capsule with 5 prominent wings, eventually dividing into 5 equal parts; seeds one per cell, reniform.

A genus of one or perhaps two species but often wrongly said to have three in some literature. Vollesen recognised two species but Thulin considers the genus monotypic (see note).

Senra incana *Cav.*, Diss. 2: 83, 104, t. 35.3 (1786); Mast. in F.T.A. 1: 194 (1868); Balf. f. in Trans. Roy. Soc. Edinb. 31: 30 (1888); Abedin in Fl. W. Pak. 130: 4 (1979); Collenette, Flowers Saudi Arabia: 365 (1985); Paul in Fl. India 3: 347, fig. 97 (1993); Vollesen in Fl. Eth. & Eritr. 2 (2): 216, fig. 82.12.1–7 (1995); Wood, Handb. Yemen Flora: 108 (1997); Thulin, Fl. Somalia 2: 56, fig. 32 (1999). Type: Arabia, opposite Socotra island, collector unknown but communicated by Joseph Banks (MA, holo.)

Annual or perennial sometimes shrubby herb 0.4–1.5 m tall; stems densely short stellate-pubescent. Leaves rounded-reniform in outline, 2–8.5 × 2–9 cm, cordate at base, lobes acute to rounded, toothed or subentire. Epicalyx bracts ovate, 1.5–3 × 1.3–2.5 cm, subacute to rounded, cordate at base; calyx 0.9–1.4 cm long, lobes narrowly ovate, distinctly 3-veined. Petals dark purple to almost black or yellow with purple base, 2–3 cm long. Capsule subglobose, 5–8 mm long; seeds 3–4 mm long, densely silky. Fig. 15, p. 90.

KENYA. Northern Frontier District: Daua R., near Ramu, 23 May 1952, *Gillett* 13286A!; Tana River District: Bura, 26 Mar. 1963, *Thairu* 92! & Tana River National Primate Reserve, 7 km N of main gate, 20 Mar. 1990, *Kabuye et al.* TPR 744!

DISTR. **K** 1, **7**; Sudan, Ethiopia, Djibouti, Somalia; Socotra, Arabia, Pakistan and India

HAB. *Acacia–Commiphora* bushland sometimes watercourse edge, scattered tree grassland with *Acacia, Terminalia, Commiphora, Dobera* and *Salvadora*; 40–300 m

NOTE. Vollesen keeps *S. zoës* Volkens & Schweinfurth (Liste Plantes-Ghika-Pays de Somalis: 8 (1897). Type: Ethiopia, Harerge, banks of torrent Faf, *Ghika-Comanesti* (B? holo. †)) separate on account of the stem indumentum of long spreading simple hairs and glands instead of dense stellate pubescence; but Thulin pointed out that more material from Somalia shows variation in indumentum. Nevertheless a few specimens from the gypsaceous areas of the Nugaal valley e.g. *McKinnon* S/185, Nogal, 5 Dec. 1938, *Peck* 8, Wadamago, 4 Aug. 1941 and *Warfa* 1001, 11 km from Garoe towards Talex, Armola, 8 Nov. 1984 have long rather sparse bristly setae on the stems and glands and are certainly distinctive.

* As Vollesen points out Cavanilles first spelt the name *Serra* correcting it later in the same paper. The reason is obscure since it is named for a botanist called Serra.

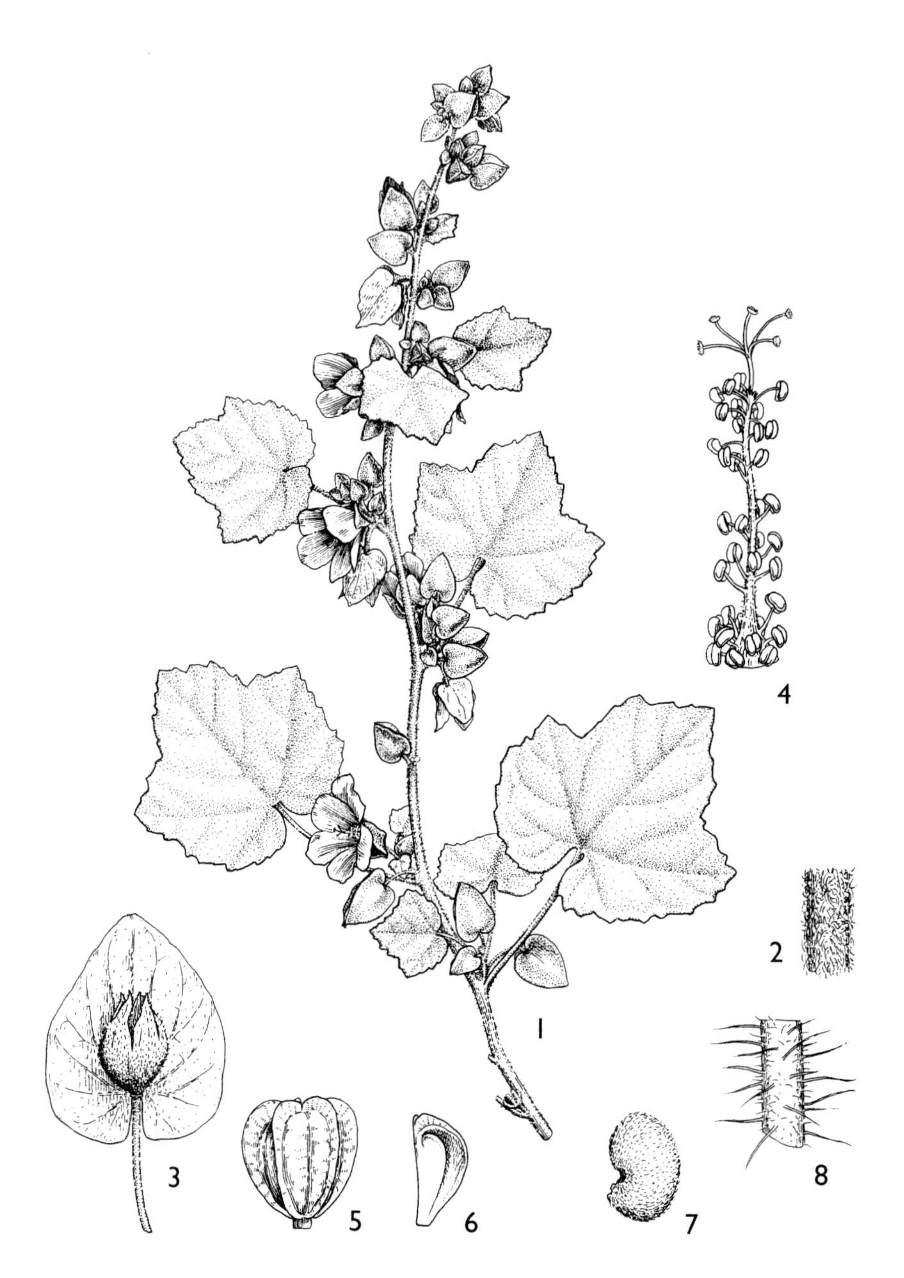

FIG. 15. *SENRA INCANA* — **1**, flowering and fruiting stem, × $^1/_2$; **2**, stem indument, × 2; **3**, epicalyx (part) and calyx, × 1.3; **4**, staminal column, × 3; **5**, capsule, × 2; **6**, segment of dehisced capsule, × 2; **7**, seed, × 4; **8**, stem indument, × 2. 1 from *Ellis* 235; 2, 5 from *Ash* 2316; 3 from *McKinnon* S/185; 4, 6 from *De Wilde et al.* 10557; 7–8 from *IECAMA* BH-43. Drawn by Eleanor Catherine, and reproduced with permission from Flora of Ethiopia and Eritrea 2, 2.

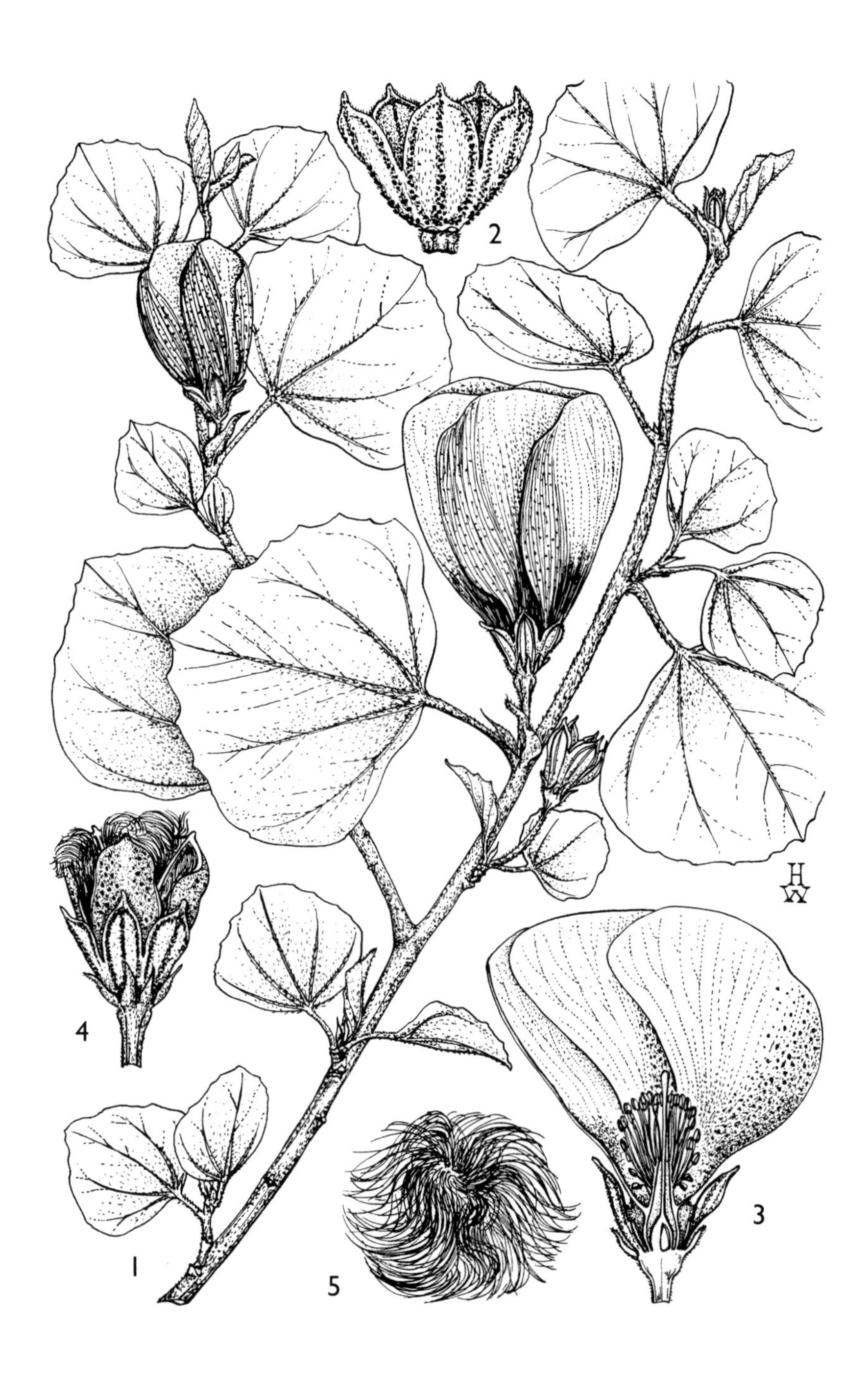

FIG. 16. *CIENFUGOSIA HILDEBRANDTII* — **1**, flowering stem, × ⅔; **2**, calyx (epicalyx removed), × 2; **3**, longitudinal section of flower, × 1; **4**, fruit, × 1; **5**, seed, × 1.4. All from *Drummond & Hemsley* 2285. Drawn by Heather Wood.

8. CIENFUEGOSIA

Cav., Diss. 3: 174, t. 72.2 (1787); Fryxell in Ann. Miss. Bot. Gard. 56: 179–250 (1969) & in Nat. Hist. Cotton tribe: 21–36, etc. (1979)

Herbs with woody rootstocks, subshrubs or shrubs, usually conspicuously gland-dotted; dioecious in one species. Leaves lobed or unlobed; stipules minute to large and leafy. Flowers solitary in leaf axils or in 2-flowered cymes. Epicalyx with (3–)9–12 lobes, subulate to spatulate, minute or equalling calyx, rarely absent. Calyx 5-lobed or splitting into only 2 lobes, with black glands in two rows along veins. Petals glandular-punctate. Ovary 3–4(–5)-locular, locules with 3–many ovules; style divided or not. Capsule not separating from receptacle, woody, glandular-punctate, dehiscing loculicidally. Seeds 3–8 per locule, densely woolly with appressed or spreading brownish hairs up to 1 cm long.

About 26 species, 8 of which occur in tropical and southern Africa, Madagascar and Arabia; the rest occur in South and Central America and southern U.S.A. The single East African species belongs to subg. *Articulata* Fryxell sect. *Garckea* Fryxell.

Cienfuegosia hildebrandtii *Garcke* in Jahrb. Königl. Bot. Gart. Berlin 2: 337 (1883); Gürke in P.O.A. C: 268 (1895); V.E. 3(2): 407 (1921); Marshall, Parson & Hutchinson in Bull. Entomol. Res. 28: figs. 1–3 (1937); T.T.C.L.: 299 (1949); Exell in F.Z. 1: 428, t. 85 (1961); Fryxell in Ann. Miss. Bot. Gard. 56: 197, fig. 17, a–c (1969) & Nat. Hist. Cotton Tribe: 26 (1978); Exell & Gonçalves, Fl. Moçambique 25: 12, t. 4 (1979). Type: Kenya, Kwale/Kilifi District: Maji ya Chumvi [Tschamtei, 'Txamtei'], *Hildebrandt* 2325 (B†, holo.); Tanzania, Lushoto District: near Mombo, *Drummond & Hemsley* 2285 (BR, neo; B, Fl, K!, isoneo., chosen by Fryxell)*

Erect branched woody herb or shrub 0.6–1.5(–3) m tall; young stems shortly stellate-pubescent, soon ± glabrous. Leaves broader than long, oblate to very broadly obovate or ± semicircular, usually unlobed or sometimes shallowly palmately 3(–5)-lobed, 2–6.5 × 2.5–8.5 cm, broadly rounded at apex, subtruncate to broadly cuneate at base, shallowly crenate-undulate and irregularly dentate especially on upper margin, densely stellate-pubescent to almost glabrous, 5–7-veined from base, the 3 central more marked veins with narrow elliptic slits or ± circular pits (nectaries) near base on lower surface; stipules obliquely lanceolate, 2.5–10 × 1–6 mm, subauriculate, acuminate at apex; sometimes also with short shoots in the axils bearing small leaves; petiole 0.8–3 cm long. Inflorescence with peduncles usually solitary, 1–2.5 cm long, jointed and bracteolate near middle or base; involucre lobes 9–10, in 3 groups, each usually with basal nectary, 1–3.5 mm long, the central one of each group the longest; calyx 7–10(–13) mm long, lobed about halfway; lobes ovate, 3.5 × 4 mm, with hyaline margins, each with 3 dark ribs joined at apex and produced into an acumen from the ± rounded lobe apex. Corolla yellow with red or dark purple centre (always?–some collectors do not mention central blotch); petals obovate, 2–4 cm long and wide, punctate; filaments and style dark red. Fruit ovoid to obovoid-oblong, 10–15(–17) mm long, 8–10(–12) mm diameter, rather obscurely punctate; seeds turbinate, 3–6 mm long with dense spreading pale rusty brown hairs, 8–10 mm long. Fig. 16, p. 91.

KENYA. Kwale District: 34 km W of Mombasa on Nairobi road, 16 Apr. 1969, *Q. Jones* 69/164! & Daruma area, Avisana, 23 Mar. 1902, *Kassner* 439!; Kilifi District: 9 km from Bamba to Karimani, 21 Nov. 1989, *Luke & Robertson* 2138!

TANZANIA. Lushoto District: 2.4 km N of Mombo, base of escarpment before Mkomazi R. flood plain, 8 Mar. 1940, *Milne* H16/40! & 4.8 km NW of Mombo, 29 Apr. 1953, *Drummond & Hemsley* 2285!

DISTR. **K** 7; **T** 3; Mozambique, Swaziland, South Africa

* *Gillett* 19550 (EA, K) is a topotype

HAB. Alluvial plains with *Acacia drepanolobium* on waterlogged soil, seasonally flooded black cotton areas with *Combretum, Euclea, Grewia* etc., salt flat fringes, *Commiphora–Acacia–Thespesia* bushland with *Aloe, Sansevieria* etc. on white gritty soil with basement rock outcrops; 100–450 m

NOTE. The generic name is quite often misspelt *Cienfugosia* in collections etc.

9. GOSSYPIUM*

L., Sp. Pl.: 693 (1753) & Gen. Pl. ed. 5: 309 (1754); Vollesen in K.B. 42: 337–349 (1987)

Shrubs or subshrubs, erect or decumbent to prostrate, or small to medium-sized trees, usually ± punctate throughout. Leaves usually entire (serrate in one species), shallowly lobulate to deeply palmately 3–7-lobed, palmately or pedately 3–7-veined; foliar nectaries 1–5, abaxially on principal veins; stipules filiform, persistent or caducous. Flowers borne singly or clustered on axillary pedicel or several on lateral inflorescences; pedicel usually surmounted by trimerous nectaries. Epicalyx of 3 bracts, usually persistent in fruit. Calyx truncate, 5-toothed, or deeply 5-lobed, often prominently punctate. Petals often large and showy. Staminal column antheriferous throughout length or only in upper half. Style clavate, stigmatic surfaces 3–5-lobed. Capsule 3–5-locular, often prominently punctate, dehiscent and opening slightly or flaring widely. Seeds 1–several per locule, usually free but sometimes ± coherent into a single unit for each locule; usually comose, sometimes minutely puberulent or glabrous, turbinate; seed hairs white to various shades of brown, copious, 1–3 cm long, usually crimped but patent in two species. Chromosome numbers: $2n = 26, 52$.

Gossypium includes about 50 species worldwide, found in the warmer parts of Africa, the Middle East, Australia, and the Americas (principally Mexico), often in arid habitats. Fourteen species occur in Africa (including cultivated species), six in the region of this flora.

The cultivated species have been included in the main sequence since apparently they occur as escapes or are naturalised.

1. Plants scandent; leaves unlobed; calyx deeply divided, the lobes narrowly triangular; petals 1.5–2 cm long, yellow throughout . . . . 1. *G. longicalyx*
   Plants erect; leaves sometimes unlobed, but usually shallowly to deeply lobed; calyx truncate or undulate to shallowly or moderately lobed; petals 2–8 cm long, often with a prominent, dark basal spot 2
2. Calyx with prominent black glands; capsules 2–6 cm long; seeds with long fibres (cotton), usually white; cultivated plants, occasionally escaped 3
   Calyx with scattered punctae or epunctate; capsules ± 1 cm long; seeds with short, appressed fibers, usually brownish; indigenous plants, not cultivated 4
3. Foliar nectary positioned within 5 mm of petiole; epicalyx bracts coarsely 5–13-dentate, the main veins terminating at margin; calyx with scattered black glands; capsules 3–4-locular, glabrous to sparsely strigose, acute or apiculate . . . . 2. *G. somalense*
   Foliar nectary positioned 10–15 mm from petiole; epicalyx bracts entire or weakly undulate, all veins anastomosing; calyx without glands; capsules 5-locular, densely strigose, cuspidate . . . . 3. *G. bricchettii*

* by Paul Fryxell and Bernard Verdcourt

4. Epicalyx bracts entire or few-laciniate, sometimes basally connate; epicalyx nectaries commonly absent; capsules elongated, flaring widely, releasing cotton .... 5. *G. arboreum*
   Epicalyx bracts prominently laciniate, distinct; epicalyx nectaries usually present; capsules globose, ovoid or elongate, usually not flaring widely at maturity 5
5. Leaves 3–5-lobed, the central lobe triangular to ovate, usually 1–1.5 times as long as wide; bracts of the epicalyx with the teeth separated by ± acute sinuses; stipules 0.5–1.5(–2) cm long; calyx truncate or with acute lobes or long-acuminate teeth, usually less than 6 mm long (excluding teeth); capsules 3–5-locular, ovoid or subglobose, 2–4 cm long, smooth .... 4. *G. hirsutum*
   Leaves 3–7-lobed, the central lobe ovate to lanceolate, usually more than 1.5 times as long as wide; bracts of the epicalyx with the teeth separated by rounded sinuses; stipules 1–5 cm long; calyx usually truncate, to 10 mm long; capsules usually 3-locular, narrowly ovoid, 3.5–6 cm long, pitted .... 6. *G. barbadense*

1. **Gossypium longicalyx** *J.B. Hutchinson & Lee* in K.B. 13: 221 (1958); Fryxell, Nat. Hist. Cotton tribe: 68, fig. 28 (1979); Vollesen in K.B. 42: 347, fig. 1E, map 4 (1987). Type: Tanzania, Dodoma District: Nondwa, *Disney* 33 (EA!, holo.; K!, iso.)

Scandent shrubs 0.6–1 m tall, usually supported by other vegetation; stems sparsely pubescent to glabrescent, prominently gland-dotted. Leaves petiolate, ovate, 4–6 × 3.5–5 cm, truncate to cordate, entire, acuminate, gland-dotted, with simple hairs on main veins beneath, otherwise subglabrous, pedately 7-veined, with an inconspicuous foliar nectary ± 5 mm from petiole on midrib beneath; stipules falcate, 4–11 × 1–4 mm, acuminate, caducous. Flowers on sympodial inflorescences; pedicel 8–20(–30) mm long, sparsely pubescent and gland-dotted; epicalyx nectaries lacking; bracts of the epicalyx ovate-cordate, 15–25 mm long, almost as broad, entire, acute to acuminate, gland-dotted; calyx 8–15 mm long, prominently gland-dotted, minutely pubescent, deeply divided, the lobes narrowly triangular; petals yellow, minutely gland-dotted, 15–25 mm long (slightly exceeding epicalyx), ciliate on claw, pubescent externally where exposed in bud, glabrous internally; staminal column pallid, glabrous, eglandular, 8–9 mm long; filaments 1–1.5 mm long; style exceeding the androecium. Capsules ovoid, 10–15 mm long, 3-locular, prominently gland-dotted, glabrous; seeds 2–3 per locule, 5–6 mm long, densely pubescent, fibres grayish, appressed.

UGANDA. Karamoja District: Bokora County, 17 July 1957, *J. Wilson* 390
TANZANIA. Dodoma District: 11 km on Bahi to Kilimatinde road, 18 Apr. 1988, *Bidgood, Mwasumbi & Vollesen* 1207! & Nondwa, 25 Mar. 1955, *Disney* 33!; Iringa District: Msembe–Causeway Track, km 9, 24 Mar. 1970, *Greenway & Kanuri* 14193! & 25 Mar. 1970, *Greenway & Kanuri* 14206!
DISTR. **U** 1; **T** 5, 7; Sudan
HAB. Seasonally wet *Acacia* bushland, *Terminalia–Acacia–Delonix–Commiphora* scrub; 800–900 m

2. **Gossypium somalense** (*Gürke*) *J.B. Hutchinson*, Evol. Gossyp.: 31 (1947); Fryxell, Nat. Hist. Cotton tribe: 65 (1979); Vollesen in K.B. 42: 347 (1987) & in Fl. Eth. & Erit.2 (2): 220, fig. 82.13.1–5 (1995); Thulin, Fl. Somalia 2: 60, fig. 34 A–F (1999). Type: Somalia, Fulla Valley, *Ellenbeck* 220 (B†, holo.)

Shrubs 0.7–2 m tall; stems stellate-tomentose. Leaves petiolate, ± rotund, 2.5–6 × 2.5–6.5 cm, cordate, shallowly 3-lobed, entire or pedately 5–7-lobed; lobes acute to obtuse (sometimes mucronate), stellate-tomentose above and beneath, clearly punctate beneath, with an obscure foliar nectary at or near the base of the midrib beneath (5 mm distant from petiole or less); stipules 6–11 mm long, subulate,

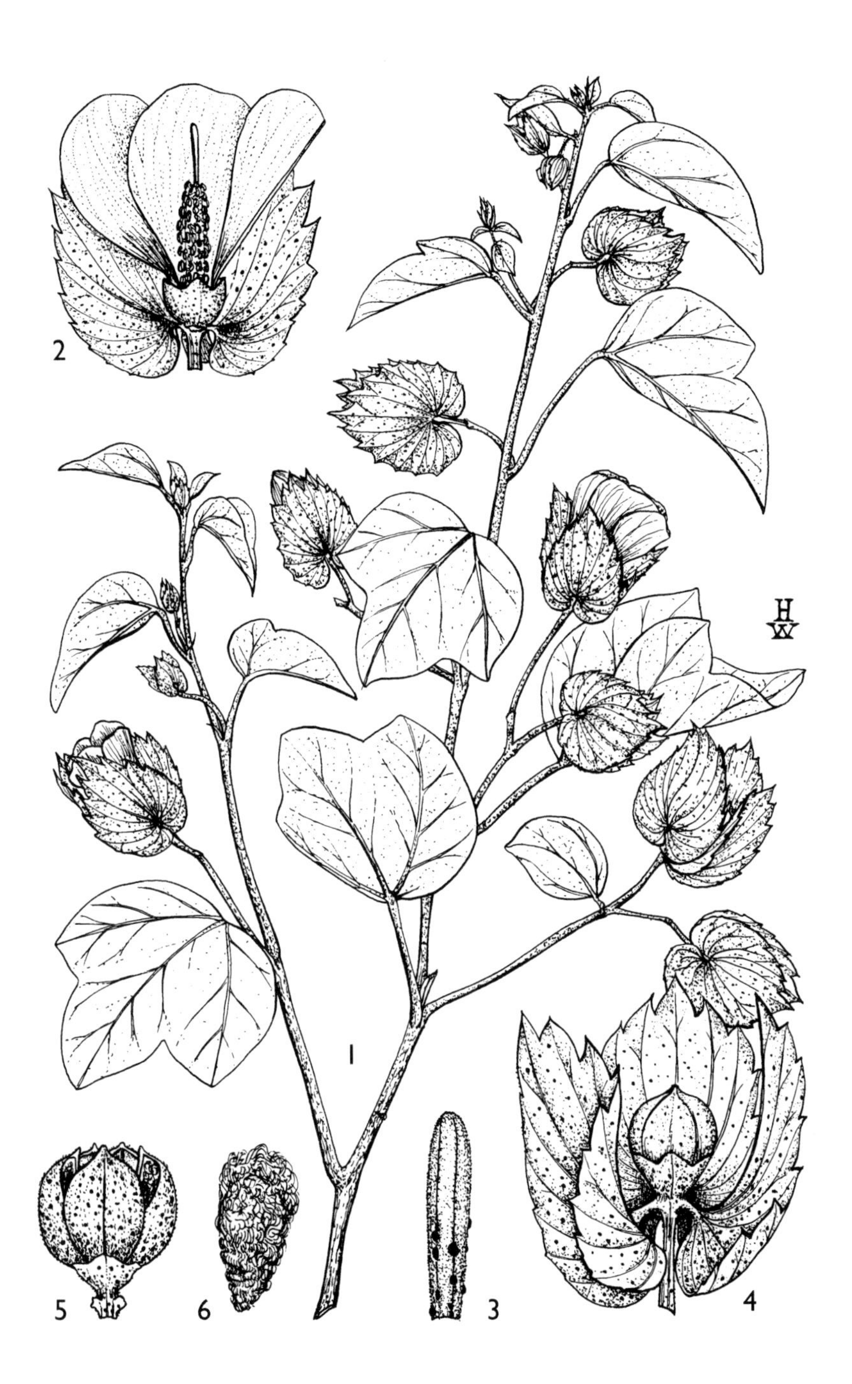

Fig. 17. *GOSSYPIUM SOMALENSE* — **1**, flowering stem, × $^2/_3$; **2**, flower (one bract and two petals removed), × 1.5; **3**, stigma, × 10; **4**, young fruit and bracts, × $1^1/_2$; **5**, fruit, × 2; **6**, seed, × 3. All from *Gillett* 13331. Drawn by Heather Wood.

caducous. Flowers on short (1–2-flowered) sympodial inflorescences; pedicel 6–12 mm long, bracteate at articulation, surmounted by 3 epicalyx nectaries; bracts of the epicalyx inserted above the nectaries, broadly ovate and deeply cordate, narrowed to a short claw, 2–3 × 3–3.5 cm, coarsely 5–13-dentate, tomentose, obscurely punctate; calyx yellow-green, 5–7 mm long, undulately 5-lobed, minutely pubescent, with scattered black glands; lobes obtuse. Corolla yellow or apricot to pink or dull red with prominent dark red centre or reddish green at base, funnelform, ± 2 cm long (subequal to or slightly exceeding epicalyx), epunctate or nearly so, pubescent externally especially where exposed in bud, glabrous internally; staminal column pallid, glabrous, epunctate; anthers subsessile or the filaments to 1 mm long; style exceeding the androecium, punctate. Capsules ovoid, beaked, 0.8–1 cm long, 3–4-locular, ± appressed-hirsute (at least distally), prominently punctate, included in persisting epicalyx; seeds solitary, ± 7 mm long, densely pubescent, fibres brownish, appressed. Fig. 17, p. 95.

UGANDA. Karamoja District: base of Turkana Escarpment, Apr. 1960, *J. Wilson* 867!
KENYA. Northern Frontier District: 42 km SE of Malka Mari, 21 Jan. 1972, *Bally & Smith* 14934!; Meru District: Meru Game Reserve, Kenmare Lodge, 24 Mar. 1966, *J. Adamson* in EA 13542!; Tana River District: Kora Game Reserve, area between Tana R. and Mwitamyisi watercourse, 26 May 1977, *Gillett* 21101!
DISTR. **U** 1; **K** 1, 2, 4, 7; Niger, Chad, Sudan, Ethiopia, Somalia
HAB. *Acacia misera–Commpihora* thicket, *Acacia–Commiphora–Delonix* bushland, dried-up river-beds on limestone; (300–)400–700 m

SYN. *Cienfuegosia somalensis* Gürke in E.J. 33: 380 (1903)
C. ellenbeckii Gürke in E.J. 33: 381 (1903). Types: Ethiopia, Boran, Tarro Gumbi, *Ellenbeck* 2069 & 2082 (B†, syn.)
Gossypium paolii Mattei in Boll. Stud. Inform. R. Giord. Palermo 2(4): 223 (1916). Type: Somalia, near Salagle, Juba [Giuba] R., Paoli *s.n.* (FT, holo.)
G. ellenbeckii (Gürke) Mauer in Trudy Sredne-Aziatsk Gosud. Univ. Lenina n.s. 18: 19 (1950)

NOTE. At least one field note points out that the corolla starts yellow but turns purple, which explains the wide variety of colours cited in other field notes.

3. **Gossypium bricchettii** (*Ulbr.*) *Vollesen* in K.B. 42: 349, fig. 2B, map 3 (1987) & in Fl. Eth. & Eritr. 2 (2): 220, fig. 82.13.6–7 (1995); Thulin, Fl. Somalia 2: 61, fig. 34G, F (1999). Type: Somalia, between Obbia and Uarandi, *Roberti-Brichetti* 238 (B†, holo.; FT, iso.)

Erect shrubs to 2.5 m tall; stems stellate-tomentose, the hairs obscuring the punctae. Leaves petiolate, broadly ovate, 2–7 × 1.5–7.5 cm, simple or shallowly trilobate, acute, basally rotund to shallowly cordate, pedately 5(–7)-veined; lobes rotund, stellate-tomentose; foliar nectary positioned 1–1.5 cm distal to petiole on the midrib beneath; petiole $^1/_2$–$^2/_3$ length of lamina, obscurely punctate; stipules 5–11 mm long, subulate, caducous. Flowers on short sympodial inflorescences; pedicel 8–12 mm long, with falcate bracts at articulation, with 3 epicalyx nectaries; bracts of the epicalyx inserted above the nectaries, broadly ovate, deeply cordate, acute to obtuse, entire or subentire, stellate-tomentose, epunctate; calyx cupuliform, pubescent, truncate to sinuate-lobate, epunctate. Corolla yellow with reddish cast, internally with dark red centre, funnelform, ± 2 cm long, subequal to or slightly exceeding epicalyx, externally stellate-tomentose. Capsules 5-locular, strigose.

KENYA. Nothern Frontier District: Yabichu, near Ramu, 23 May 1952, *Gillett* 13286!
DISTR. **K** 1; Ethiopia, Somalia
HAB. Open *Commiphora-Acacia* scrub on pale limestone soils; ± 360 m

SYN. *Cienfuegosia bricchettii* Ulbr. in E.J. 48: 378 (1912)
Gossypium benadirense Mattei in Bol. Stud. Inform. R. Giord. Colon. Palermo 2 (4): 223 (1916); Vollesen in K.B. 42: 349, fig. 2A, map 4 (1987) & in Fl. Eth. & Eritr. 2 (2): 220, fig. 82.13.11 (1995). Type: Somalia, Lugh area, 26 Oct 1913, *Paoli* s.n. (FT, holo.)

4. **Gossypium hirsutum** *L.*, Sp. Pl. ed. 2: 975 (1763); T.T.C.L.: 299 (1949); Keay, F.W.T.A. ed. 2, 1(2): 349 (1958); Fryxell, Nat. Hist. Cotton tribe: 68 (1979); Vollesen in Fl. Eth. & Eritr. 2 (2): 222 (1995); Thulin, Fl. Somalia 2: 61 (1999). Type: specimen cultivated at Chelsea Physic Garden, *Miller* s.n. annotated "Gossypium hirsutum LM" in Miller's hand (BM!, lecto.)

Shrubs 1–2 m (or more) tall, usually widely branching, ± stellate-pubescent, sometimes densely so. Leaf blades mostly 4–10.5 cm long, cordate, weakly to distinctly 3–5-lobed, the lobes broadly triangular to ovate, acute to acuminate, densely pubescent to glabrous, foliar nectaries present; petiole $^1/_2$–1 times length of blade; stipules 0.5–1.5(–2) cm long, subulate or falcate. Flowers usually in sympodial inflorescences; pedicel 2–4 cm long, surmounted by 3 nectaries; bracts of the epicalyx 2–4.5 cm long, each inserted above a nectary, foliaceous (enclosing the bud), ovate, cordate, 3–19-laciniate, persistent; calyx 5–6 mm long (excluding teeth), truncate or 5-toothed. Corolla cream-coloured or pale yellow, with or without a dark spot at base, the petals 2–5 cm long; staminal column 1–1.5 cm long; style ± enclosed by androecium or somewhat exceeding androecium. Capsules broadly ovoid or subglobose, 2–4 cm long, 3–5-locular, glabrous, smooth; seeds 8–10 mm long, several per locule, lanate, the seed hairs (cotton) white, tan, or red-brown.

KENYA. Northern Frontier District: Moyale, 27 Apr. 1952, *Gillett* 12946! & 4 July 1952, *Gillett* 13497!; Turkana District: Lake Rudolf, 1899, *Wellby* s.n.!; Kilifi District: Gedi, Watuma, *Greenway* 10455!

TANZANIA. Mwanza District: ?Ukerewe Is., *Conrads* in EA 13347!; Rufiji District: Mafia I., Nyororo Is., 6 Oct. 1937, *Greenway* 5394!; Pemba, Vitongoje, 15 Feb. 1929, *Greenway* 1453!

DISTR. **K** 1, 2, 7; **T** 1, 6; **P**; indigenous to Central America, Caribbean and certain Pacific islands (Marquesas, Samoa etc.); now cosmopolitan in cultivation

HAB. Cultivated in gardens and sometimes occurs as an escape in coconut plantations; 0–1150 m

SYN. *G. barbadense* L. var. *hirsutum* (L.) Triana & Planchon in Ann. Sci. Nat. Bot. ser. 4, 17: 171 (1862), *non* var. *hirsutum* Hooker & Bentham (1849)
G. herbaceum L. var. *hirsutum* (L.) Masters in Fl. Brit. India 1: 347 (1874)

5. **Gossypium arboreum** *L.*, Sp. Pl. 693 (1753); Mast. in F.T.A. 1: 211 (1868); U.O.P.Z.: 278 (1949); Keay, F.W.T.A. ed. 2, 1(2): 349 (1958); Fryxell, Nat. Hist. Cotton tribe: 62 (1979); Vollesen in Fl. Eth. & Eritr. 2 (2): 222 (1995); Thulin, Fl. Somalia 2: 59 (1999). Type: "India", Linn. Herb. 874.3 (LINN!, holo.)

Shrubs or subshrubs up to 2 m tall, highly variable in morphology and pubescence; stems usually puberulent to pubescent, the young stems punctate. Leaves petiolate, moderately to deeply palmately 3–7-lobed, 2–5 × 2–4.5 cm, pedately 7–9-veined, sometimes with small accessory lobes in the sinuses; lobes broadly ovate to narrowly lanceolate and ± constricted basally, entire, acute or acuminate (sometimes obtuse), the foliar nectary inconspicuous or absent; petiole punctate; stipules 5–15 mm long, linear to falcate, caducous. Flowers borne on sympodial inflorescences (sometimes solitary); pedicel 10–60 mm long, the epicalyx nectaries commonly absent; bracts of the epicalyx broadly ovate, 15–35 mm long, almost as wide, ± connate basally forming a cup, entire or 3–7-laciniate, pubescent externally, glabrous internally; a trio of nectaries sometimes present within the epicalyx at the juncture of epicalyx and calyx alternate with the bracts; calyx 7–12 mm long, glabrous, prominently punctate, truncate to distinctly 5-lobed. Petals usually yellow with a purplish basal spot, sometimes white or flushed purplish or with basal spot absent, 25–45 mm long, minutely punctate, ciliate on claw, pubescent externally where exposed in bud, glabrous internally; staminal column glabrous, pallid, usually punctate, $^1/_2$ the length of petals or less, the filaments 2–4 mm long; style only slightly exceeding androecium. Capsule ovoid to elongate, usually 3-locular, usually beaked, glabrous, punctate, the glands sunken, flaring widely at maturity; seeds bearing white or tan fibers.

UGANDA. 'Nile Province', in native gardens, no date, *Dawe* 884!
TANZANIA. Morogoro District: Kissaki steppe, 1890, *Goetze* 75!; Rufiji District: Mafia Island, Maandaa, Juani Isl., Sep. 1937, *Greenway* 5292!
DISTR. **U** 1; **T** 6; not certainly known from the wild; cultivated principally in southern Asia, sometimes elsewhere, occasionally escaped
HAB. Secondary bushland; 3–?1000 m

SYN. *G. transvaalense* sensu T.T.C.L.: 300 (1949), *non* Watt

6. **Gossypium barbadense** *L.*, Sp. Pl. 693 (1753); Mast. in F.T.A. 1: 210 (1868); Oliv. & Grant in Trans. Linn. Soc. Lond. 29: 38 (1873); T.T.C.L.: 299 (1949); U.O.P.Z.: 278 (1949); Keay, F.W.T.A. ed. 2, 1(2): 349 (1958); Fryxell, Nat. Hist. Cotton tribe: 70 (1979); Vollesen in Fl. Eth. & Eritr. 2 (2): 222 (1995); Thulin, Fl. Somalia 2: 61 (1999). Type: Plukenet, Phytograph.: 2, t. 188, fig. 1 (lecto.)

Shrubs 1–3 m tall, sometimes arborescent, sparsely stellate-pubescent to glabrate. Leaf blades mostly 8–20 cm long, 3–7-lobed (less than half-divided), pedately 5–9-veined; lobes ovate to lanceolate; petiole $^1/_3$–$^3/_4$ the length of the blades, punctate; stipules 1–5 cm long, often prominent, subulate to falcate. Flowers solitary in leaf axils or in sympodial inflorescences; epicalyx nectaries generally prominent; bracts of the epicalyx broadly ovate, 4–6 cm long, cordate, foliaceous, 5–17-laciniate, persistent; calyx 8–10 mm long, truncate, punctate. Corolla usually yellow with a dark red or purple centre, the petals up to 8 cm long; staminal column 2.5 cm long; style exceeding androecium. Capsules usually narrowly ovoid to elongate, 3.5–6 cm long, 3-locular, glabrous, pitted; seeds 8–10 mm long, several per locule, free or fused together, lanate, the seeds hairs (cotton) usually white.

UGANDA. Karamoja District: Dodoth, Kasile, Magos Tsetse Post, Feb. 1964, *Wilson* 1526!; Toro District: Bwamba, 18 Jan. 1932, *Hazel* 145! & Bwamba, 27 Sep. 1932, *A.S. Thomas* 717!
KENYA. West Suk District: Sigor, 2 Dec. 1959, *Bogdan* 4956!; Kitui District: Kilimakei, 16 Apr. 1902, *Kassner* 611! Kwale District: Shimba Hills, Kwale–Mombasa road, 4 May 1965, *Magogo & Glover* 985!
TANZANIA. Buha District: Kasulu, 1931, *Rounce* A1!; Rufiji District: Mafia I., Upanga, 9 Aug. 1937, *Greenway* 5041!; Mbeya District: Mbeya, seeds from there grown at Agric. Dept. Morogoro, 13 Oct. 1938, *Greenway* 5809!
DISTR. **U** 1, 2; **K** 1, 2, 4, 7; **T** 4, 6, 7; originally South American, now widely distributed in cultivation in many parts of the world
HAB. Gardens and experimental plots, and as an escape on middens; 5–1200 m

SYN. *G. peruvianum* Cav., Diss. 6: 313, t. 168 (1788). Type: from Peru, cult. at Etche in Spain, Herb. Madrid 476, 755 (MA, holo.)
G. lapideum Tussac, Fl. Antilles 2: 67 (1818); T.T.C.L.: 299 (1949). Type: Guyana, Cayenne (no material or collector cited)
G. brasiliense MacFadyen, Fl. Jamaica 1: 72 (1837). Type: Jamaica (without locality or collector)
G. microcarpum sensu T.T.C.L.: 299 (1949), ?*non* Tod.

NOTES. Percy & Wendel (1990) have concluded that Northwestern South America is the "ancestral home" of the species and have used evidence from allozyme variability to trace its dispersal to other parts of South America and other parts of the world, and have detected appreciable introgression from *G. hirsutum* in certain parts of this range.

10. **GOSSYPIOIDES**

J.B.Hutch.* in New Phytol. 46: 131 (1947); Fryxell, Nat. Hist. Cotton tribe: 36 (1979)

Subscandent or climbing shrubs. Leaves 3–5-lobed; foliar nectaries 1–5 on principal veins beneath; stipules falcate or auriculate-clasping, usually persistent. Flowers axillary or in 1–4-flowered sympodial inflorescences. Epicalyx bracts 3, incised-lobed, persistent. Calyx short, subtruncate or very shortly 5-lobed or -undulate. Staminal column bearing numerous anthers on short filaments. Ovary 3–5-locular with 2–9 ovules in each locule; styles united below but ± free above forming 3–5 lobes; stigmas distinct, clavate. Capsule 3–5-locular, loculicidally dehiscent. Seeds ellipsoid, glabrescent or densely lanate.

Two species, the other in Madagascar. Exceedingly similar in facies to *Gossypium* and I have some doubts about separating it; the chromosome number is however 2n = 24 (not 26 or 52).

Gossypioides kirkii (*Mast.*) *J.B. Hutch.* in New Phytol. 46: 132 (1947); Exell in F.Z. 1: 432, t. 87 (1961); Fryxell, Nat. Hist. Cotton Tribe: 37, fig. 14 (1979); Vollesen, Opera Bot. 59: 35 (1980); K.T.S.L: 171, fig. (1994). Type: Tanzania, Uzaramo District: Dar es Salaam, *Kirk* s.n. (K!, syn.)**

Scandent or subscandent shrub 1.2–5 m tall with ribbed, angled or ± winged glabrous or pubescent gland-dotted branches. Leaves ± deeply 3–5-lobed, 5–14 × 5–16 cm, cordate at base, glabrous to densely pubescent, gland-dotted (green in life), lobes ovate to lanceolate, acuminate, narrowly acute at apex but lowest lobes often quite rounded, margin entire; petiole up to 8 cm long; stipules obliquely ovate-falcate, up to 15 × 5 mm. Epicalyx bracts ovate, 2.5–3.5 × 2–3.5 cm, incised-lobed, gland-dotted; calyx cupuliform, 6 mm long, 5-dentate, gland-dotted. Corolla yellow flushed mauve outside, blood-red or ± black at base, ± 5 cm wide; petals obliquely obovate, 2.5–3 cm long and wide, stellate-pubescent outside, at first connivent into a tube but spreading eventually; staminal column and anthers yellow, free filaments maroon, 1–1.5 mm long, column 6–7 mm; ovary 3–4-locular with 2 ovules per locule. Capsule subglobose, 1.2–1.5 cm diameter, deeply glandular-punctate; seeds ellipsoid, either all black or alternately striped black and white (due to fine pubescence), 6–7 × 3.5 mm, embedded in a dull brown floss 5–10 mm long, which partly separates from the seeds early and only a few apical attached tufts remain visible if any and partly is derived from internal capsule sutures (see note). Fig. 18, p. 100.

KENYA. Northern Frontier District: Boni Forest, 2 Oct. 1947, *J. Adamson* 409 in *Bally* 5895!; Kwale District: Mrima Hill, 4 Sep. 1957, *Verdcourt* 1871!; Kilifi District: Marafa, 40 km NW of Malindi, 22 Nov. 1961, *Polhill & Paulo* 842!

TANZANIA. Tanga District: 8 km SE of Ngomea, 31 July 1953, *Drummond & Hemsley* 3580!; Uzaramo District: Banda Forest Reserve, near Mfyoza Village, 13 Aug. 1968, *Shabani* 176!; Rufiji District: Mafia 1., Chungarumo, 1 Oct. 1937, *Greenway* 5359!; Pemba, Mkoani, 7 Sep. 1929, *Vaughan* 636!

DISTR. **K** 1, 7; **T** 3, 6, 8; **Z**; **P**; Mozambique, South Africa

HAB. Bushland, bushland with scattered trees, very mixed coastal forest with *Sterculia appendiculata, Gyrocarpus* etc., forest edges, *Brachystegia* woodland, abandoned cultivations; usually on sand; near sea level–450 m

* Based on Skovsted in Journ. Genet. 31: 287 (1935) where it was published without a Latin diagnosis.
** There are three Kirk specimens at Kew, labelled Province Zanguebar 1869, Dar es Salaam, June 1879 and a duplicate of it; the first is not strictly a syntype and might be from Zanzibar island.

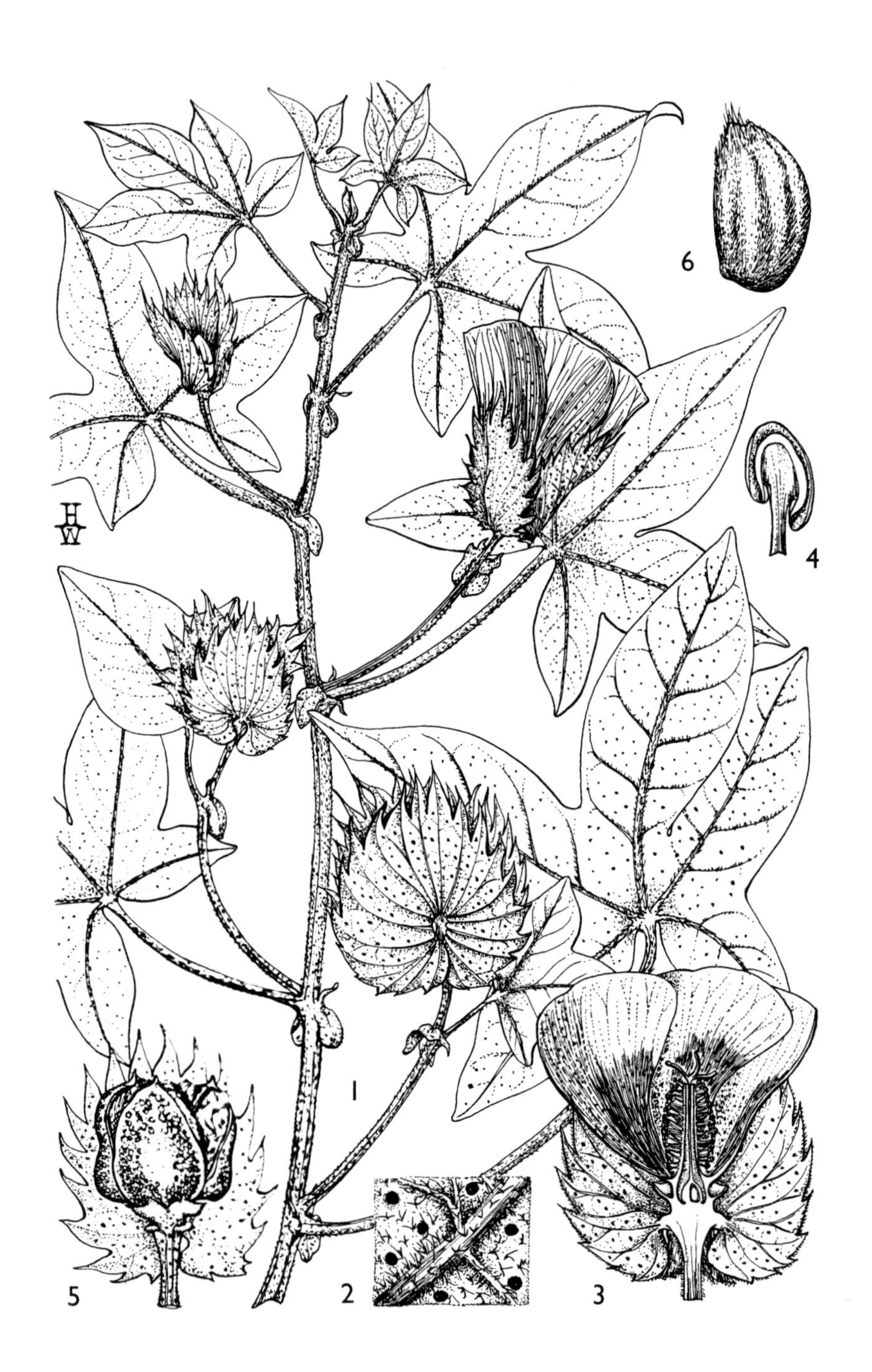

FIG. 18. *GOSSYPIOIDES KIRKII* — **1**, flowering stem, × $^{2}/_{3}$; **2**, undersurface of leaf, × 3; **3**, longitudinal section of flower, × 1; **4**, stamen, × 15; **5**, fruit (two bracts removed), × 1; **6**, seed, × 4. 1–4 from *Polhill & Paulo* 842; 5–6 from *Tweedie* 2383. Drawn by Heather Wood.

SYN. *Gossypium kirkii* Mast. in J.L.S. 19: 214 (1882); P.O.A. C: 268 (1895); G. Watt, Wild & Cult. Cottons: 317, t. 51* (1907); U.O.P.Z.: 279 (1949); T.T.C.L.: 229 (1949)

G. kirkii Mast. subsp. *scandens* Roberty in Candollea 13: 31 (1950). Type: Tanzania "région de lacs" fide Roberty but actually Songea District: Kwa Likumbi, *Busse* 2991 (ubi?, holo.) (*G. bussei* Gürke MS)

NOTE. Roberty records subsp. *scandens* from Madagascar which would extend the distribution. Exell, J.B. Hutchinson, Roberty and Fryxell have all commented on the fact that there are two well matched varieties or subspecies, glabrous and tomentose but study of material does not suggest any geographical or other correlation which would make recognition of the two taxa acceptable. J.B. Hutchinson states that the brown floss originates from the capsule and Fryxell uses this in his key whereas Watt and Brenan indicate it originates from the seed as one would expect but is very easily separable. I have noted that at least some hairs are attached to the apex of the seed but the majority do originate from the capsule sutures. This needs confirming from fresh material in the field. *Greenway* (3292 and 4988) states that the floss is white or said to be white but in both cases I think due to confusion.

11. **THESPESIA**

Correa in Ann. Mus. Paris 9: 290, t. 8, fig. 2 (1807); Fryxell, Nat. Hist. Cotton Tribe: 84–102 (1979), *nom. conserv.*

Azanza Alef. in Bot. Zeit. 19: 298 (1861); Exell & Hillcoat in Contr. Conhec. Fl. Moçambique 2: 58 (1954)

Shantzia Lewton in J. Wash. Acad. Sci. 18: 10, fig. 1–2 (1928)

Thespesiopsis Exell & Hillcoat in Contr. Conhec. Fl. Moçambique 2: 55 (1954)

Trees or less often shrubs; indumentum of stellate hairs or lepidote. Leaves petiolate, lobed or unlobed, cordate to ± truncate at base, entire, palmately or pedately 5–9-veined or rarely penniveined; stipules subulate or filiform, rarely absent. Flowers showy, solitary and axillary or sometimes aggregated terminally; pedicel sometimes subtended by a pair of tiny bracts. Epicalyx of 3 to many bracts, either whorled or spirally inserted, often caducous. Calyx truncate to 5-lobed, persistent (save in one species). Petals white, yellow, pink or ± purple often with a dark red or purple base. Stamens numerous. Ovary 5-locular, each with 1–4 ovules. Styles not branched, the stigmatic surfaces born on the clavate apex. Fruit (3–)5-locular, dehiscent or indehiscent, woody, coriaceous, or somewhat fleshy. Seeds ± angular, glabrous or pubescent.

As circumscribed by Fryxell, a pantropical genus of about 16 species. Exell & Hillcoat (see above) and Exell (F.Z. 1: 420 (1961)) keep up three genera separating them as follows:

- Fruit indehiscent, not lobed; ovules 3–4 per locule; epicalyx of 3–5 bracts . . . . . *Thespesia*
- Fruit indehiscent, (4–)5-lobed; ovule 1 per locule; epicalyx of 3 bracts . . . . . . *Thespesiopsis*
- Fruit dehiscent; ovules 2–several per locule, epicalyx of 5–15 bracts . . . . *Azanza*

There are however, partially dehiscent fruits in *Thespesia* and other difficulties and I have followed Fryxell. It is, however, very much a matter of opinion and the alternative names are extensively used.

T. lampas (Cav.) Dalz. & Gibs. (*Azanza lampas*) (Cav.) Alef. has been cultivated in Tanzania (Lushoto District: Amani, 9 Jan. 1934, *Greenway* 3695) (T.T.C.L.: 307 (1949). A native of tropical Asia it is a shrub 1.8–3 m tall with 3–5-lobed leaves (or the upper sometimes entire, ovate-triangular), 7–15 × 4.5–14 cm, glabrous to sparsely

* This is based on *Kassner* 28 (Tanzania, Tanga District: Bote on Umba R. 24 Feb. 1902); part of Matilda Smith's original drawing is attached on the sheet (K).

stellate-scabrid or -pubescent above, more densely stellate-tomentose beneath, the lobes usually acuminate; epicalyx bracts 3–5(–8), linear, 1–3 mm long, soon deciduous; calyx usually distinctly toothed or lobed with linear, triangular or subulate teeth 1–6 mm long (8–10 in var. *longisepala* Borss.); corolla yellow with purple or crimson centre, 7–12 cm long. Fruit ellipsoid, 2.5 cm long, 2 cm diameter, woody, dehiscent. Sivarajan & Pradeep, Malvaceae S. Pen. India: 43 (1996) state "also distributed in East Africa ...." but it is only rarely cultivated. It has been grown for fibre in West Congo-Kinshasa.

1. Indumentum lepidote; fruit indehiscent or only exocarp dehiscent 2
   Indumentum never lepidote; fruit dehiscent 3
2. Epicalyx bracts inserted at base of calyx, quickly falling; petals 4–10.5 cm long; leaves acuminate at apex .... 1. *T. populnea*
   Epicalyx bracts inserted on calyx, persistent even in fruit; petals 2–3.5 cm long; leaves rounded to ± acute at apex ... 2. *T. danis*
3. Calyx truncate with only minute teeth if present; native .... 3. *T. garckeana*
   Calyx usually distinctly lobed with teeth 1–10 mm long, less often ± truncate; cultivated .... *T. lampas*

1. **Thespesia populnea** (*L.*) *Correa* in Ann. Mus. Paris 9: 290, t. 8, fig. 2 (1807); Mast. in F.T.A. 1: 209 (1868); P.O.A. C: 268 (1895); T.T.C.L.: 307 (1949); U.O.P.Z.: 468 (1949); Keay, F.W.T.A. ed. 2, 1(2): 342 (1958); K.T.S.: 363 (1961); Exell in F.Z. 1: 421 (1961); Borss. in Blumea 14: 106 (1966); Fryxell, Nat. Hist. Cotton Tribe: 86, fig. 33 (1978); Abedin, Fl. W. Pak. 130: 32 (1979); Fosberg & Renvoize, Fl. Aldabra: 66, fig. 6/3–5 (1980); Blundell, Guide Wild 71, E. Afr.: 80, fig. 302 (1982); K.T.S.L.: 174 (1994); Vollesen, Fl. Eth. 2 (2): 222, fig. 82.14.13–14 (1995); Sivarajan & Pradeep, Malvaceae S. Pen. India: 37 (1996); Thulin, Fl. Somalia 2: 62, fig. 35 D, E (1999). Type: Sri Lanka, Herb. Hermann vol. 4. folio 34 Linn. no. 258 (BM, lecto.)

Shrub or small tree 0.3–10(–15) m tall with smooth brown bark; branchlets densely lepidote. Leaves ovate, 6–15 × 5–12 cm, acuminate at the apex, cordate at base, entire, ± covered on both sides with minute sessile peltate scales, otherwise smooth and glabrous; petiole 4–11 cm long. Flowers erect or pendent, solitary in leaf axils or a few aggregated in terminal inflorescences; pedicel 2–12 cm long; bracts of epicalyx 3–5, narrowly triangular or oblong-lanceolate, 2–7(–12) mm long, quickly falling; calyx 0.7–1.5 × 0.7 cm, densely lepidote, ± entire or with minute teeth about 1 mm long. Petals yellow turning red with age, obovate, 4–10.5 × 3.5–7 cm, lepidote outside. Fruit depressed globose, 2.5–3 cm long, 3–4.5 cm diameter, lepidote; seeds angular-ovoid, 10–15 × 6–9 mm, rusty brown pubescent or ± glabrous.

a. var. **populnea**

Fruit completely indehiscent, the exocarp not splitting; seeds with long soft hairs.

KENYA. Lamu District: Witu, *Thomas* s.n.! & Kipini, E of Osi on N bank of Tana R., 8 Nov. 1957, *Greenway & Rawlins* 9484!

TANZANIA. Tanga District: Moa, June 1983, *Holst* 3034!; Bagamoyo District: Ruvu mouth, 14 Aug. 1967, *Mwasumbi* 10248!; Mikindani District: Mtwara, 7 Mar. 1963, *Richards* 17777!; Zanzibar: Mkokotoni, 15 Aug. 1961, *Faulkner* 2886!

DISTR. **K** 7; **T** 3, 6, 8; **Z**; **P**; Senegal to Nigeria, Eritrea, Ethiopia, Mozambique, South Africa; pantropical

HAB. Mangrove swamps, borders of *Avicennia* saltmarsh, *Tamarindus*, *Mimusops*, *Sapindaceae* bushland often with *Hibiscus tiliaceus* and *Sophora*; 0–6 m

SYN. *Hibiscus populneus* L., Sp. Pl.: 694 (1753)

NOTE. I have included citations of specimens for which no fruits are available; they might be var. *populneoides*, in fact using the leaf-base characters and pedicel length considered valid by some authors, several would be referred to that but these characters do not work (see general notes).

b. var. **populneoides** (*Roxb.*) *Pierre*, Fl. Forest. Cochinch. 3: t. 173B (1888). Type: India, gardens near Samulcotta, *Roxburgh* in *Wallich Herb.* 1888F (K–W!, lecto.)

Fruit with exocarp dehiscent; seeds with short clavate or bulbous hairs.

TANZANIA. Zanzibar, May 1868, *Kirk* s.n.! & Fumba, 18 July 1972, *Robins* 12!
DISTR. **Z**; Mozambique, coasts of Indian Ocean and islands, also reported to be introduced into West Indies and cultivated in Brazil and Guyana
HAB. As for var. *populnea*

SYN. *Hibiscus populneoides* Roxb., Fl. Ind. ed. Carey 3: 191 (1832)
Thespesia populneoides (Roxb.) Kostel., Allg. Med. Pharm. 5: 1861 (1836); Fosberg & Sachet in Smith. Contrib. Bot. 7: 10 (1972); Fryxell, Nat. Hist. Cotton Tribe: 87, fig. 34 (1978); Abedin, Fl. Pakistan 130: 32 (1979); Fosberg & Renvoize, Fl. Aldabra: 67, fig. 6/6–8 (1980); Paul in Fl. India 3: 353 (1993); Sivarajan & Pradeep, Malvaceae S. Pen. India: 39 (1996)

NOTE. Sivarajan & Pradeep, Malvaceae S. Pen. India: 37 (1996), Abedin, Fl. W. Pakistan 130: 32, fig. 6 (1979) and Fryxell, Nat. Hist. Cotton Tribe: 86 (1979) all keep *T. populnea* and *T. populneoides* (Roxb.) Kostel. as separate species, the former distinguishing the species by "Capsules with a hard fibrous indehiscent endocarp and exocarp dehiscing into 4–5 pieces – *populneoides*; capsules without a hard fibrous endocarp, both endocarp and exocarp indehiscent – *populnea*". The latter uses "Leaves copper-coloured, shallowly cordate, pedicel often drooping, 5–12 cm long, without bracteate articulation – *populneoides*; leaves green, deeply cordate, pedicel erect, 1–5 cm long with a bibracteate articulation near base – *populnea*". The leaf-base shape is certainly of no use for most of the material from Australia and Pacific islands etc. which is undoubtedly true *T. populnea*, has subcordate to truncate leaf-bases. Equally useless is whether the pedicel are erect or drooping and the articulation character is difficult to see. There is certainly something in the length of the pedicel and the dehiscence of the fruit but none of the characters correlates well. Philcox has discussed the matter (Rev. Fl. Ceylon 11: 314–315 (1997)) and concluded as did Borssum Walkes (Blumea 14: 107 (1966)) and Nicolson *et al.*, (Interpretion Rheede's Hort. Malab.: 174 (1988)) that the two cannot be kept as separate species; Merrill (Sp. Blanc.: 255 (1918)) commented on these being two forms differing in pedicel but was not convinced two taxa were involved. More investigation could be done on the East African coast by tagging plants and correlating fruit, seed, pedicel and leaf and flower characters. Much of the material I have cited under var. *populnea* has long pedicels but without fruits I have left them under that variety. *Faulkner* 2250 (Zanzibar, Massazine, 5 May 1959) has a 9 cm long pedicel but indehiscent exocarp or perhaps, the fruits are not mature enough. Some specimens are typical var. *populnea* without doubt e.g. *Greenway* 1463 (Pemba, Verani, 18 Feb. 1929) with short pedicel, indehiscent exocarps, more ovoid seeds with a tuft of ferruginous hairs 2 mm long at the narrow end. I suspect corolla size might be worth investigating. *Faulkner* 2292 (Zanzibar, Massazine, 26 June 1959) has a corolla 10 cm long whereas in *Faulkner* 2250 mentioned above it is only 6 cm. Sivarajan and Pradeep have dismissed leaf base and pedicel length as useless characters. Fosberg & Sachet suggest that the numerous intermediate specimens could be of hybrid origin. It certainly seems possible that '*populneoides*' could have evolved as a distinct species in some area of the Indian Ocean but has since hybridised with *T. populnea* so that the possible distinctions it once possessed have become blurred. Until the problem is resolved by modern methods I have treated it as a variety. Pierre, Fl. Forest. Cochinch. 3: t. 173 A & B (1888) gives excellent drawings of the two varieties but uses an illegitimate name var. *rheedii* instead of var. *populnea*.

2. **Thespesia danis** *Oliv.* in Hook. Ic. Pl. 14: 26, t. 1336 (1881); P.O.A. C: 268 (1895); V.E. 3 (2): 405 (1921); T.T.C.L.: 307 (1949); K.T.S.: 263 (1961); Fryxell, Nat. Hist. Cotton Tribe: 89 (1978); Vollesen in Op. Bot. 59: 36 (1980); Paul in Fl. India 3: 349 (1993); K.T.S.L.: 173, fig. (174) (1994); Vollesen in Fl. Eth. & Eritr. 2 (2): 222, fig. 82.14.10–12 (1995); Sivarajan & Pradeep, Malvaceae S. Pen. India: 37 (1996); Thulin, Fl. Somalia 2: 61 fig. 35, A–C (1999). Type: Kenya, Kilifi District: Ribe, *Wakefield* s.n. (K!, lecto. & isolecto!) (see note)

Much branched or rambling shrub or small tree 0.9–12 m tall with smooth grey bark; stems etc. sparsely to densely lepidote. Leaves reniform to rounded, 1.5–9 × 2–11 cm, rounded to subacute at the apex, cordate at base, not lobed, entire at margin, covered with peltate scales, particularly beneath, otherwise smooth and glabrous; petiole 0.5–4.5 cm long. Flowers solitary in the leaf axils, or can appear terminal at leafless apices of branches; pedicel 0.5–5.5 cm long; epicalyx bracts 3, attached on sides of calyx, very variable, broadly ovate to ovate-lanceolate, 0.5–2 × 0.2–1 cm, persistent in fruit; calyx 5–10 mm long, truncate or with minute teeth 1 mm long, lepidote. Petals yellow or orange-yellow with claret or purple base, turning entirely salmon on fading, 2–3.5 cm long, conspicuously gland-dotted; ovary 8-locular. Fruit rose-pink when ripe, subglobose, coriaceous but eventually soft, 1–2 cm diameter, acute or apiculate, glabrous.

KENYA. Northern Frontier District: Sala (not traced), 5 July 1951, *Kirrika* 139!; Kwale District: between Mariakani and Samburu, about 40 km from Mombasa, 22 Sep. 1961, *Verdcourt* 3217!; Kilifi District: Malindi, Oct. 1951, *Tweedie* 1007!

TANZANIA. Tanga District: 6.4 km N of Amboni, 5 Dec. 1935, *B.D. Burtt* 5362!; Pangani District: Bushiri, 8 Feb. 1950, *Faulkner* 710!; Uzaramo District: Dar es Salaam, University College, near lagoon, 10 Dec. 1965, *Mwasumbi* 10030!

DISTR. **K** 1, 7; **T** 3, 6, 8 (fide Vollesen); Ethiopia, S Somalia; India (see note); cultivated in Nairobi

HAB. Alluvial flood-plains with *Cordia*, *Hyphaene*, *Borassus* etc., riverine grassland, coastal bushland and thicket with *Rhoicissus*, *Cissus rotundifolia*, *Ziziphus* etc.; sometimes on termite mounds; extends down to highwater mark but never in mangrove swamps; 0–450 m

SYN. *T. danis* Oliv. var. *grandibracteata* Chiov. in Fl. Som. 2: 28 (1932). Types: Somalia, Dibbei, *Senni* 537 (FT, syn.); from Anole to Cu Daio, *Senni* 99 (FT, syn.)

T. danis Oliv. var. *somalica* Chiov. in Fl. Som. 2: 28 (1932). Types: Somalia, many syntypes, *Senni* 5, 14, 32 & 177, *Gorini* 28, 86 & 156 (FT, syn.)

NOTE. When Oliver described the species he mentioned "also in Galla country where it is considered sacred and called 'danis'"; This is taken directly from the label attached to one of the Ribe specimens and there is no material from Galla. Oliver also cites *Kirk* s.n. and *Hildebrandt* 1929 from 'Zanzibar Coast'. The *Hildebrandt* specimen clearly says 'Insel Mombasa' but I can find no Kirk specimen annotated by Oliver; but there is one labelled "Mombasa 300'" collected in Sep. 1873 which he should have seen. Since there is no locality on the island at that altitude, it was obviously collected some distance inland, perhaps at Rabai or Ribe. It has been recorded from India but if correct presumably introduced (see Sivarajan & Pradeep cited above; they were unable to confirm its occurrence). The fruits are edible and the wood is used for making bows.

3. **Thespesia garckeana** *F. Hoffm.*, Beitr. Fl. Centr.-Ost-Afr.: 12 (1889); P.O.A. C: 268 (1895); T.T.C.L.: 302 (1949); F.P.S. 2: 42, fig. 18 (1952); Fryxell, Nat. Hist. Cotton Tribe: 93, fig. 42 (1978). Type: Tanzania, Tabora District: Gonda, *Böhm* 145 (B†, holo.; K!, Z, iso.)

Small much branched spreading tree or shrub 2–10.5(–12) m tall with pale grey rough bark or greenish brown smooth bark; branchlets stellate-tomentose, ± floccose when young but ultimately glabrescent. Leaves ± round in outline, 3–20 × 4–20 cm, palmately 3–5-lobed or rarely unlobed, cordate at base, 5–7-veined, stellate-pubescent to almost glabrous above, stellate-pubescent to tomentose beneath; lobes rounded or subacute at the apex; median vein often with a short longitudinal fissure beneath near base; petiole 1.5–13 cm long; stipules linear to ligulate, 5–7 mm long. Flowers solitary in the leaf axils; peduncles 2–7 cm long, articulated above the middle; epicalyx base cupuliform, fused with the calyx; bracts (7–)9–10, linear to lanceolate 7–16(–22) × 2–3(–5) mm, deciduous, leaving abscission scars on the rim of the persistent base; calyx cupuliform, 7–10 mm long, stellate-tomentose, truncate or very shortly 5-toothed. Petals yellow to purplish with dark purple or dark red base,

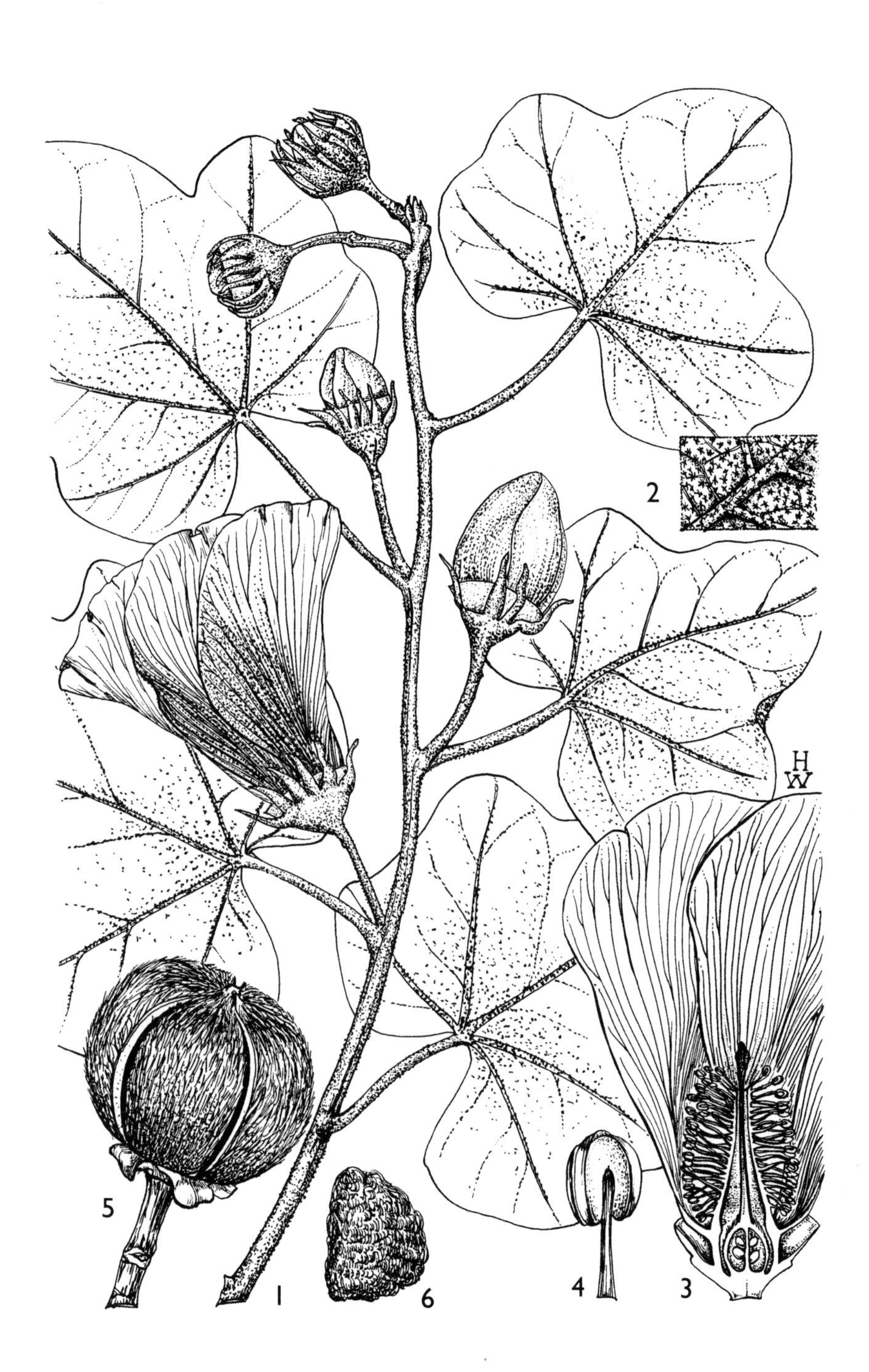

FIG. 19. *THESPESIA GARCKEANA* — **1**, flowering stem, × $^2/_3$; **2**, undersurface of leaf, × 4; **3**, longitudinal section of flower, × 1; **4**, stamen, × 5; **5**, fruit, × 1; **6**, seed, × 2. 1–2 from *Greenway & Kanuri* 12032;3–4 from *Richards* 18423; 5 from *Boaler* 914; 6 from *Milne-Redhead & Taylor* 7625. Drawn by Heather Wood.

fading to dull red, obovate, 5.5–6.5 × 3.5–4 cm, stellate-pubescent outside; style apex bearing stigmatic surfaces clavate, 3.5 mm long. Capsule red, subglobose or ovoid, 2.5–4 cm long, 3–3.8 cm diameter, loculicidally dehiscent by 5 valves, with sugary mucilage beneath the exocarp; seeds hemispherical, ± 10 × 7 mm, with dense long pale hairs.

a. var. **garckeana**

Leaves 3–5-lobed, the lobes rounded to subacute. Fig. 19, p. 106.

KENYA. Meru District: just outside Meru National Park, Kina, 18 Mar. 1968, *Fosberg & Mwangangi* 49926!; Machakos District: Emali Valley, 7 Mar. 1940, *V.G.L. van Someren*! 17; Kitui, May 1877, *Hildebrandt* 2765!

TANZANIA. Masai District: Monduli area, Arusha–Dodoma road, about km 99, 8 June 1965, *Greenway & Kanuri* 12032!; Iringa District: Lukosi R., about 10 km from base of Kitonga gorge, 1 Feb. 1982, *Lovett & Congdon* 1413!; Lindi District: Nachingwea, 28 Mar. 1955, *Anderson* 1025!

DISTR. **K** 4; **T** 1–8; Nigeria (see note), Congo-Kinshasa, Burundi, Sudan, Zambia, Malawi, Mozambique, Zimbabwe, Botswana, Namibia

HAB. Mainly in wooded grassland or bushland with *Combretum, Cordia, Commiphora, Acacia* etc., occasionally codominant, also *Brachstegia-Julbernardia* and *Terminalia* woodland, sometimes on termite mounds; 900–1950 m

SYN. *T. trilobata* Baker f. in J. Bot. 35: 52 (1897); T.T.C.L.: 308 (1949); Fryxell, Nat. Hist. Cotton Tribe: 93 (1978). Type: Tanzania, Mpanda District: E shore of Lake Tanganyika, near Karema [Kareni], *Scott Elliot* 8356 (BM, holo., K!, iso.)

T. debeerstii De Wild. & Durand in Ann. Mus. Congo Belge sér. Bot. 2, 1 (2): 6 (1900). Type: Congo-Kinshasa, upper Marungu, *Debeerst* s.n. (BR, holo.)

T. hockii De Wild. in B.J.B.B. 3: 266 (1911). Type: Congo-Kinshasa, Shaba [Katanga], Luembe Valley, *Hock* s.n. (BR, holo.)

T. rogersii S. Moore in J. Bot. 56: 5 (1918). Type: Zimbabwe, Bulawayo, *Rogers* 5839 (BM, holo.)*

Shantzia garckeana (F. Hoffm.) Lewton in J. Wash. Acad. Sci. 18: 13, t. 1, 2 (1928)

Azanza garckeana (F. Hoffm.) Exell & Hillcoat in Contr. Conhec. Fl. Moçambique: 59 (1954); Jex Blake, Gard. E. Afr. ed. 4: 220, t. 8 (1957); Keay, F.W.T.A. ed. 2, 1(2): 342 (1958); K.T.S.: 262, fig. 53 (1961); Exell in F.Z. 1: 432, t. 88 (1961); Hauman in F.C.B. 10: 150, t. 15 (1963); K.T.S.L.: 170, fig. (171) (1994)

NOTE. Fryxell kept *T. trilobata* separate and Gillett put a note on the covers at Kew that he thought it should be retained as a variety. At first I thought it could be retained as a glabrescent subspecies in Western Tanzania with thinner leaves exemplified by *Pirozynski* 329 (Buha District: Kakombe Valley, 6 Feb. 1964), *Semsei* 75 (in F.D. 2507) (Mpanda District: Kabungu, 15 July 1948), *Siame* 134 (Ufipa District: Rukwa Valley, Tumba, 12 Feb. 1952) and other similar specimens, but a survey of the whole material available from Africa shows that similar specimens occur in East Tanzania, North Malawi, Zambia (Ndola) etc. I suspect that these are all ecological variants, the indumentum and leaf texture being associated with a wetter environment, edges of flood plains etc. Material from much drier areas can have a dense tomentum of stellate hairs. I have therefore refrained from treating it even as a variety. *T. rogersii* is a densely velvety form.

A specimen from Nigeria (Bauchi Plateau, *Kennedy* FHI 6276) is described as a tree to 18 m with a girth up to 90 cm; the epicalyx bracts are 2.2 cm long but it appears to be only an extreme variant.

b. var. **schliebenii** *Verdc.* **var. nov**. a var. *garckeana* foliis cordatis haud lobatis apice anguste acuminatis differt. Typus: Tanzania, Lindi District: Nambiranji, *Schlieben* 6009 (BR!, holo.; K!, iso.)

Small tree 6–12 m tall. Leaves cordate, not lobed, narrowly acuminate at the apex, with very sparse scattered minute stellate hairs, almost glabrous.

* *Rogers* 5389 from Victoria Falls has been labelled as a type at K but S. Moore stated only probably the same.

TANZANIA. Lindi District: Nambiranji, 17 Feb. 1935, *Schlieben* 6009!
DISTR. **T** 8; not known elsewhere
HAB. Grassland with scattered trees ('Parklandschaft'); ± 200 m

NOTE. Schlieben states frequent (haüfig) so it is strange that no further material appears to have been collected by others who have visited the area. Fryxell annotated the Kew specimen *T. trilobata* Baker f.; it is not that but similarly glabrescent. No fruits are known. I had at first thought it was a new species but until more material is available I have preferred to treat it as a variety since there are no floral characters to distinguish it.

12. **MALVA**

L., Sp. Pl.: 687 (1753) & Gen. Pl., ed. 5: 308 (1754)

Annual or perennial herbs; indument stellate. Leaves unlobed to deeply palmately divided, often with angular or broadly rounded lobes; margins crenate; stipules triangular to falcate, ± persistent. Flowers solitary in leaf axils or fasciculate; sometimes the clusters forming narrow interrupted panicles. Epicalyx of 3 completely free bracts or adnate to calyx. Calyx 5-lobed. Petals usually pink to purple or white, often with dark veins, 5, free, mostly broadly obovate. Stamens numerous. Ovary with 8–15 carpels each with a single ascending ovule; styles the same number as the carpels. Fruit a schizocarp; mericarps separating, indehiscent, smooth to variously sculptured, glabrous or hairy.

About 30 species in the Old World, mostly in temperate or warmer areas, but several species are now widespread cosmopolitan weeds.

1. Mature mericarps with the back smooth or weakly ribbed near edges which are ± rounded; sides weakly ribbed; leaves larger, 4–24(–30) cm long and wide . . . . 1. *M. verticillata*
   Mature mericarps with the back strongly reticulate with edges angular; sides strongly ribbed; leaves usually smaller, 2.5–10 cm long and wide . . . . 2. *M. parviflora*

1. **Malva verticillata** *L.*, Sp. Pl.: 689 (1753); Mast. in F.T.A. 1: 177 (1868); A.V.P.: 128 (1957); Exell in F.Z. 1: 502 (1961); Hauman in F.C.B. 10: 153 (1963); Riedl in Flora Iranica 120: 22 (1976); Abedin, Fl. W. Pak. 130: 43, fig 9A (1979); Vollesen in Fl. Eth. & Eritr. 2(2): 237, fig 82.19.1–4 (1995); U.K.W.F., ed. 2: 102, t. 29 (1994). Type: China, Syria, *Herb. Linn.* 870.26 (LINN, lecto.)

Annual or biennial herb 0.5–2.4 m tall, erect, decumbent or ascending, pubescent; stems somewhat woody at base; the 'stellateness' of the hairs on foliage and calyx etc. varies considerably, the hairs being almost simple to very distinctly stellate. Leaves cordate to reniform in outline, 4–24(–30) cm long and wide, shallowly 5–7-lobed, the lobes triangular, acute to rounded, crenate to denticulate; petiole 3–20 cm long. Flowers in dense axillary clusters merging into interrupted panicles; pedicel 0–7(–18 in fruit) mm long(–30 fide Exell); epicalyx lobes lanceolate, 5 × 2 mm; calyx 4–5 mm long, later accrescent, scarious and reticulate attaining 12 × 15 mm in fruit; lobes triangular, acute. Petals pink, mauve, pale bluish lilac or white with pink tips, 6–10 mm long. Mericarps 10–13, 1.5–2 mm long, usually glabrous (see note); back smooth or slightly ribbed near edges; sides with stronger ridges. Fig. 20, p. 108.

UGANDA. Karamoja District: Mt Morongole, Apr. 1960, *J. Wilson* 981!; Kigezi District: Kachwekano Farm, Dec. 1949, *Purseglove* 3146!; Mbale District: Bugishu, Bulambuli, 4 Sep. 1932, *A.S. Thomas* 584!

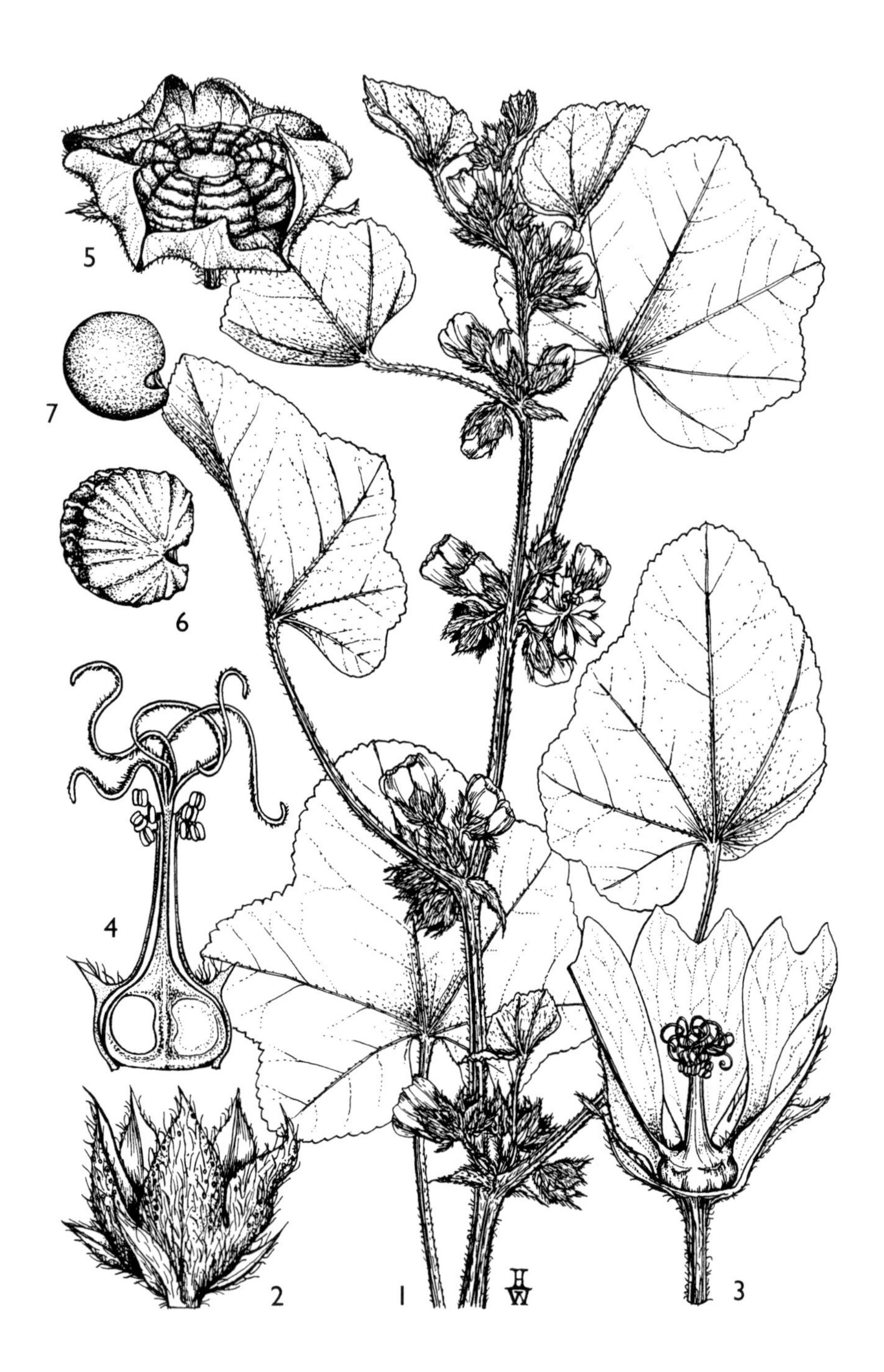

FIG. 20. *MALVA VERTICILLATA* — **1**, flowering stem, × 1; **2**, epicalyx and calyx, × 4; **3**, flower (half the calyx and petals removed), × 4; **4**, longitudinal section of gynoecium and androecium, × 6; **5**, fruit, × 4; **6**, mericarp, × 10; **7**, seed, × 10. 1–4 from *Tweedie* 364; 5–7 from *Bogdan* 5718. Drawn by Heather Wood.

KENYA. Northern Frontier District: Maralal, 1.6 km N of Muramur dam, 18 May 1968, *Nesbit-Evans* 61!; Elgeyo District: North Cherangani Hills, Chepkotet, 13 Aug. 1968, *Thulin & Tidigs* 241!; Kiambu District: Limuru, 30 Sep. 1951, *Verdcourt* 623!

TANZANIA. Masai District: Ngorongoro Conservation Area, Empakai Crater, 6 July 1973, *Frame* 211!; Kilimanjaro, Leranjwe, 31 July 1993, *Grimshaw* 93461!; Lushoto District: Jaegertal, 20 Feb. 1984, *Kisena* 166!

DISTR. **U** 1–3; **K** 1, 3, 4, 6, 7; **T** 2, 3; now widespread in tropical, subtropical and warm temperate areas of the Old World

HAB. Bushland, forest, upper edges of bamboo zone, ericaceous and alpine zones, banks of mountain streams, cultivations, waste and fallow ground, roadsides; 1200–4050 m

NOTE. This must have been widespread at an early date – already in West Uganda during last decade of 19th Century. It penetrates right into alpine zone e.g. *Hedberg* 894 from northern slope of Koitobos, Mt Elgon. Has been used as a pot herb and vegetable. *Grimshaw* 93757 (West Kilimanjaro, 24 Sep. 1993) is a distinctive form with pubescent mericarps and *Hepper & Jaeger* 7049 (Mt Kulal, Gatab, 24 Nov. 1978) appears to be the same and also probably *Bally* 5555 (also from Mt Kulal in forest glade near waterhole at 1905 m, 9 Oct. 1947). Bally had named his plant *M. verticillata*; Gillett had named it *M. parviflora*, Abedin cf. *M. rotundifolia* and Krebs cf. *pusilla* demonstrating how difficult these weedy *Malva*'s are. The mericarps of *Bally* 5555 are not rounded at the edges. More extensive and better material with corollas and fruits may solve the true identity of these plants. *M. neglecta* Walls. has mericarps dorsally ± smooth and pubescent but is much less robust with much more slender stems.

2. **Malva parviflora** *L.*, Demonstr. Pl.: 18 (1753) & Am. Acad. 3: 416 (1756) & Spl. Pl. ed. 2: 969 (1763); Exell in F.Z. 1: 501 (1961); Cullen in Fl. Turkey 2: 407 (1967); Meikle, Fl. Cyprus: 307 (1977); Abedin, Fl. W. Pak. 130: 42, fig. 7/C–F (1979); Townsend in Fl. Iran 4: 239, t. 44, 4–9 (1980); U.K.W.F. ed. 2: 102, t. 30 (1994); Vollesen in Fl. Eth. & Eritr. 2 (2): 387, fig. 82.19.7 (1995); Thulin, Fl. Somalia 2: 74, fig. 45 (1999). Type: not cited but presumably plant growing in Hort. Upsal. (In 1763 Linnaeus gave locality as in Barbaria i.e. North Africa, W of Egypt) *Linnean Herb.* 870. 17 (LINN, lecto.)*

Annual herb 5–60 cm, tall with erect, prostrate or ascending ± pubescent branches. Leaves reniform to round in outline, 2.5–10 cm long and wide, cordate at base, ± unlobed to 5–7-lobed, the lobes broadly rounded, crenate-dentate (or in some varieties deeply lobed and sharply acutely dentate), ± stellate-pubescent; petiole 3–7(–17) cm long; stipules lanceolate to ovate, 2–5(–12 fide Townsend) mm long. Flowers in axillary fascicles; pedicel up to 2.5 cm long; epicalyx bracts linear, 3.5–7 mm long; calyx 3–5 mm long but distinctly accrescent in fruit to 1 cm, glabrous or shortly pubescent; lobes broadly ovate, acuminate. Petals white to pinkish with darker lines, obovate, 4.5 mm long. Mericarps 9–11, 1.5–2 mm long, usually glabrous but sometimes shortly pubescent, conspicuously reticulate dorsally with raised walls; margins carinate to deeply denticulate or ± serrate or forming dentate crests at the sutures; lateral faces with radiating ribs; seeds brown, 1.4–1.8 mm long, minutely rugulose.

var. **parviflora**; Meikle in Fl. Cyprus: 307 (1977)

Leaves with broad blunt crenate or shallowly toothed lobes.

KENYA. Northern Frontier District: Moyale, 26 Oct. 1952, *Gillett* 14103!; Nakuru District: Elmenteita, 18 June 1964, *Bogdan* 5713!; Naivasha District: Lake Naivasha, Crescent I., 11 Oct. 1976, *Hayes* 139!

TANZANIA. Musoma District: Seronera Plains, Simba Kopjes, 22 Dec. 1962, *Greenway* 10903!; Masai District: Ngorongoro Crater, 17 Jan. 1965, *Goddard* 45! & 27 km from Loliondo on road to Seronera, 26 Aug. 1968, *Carmichael* 1489!

* Abedin considered this to be the holotype and this is equivalent to lectotypification – it is possible it is the holotype.

DISTR. **K** 1, 3; **T** 1, 2; Ethiopia, Somalia, Sudan, Zimbabwe; native of Europe and Asia but now a cosmopolitan weed

HAB. Roadsides, grassland with lava boulders, muddy edge of rain ponds in *Pennisetum*, *Digitaria*, *Cynodon* grassland; 1100–1900 m

NOTE. I have had considerable difficulty in deciding whether the weedy species of *Malva* with reticulate mericarps occurring in East Africa should be called *M. parviflora* L. or *M. pusilla* Sm. (Type: England, Kent, Hythe, *Sherard & Rand* s.n. (ubi: not found in Herb. Smith) (= *M. rotundifolia* L. nom. rejic., *non* auctt.) or if some hybridization has taken place. G. Krebs annotated *Gillett* 14103 as *M. pusilla* Sm. and *Bogdan* 5713 as *M. parviflora*. After examination of much material from Europe and the Middle East and also from throughout the world where they have been introduced, it is clear that typical material of each is distinguishable. *M. parviflora* has the mature calyx accrescent with rather broad lobes, ± scarious and reticulate, glabrous or almost so 10–15 mm across when spread out; the mericarps are strongly ridged-reticulate, glabrous or slightly pubescent; whole fruit 6–8 mm across with mericarps with raised or ± winged, often ± serrated edges; the young sepals can be slightly pubescent. *M. pusilla* has the mature calyx not accrescent, the lobes more triangular, green, long ciliate, ciliation is particularly noticeable; calyx 9–10 mm flattened out; fruits 5–6 mm across the mericarps, less strongly ridged-reticulate with only slightly raised edges. The East African material is largely intermediate and it is not possible to distribute the material seen between two species. Examination of extensive material from Iraq, Cyprus and other places in the Middle East makes it clear that accrescence and ciliation of the calyx varies. Cullen (Fl. Turkey 2: 405 (1966)) does not use the ciliation of the calyx character at all, only non-accrescent sepals and non-serrate or cristate mericarp margins. I have concluded that the treatment of *M. parviflora* and *M. pusilla* as distinct species needs experimental study. I have treated all the East African material as *M. parviflora*.

13. MALVASTRUM

A. Gray in Mem. Amer. Acad. Arts N.S. 4: 2 adnot. (1849); S.R. Hill in Rhodora 84: part I, 1–83; part II, 159–264; part III, 317–409 (1982), *nom. conserv.*

Annual or perennial herbs or subshrubs with usually stellate indumentum but less often also simple hairs and scales. Leaves serrate, petiolate; stipules linear to triangular or lanceolate. Flowers solitary in leaf axils or in terminal raceme-like or spike-like inflorescences, pedicellate or subsessile. Epicalyx of 3 linear to subcordate bracts. Calyx often strongly 5-angular; lobes united for $^1/_5$–$^3/_5$ of their length. Petals 5(–6), yellow to orange, obovate, usually emarginate. Ovary with 5–18 carpels in a single discoid whorl, each with a single ovule; style with 5–18 branches from about the middle; stigmas linear to subglobose. Fruit a schizocarp; mericarps 5–18, free from each other, indehiscent or partly to completely dehiscent, narrowly wedge-shaped, very compressed laterally, glabrous to densely pubescent, with conspicuous proximal notch, lateral faces smooth to radially ribbed, unornamented or more usually with 1–3 mucros or cusps.

Genus of 14 species native to America from U.S.A. to Argentina and West Indies; several species introduced into the Old World, one occurring extensively in the tropics.

Malvastrum coromandelianum (*L.*) *Garcke* in Bonplandia 5: 295 (1857); Keay, F.W.T.A. ed. 2, 1(2): 350 (1958); Exell in F.Z. 1: 503 (1961); Abedin, Fl. W. Pak. 130: 90, fig. 9, C–D (1979); U.K.W.F. ed. 2: 102 (1994); Vollesen in Fl. Eth. & Eritr. 2 (2): 239 adnot. (1995); Philcox in Rev. Fl. Ceylon 11: 334 (1997). Type: Hortus Upsaliensis, *Linnean Herb.* 870.3 (LINN, lecto.) (fide Abedin)

Annual or perennial herb 0.2–1 m tall but sometimes flowering when only 6 cm tall; stems with appressed stellate hairs. Leaves ovate to lanceolate, 1.5–7.8 × 0.6–5.5 cm, unlobed or sometimes shallowly 3-lobed, acute at the apex, truncate or rounded to cuneate at base with simple and stellate hairs on both surfaces; petiole 0.5–2(–4) cm

FIG. 21. *MALVASTRUM COROMANDELIANUM* — **1**, habit × $^{1}/_{4}$; **2**, flower, × 2; **3**, petal, × 2; **4**, stamens, × 10; **5**, gynoecium, × 10; **6**, mericarp, side view, × 6; **7**, mericarp, front view, × 6; **8**, seed, × 6. From Flora Brasiliensis 12, 3, t. 53.

long; stipules 2–6(–8) mm long. Flowers solitary, axillary, sometimes at maturity congested at apices or rarely on reduced 2–3-flowered axillary branches near apex; pedicel 1–2 mm long becoming 3–5 mm in fruit; epicalyx bracts 4–6 mm long; calyx 5–7 mm long, up to 11 mm long in fruit; lobes triangular, 4–5 × 2–2.5 mm (6–7 × 5–6 mm in fruit). Corolla yellow or orange, the petals 0.6–1.3 cm long; style branching 2 mm above its base. Mericarps 9–14(–15) with a single cusp on upper margin near proximal end and 2 apical cusps at distal end, upper margin with dense crest of hairs; lateral faces conspicuously ribbed especially at margin. Fig. 21, p. 111.

subsp. **coromandelianum**; S.R. Hill in Rhodora 84: 324, figs. 63, 64, 65 (1982)

Annual or perennial herb generally with several main stems 0.2–1 m tall; flowers solitary, axillary, somewhat congested apically with age; leaves usually with simple hairs above, less often stellate; mericarps 2.4–4.5 × 3–4 mm.

KENYA. Nairobi, Boulevard Hotel to Museum Bridge, 11 Jan. 1983, *Mungai* 3183! & Nairobi, *Lambinon* 75/262 (MO fide Hill)
TANZANIA. Moshi, opposite Tanesco office, 16 Nov. 1970, *Batty* 1146!; Tanga District: Muhesa, Police Station, 20 Nov. 1972, *Archbold* 1538B; Kilosa District: Ilonga, 1 Jan. 1967, *Robertson* 664!
DISTR. **K** 4; **T** 2, 3, 6; native of U.S.A. to Argentina and West Indies now widespread in Old World tropics and subtropics
HAB. Roadsides, pavements, near latrines and weed of cultivation; 0–1650 m

SYN. *Malva coromandeliana* L., Sp. Pl.: 687 (1753)

NOTE. The other two subspecies are known only from South America.

14. WISSADULA

Medik., Künstl. Geschlecht. Malv.-Fam.: 24 (1787)

Perennial herbs or shrubs with stellate indumentum. Leaves usually unlobed, acuminate at the apex, usually cordate at base; stipules setaceous. Flowers in lax terminal panicles. Epicalyx absent. Calyx shallow, 5-lobed. Corolla orange, yellow or cream. Ovary with 3–5 free 3-ovulate carpels around a central torus; style-branches 3–5; stigmas capitate. Fruit of 3–5 mericarps, each transversely divided by a septum formed an oblique constriction of the lateral walls, eventually dehiscing. Seeds 1–3 per mericarp, ± globose-reniform, pubescent.

About 40 species, mostly in tropical America but 3 occurring in the Old World.

Wissadula rostrata (*Schumach. & Thonn.*) *Hook. f.* in Niger Fl.: 229 (1849); Mast. in F.T.A. 1: 182 (1868); Exell in Bull. I.F.A.N. 21, Sér A: 452 (1959) & in F.Z. 1: 499, t. 95 (1961); Vollesen in Op. Bot. 59: 36 (1980); U.K.W.F. 103 (1994); Vollesen in Fl. Eth. & Eritr. 2 (2): 239, fig. 82.20 (1995); Wood, Handb. Yemen Flora: 106 (1997); Krapovickas in Bonplandia 9: 94 (1996). Type: Ghana, Keta [Quitta], *Thonning* 121 (C, holo.; S, iso.)

Erect subshrubby herb 0.6–2.4 m tall; stems etc. minutely tomentose and stellate-pubescent, eventually glabrescent and developing thin greyish bark. Leaves cordate, 2.5–19 × 1.5–14.5 cm, long-acuminate at the apex, truncate to deeply cordate at base, discolorous, green and minutely puberulous or ± glabrous above, whitish grey and densely finely velvety tomentose beneath and with more scattered coarser stellate hairs; petiole 5–18 cm long, stellate-pubescent; stipules ± 6 mm long. Flowers in large lax terminal panicles; pedicel 1–3.5 cm long, elongating to 6–7 cm long; calyx 3–4 mm long, puberulous. Corolla yellow or orange; petals 4–7 mm long. Mericarps 7–10 mm long including the beak, ± 3 mm wide, puberulous, spreading above so that fruit is obconic; seeds dark, 3 mm long, pubescent. Fig. 22, p. 113.

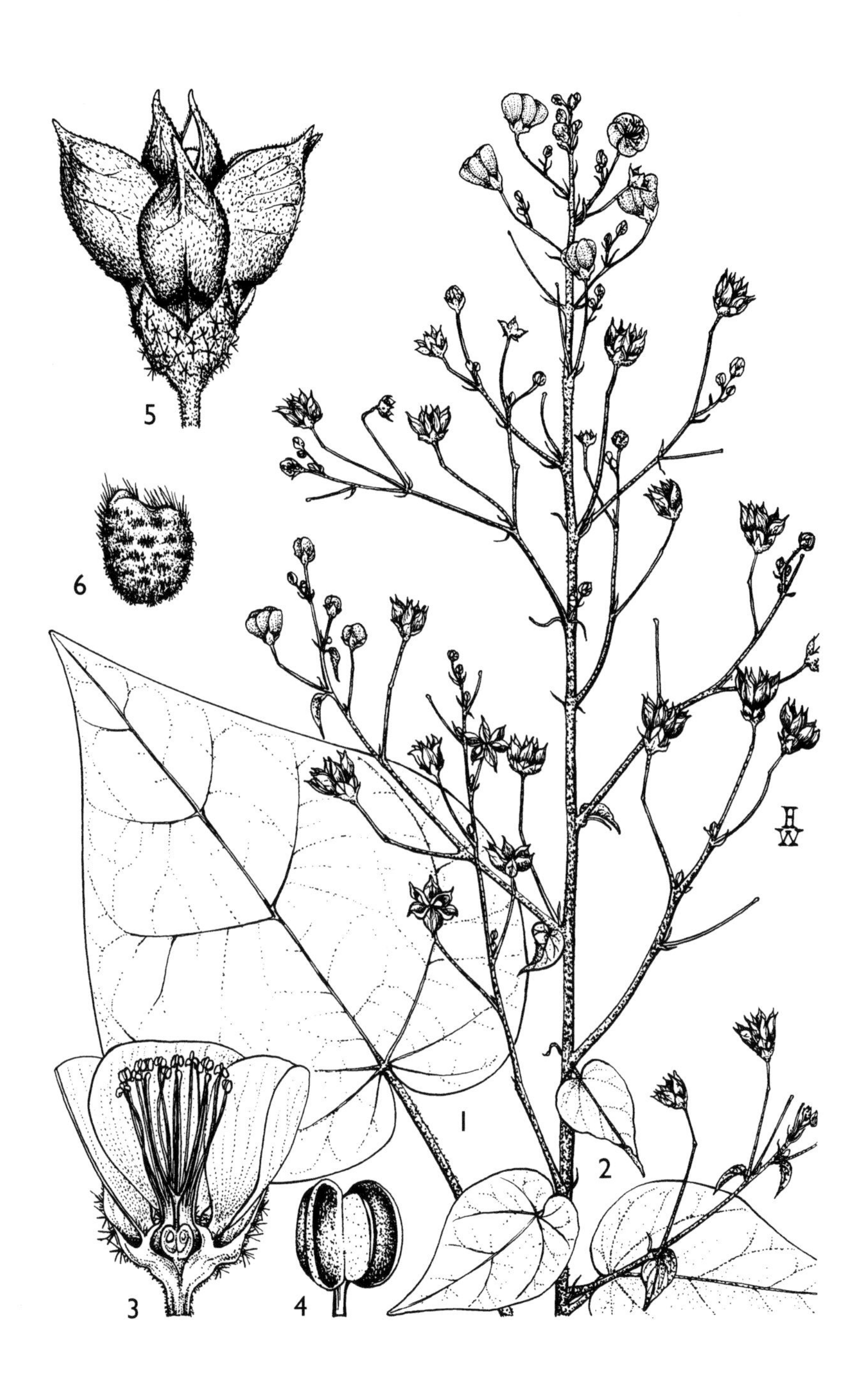

FIG. 22. *WISSADULA ROSTRATA* — **1**, leaf, × ⅔; **2**, flowering and fruiting stem, × ⅔; **3**, longitudinal section of flower, × 4; **4**, anther, × 15; **5**, fruit, × 4; **6**, seed, × 5. 1 from *Purseglove* 3519; 2 from *Purseglove* 1063; 3–4 from *Haerdi* 537/0; 5–6 from *Maitland* 402. Drawn by Heather Wood.

UGANDA. Karamoja District: W of Mt Debasien, Namalu, Feb. 1963, *J. Wilson* 1358!; Bunyoro District: Rabongo Forest, 14 May 1993, *Sheil* 1652!; Busoga District: Jinja, Aug. 1904, *E. Brown* 85!
KENYA. Kisumu-Londiani District: Kisumu, May 1961, *Tweedie* 2149!; Magadi District, 'MAG' fide UKWF
TANZANIA. Mwanza, 30 May 1937, *B.D. Burtt* 6544!; Kigoma District: Bolimba [Bulimba], 25 May 1975, *Kahurananga et al.* 2797!; Morogoro District: Turiani–Mhonda Mission road, 15 July 1957, *Semsei* 2679!
DISTR. **U** 1–4; **K** 5, 6; **T** 1, 2, 4, 6–8; widespread in tropical Africa extending to South and SW Africa; also in the Yemen
HAB. Grassland, bushland, thicket and woodland, often on grey or black soils, often riverine, sandy lake shores; (9 m, Mafia I.–)500–1650 m

SYN. *Sida rostrata* Schumach. & Thonn., Bescriv. guin. Pl.: 306 (1827) & in Kongl. danske Vidensk. Selsk. naturvidensk. math. Afhandl. 3: 80 (1829)
Wissadula amplissima (L.) R.E. Fr. var. *rostrata* (Schumach. & Thonn.) R.E. Fr. in Kungl. Svensk. Vetenskapsakad. Handl. Ny Földj. 43, 4: 51, t. 6, fig. 13–14 (1908); C.F.A. 1: 150 (1937); T.T.C.L.: 308 (1949); Keay in F.W.T.A. ed 2, 1: 336 (1958); Hepper, W. Afr. Herb. Isert & Thonning: 73 (1926)
W. hernandioides (L'Hérit.) Garcke var. *rostrata* (Schumach. & Thonn.) R.E. Fr. in Wiss. Ergebn. Schwed.-Rhod.-Kongo-Exped. 1: 143 (1914); F.P.N.A. 1: 584 (1948)
W. periplocifolia sensu Cufod., E.P.A: 540 (1959) pro parte, and numerous determinations in herbaria, *non* (L.) Thwaites

NOTE. *Hosegawa* 18 (Mahali Mts, Apr. 1981) is described as a 'small tree' so it may sometimes exceed 2.5 m and be somewhat woody. Many specimens have been determined as *W. periplocifolia* (based on a Hermann specimen from Sri Lanka but usually considered to extend to the New World or even to have been introduced from it) but that has much narrower leaves and I have seen nothing from East Africa resembling it. On the other hand it is possible that the African material is a subspecies of *W. hernandioides*; but I have followed Exell, Vollesen and Krapovickas. The latter author has shown that *W. amplissima* (L.) R.E. Fr. and *W. hernandioides* (L'Hérit.) Garcke are separate species and clarified the nomenclature (see Bonplandia 9: 89–94 (1996)).

A large number of herbarium specimens has the authority Planchon alone instead of Hook. f.; it was certainly Planchon who first indicated on early specimens from West Africa the combination *W. rostrata* so Planchon ex Hook. f. is correct.

15. **ABUTILON**

Mill., Gardn. Dict. abr. ed. 4 (1754); Adans., Fam. 2: 398 (1763); Baker f. in J. Bot. 31: 71–76, 267–272, 334–338, 361 (1893); Hochr. in Ann. Conserv. Jard. Bot. Genève 7: 24 (1902); Mattei in Boll. Ort. Soc. Palermo N.S. 1: 71–102 (1915); Ulbrich in V.E. 3 (2): 370–374 (1921); Fryxell in Lundellia 5: 79–118 (2002)

Herbs or shrubs or rarely small trees, mostly erect but a few prostrate or subscandent; indumentum stellate and often simple as well. Leaves petiolate or less often subsessile; lamina mostly cordate at base. Flowers solitary in leaf axils or fascicled or less often 2–6 on a common peduncle, sometimes forming terminal leafy or leafless panicles; peduncles usually articulated near apex (the upper part perhaps to be looked on as a pedicel?). Epicalyx absent. Calyx 5-lobed. Corolla yellow or orange (sometimes with a red to purple centre), or white (often with a dark centre) or pink, mauve or purple; petals 5, joined at the extreme base and adnate to base of the staminal tube, longer than the calyx. Staminal tube usually stellate-pubescent. Carpels 5–± 40, 2–9-ovulate, joined to form a subglobose or broadly cylindrical ovary; style-branches as many as the carpels. Fruits subglobose, broadly cylindrical or barrel-shaped, often deeply depressed in centre; mericarps laterally much compressed, apiculate or awned and usually with retrorse tooth on the ventral margin, usually dehiscing apically, (1–)2–3(–9)-seeded. Seeds usually irregularly reniform, glabrous to tomentose, smooth or variously papillate or pitted.

About 150 species, pantropical but also in subtropics and warm temperate regions. Many are very distinct but there are several very difficult complexes, the taxonomy of which is far from resolved.

Several species are widely cultivated but I have seen only three specimens from East Africa namely *A.* × *hybridum* Hort., *Gitonga* 114, Njoro Plant Breeding Station, 20 June 1971 and *Graham Bell* 6, Closeburn, Nairobi, 3 Mar. 1952. Jex-Blake (Gard. E. Afr. ed. 4: 102 (1957)) mentions *A. darwinii* Hook. f., *A. hildebrandtii* (see page 137), *A. insigne* Planch., *A. megapotamicum* (Spreng.) St. Hil. & Naudin, *A. striatum* Lindl. (now known as *A. pictum* (Hook. & Arn.) Walp.), *A. venosum* Walp., [*Gillett* 22753, Closeburn Nurseries, Nairobi, 15 Dec. 1979] and *A. vitifolium* (Cav.) Presl (sometimes placed in a separate genus as *Corynabutilon vitifolium* (Cav.) Kearney.

There are many hybrids now available. The species mentioned above may be separated as follows. All are natives of South America.

1. Staminal column exceeding the petals; flowers solitary in the leaf axils 2
   Staminal column shorter than petals or if ± equalling petals then peduncles several-flowered and bracteate 4
2. Leaves deeply palmately 5–7-lobed; calyx to 3.5 cm with lobes much longer than tube .... *A. venosum*
   Leaves not lobed or not so deeply lobed; calyx lobes shorter than or ± equalling tube 3
3. Leaves narrowly triangular-ovate, to 8 cm long, calyx-tube narrowly 5-winged, the teeth ± $^1/_4$ the length of the tube .... *A. megapotamicum*
   Leaves and calyx not as above; leaves broadly ovate, 3–7-lobed; calyx lobes ± equalling the tube .... *A. pictum*
4. Petals ± 4 cm long, about 4 times as long as the staminal column; flowers solitary or in 2–6-flowered cymes .... *A. vitifolium*
   Petals 3–4 cm long almost equalling or not so much longer than the staminal column 5
5. Upper leaves unlobed and lower only shallowly 3-lobed; flowers in 3–7-flowered axillary cymes with some leaf-like bracts; petals white or rose with dark crimson veins and margins .... *A. insigne*
   Upper leaves more distinctly 3–5-lobed or almost unlobed in *A. hybridum*; peduncles 1-flowered 6
6. Peduncles 1-flowered, 1–3 per axil; corolla dark orange-red with blood-red veins; leaves densely velvety hairy .... *A. darwinii*
   Flowers usually solitary in leaf axils but may form terminal panicles; corolla variously coloured, white, apricot, yellow, orange or red variously veined or marbled; leaves pubescent or glabrescent (very numerous cultivars) .... *A.* × *hybridum*

For the key to native species, see next page

MAIN KEY

[Note: 7. *Abutilon* sp. from **T** 3, Lushoto, has not been keyed out, as only immature fruits are known; 8. *Abutilon* sp. from **U** 3 and **K** 5 also lacks data]

1. Prostrate or scandent subshrubby herbs; leaves up to 3.3 × 2 cm; flowers solitary in leaf axils with very slender pedicels 3.5–4.5 cm long; petals apricot yellow, ± 8 mm long; carpels 7; **K** 1, Mandera (ripe fruit not known but immature mericarps shortly beaked) . . . . 10. *A. subprostratum* (p. 127)
   Erect herbs or shrubs not as above or if prostrate then a plant of coastal dunes 2
2. Flowers pink, lilac or mauve, often with a darker centre, rarely white and never yellow; trichomes of staminal column simple or 2–3-fid, often flattened when dry or petals with dense long hairs at top of claw 3
   Flowers yellow or orange but sometimes with purple centre or in some cases drying pinkish or bluish; staminal column glabrous to densely covered with stellate hairs 4
3. Staminal column with trichomes simple or 2–3-fid, often flattened when dry; mericarps rounded above (very widespread) . . . . 1. *A. longicuspe* (p. 118)
   Staminal column without these characteristic hairs but petals with dense long hairs at top of claw; mericarps shortly awned; **U** 1 . . . . 2. *A. eggelingii* (p. 120)
4. Calyx densely long-pilose, the lobes narrowly triangular; fruits smoothly rounded above when young but mericarps shortly acute when ripe; **K** 1, Garissa to Wajir area . . . . 9. *A. pilosicalyx* (p. 127)
   Calyx not densely long-pilose or if somewhat so, fruit and geography different 5
5. Mericarps rounded above or at the most with a slight angle or short point scarcely 0.5 mm long 6
   Mericarps distinctly angled to awned, the awn a narrowly triangular projection 1–7 mm long 10
6. Stems almost always distinctly angular or ridged below nodes (not in a few specimens from Kenya and Tanzania) (not glandular); flowers in terminal lateral panicles; calyx small with small lobes 3 × 4 mm (save in var. in **T** 8); stems and undersides of leaves very finely tomentose 3. *A. angulatum* (p. 121)
   Stem not angular and other characters not combined 7
7. Petals usually 5–12 mm long (rarely to 17); mericarps 9–10, 5–9 × 3–4 mm, subacute to acute but not awned . . . . 11. *A. fruticosum* (p. 128)
   Mericarps more than 13 or if 10 then wider than 5 mm 8
8. Whole plant sticky or at least inforescences densely glandular 9
   Plant neither sticky nor glandular or with only a few glands; stems usually velvety-tomentose with or without long hairs or densely pubescent . . . . 6. *A. pannosum* (p. 123)

9. Whole plant sticky; corolla yellow to orange, usually with reddish maroon or purple centre; calyx larger, 0.9–1.8 cm long with lobes ovate, ± 7 × 6 mm (widespread) . . . . 5. *A. hirtum* (p. 122)
   Inflorescence axes densely glandular but rest of plant not sticky; corolla yellow to orange; calyx 5–9 mm long with rounded ovate lobes 2.5–4 × 4 mm; flowers typically in leafless panicles; **K** 1 . 4. *A. anglosomaliae* (p. 122)
10. Flowers in (1–)2–6-flowered umbel-like cymes (sometimes forming terminal panicles); mericarps 6–8(–10); petals 4–7 × 3 mm; mericarp awns spine-like, curved, 2–3 mm long . . . . 21. *A. ramosum* (p. 143)
   Flowers solitary (sometimes forming terminal or axillary panicles); mericarps 10–40, rarely 8–9 11
11. Prostrate or spreading shrubby herb of coastal dunes and coral scrub etc.; leaves often subsessile but petiole can be up to 6 cm long; petals 7–10 mm long; mericarps 9–10(–12); **K** 7, Lamu . . . . 12. *A. rotundifolium* (p. 129)
   Plant erect and of different habitat or if on coastal dunes etc. then other characters not agreeing 12
12. Leaves distinctly but not deeply 3-lobed, stem with short and long spreading hairs; calyx 0.9–1.5 cm long; corolla yellow often with purple centre; petals 1.3 cm long; mericarps 9–11(–17) with awn 2–5 mm long; **K** 7; **T** 6–8 . 13. *A. wituense* (p. 130)
   Leaves not 3-lobed (or only with slight traces) 13
13. Leaves green on both sides, only thinly pubescent, the upper subsessile and much narrower than the distinctly petiolate cordate lower ones; mericarps distinctly veined when dry, green to brown; **T** 8 . . . . 20. *A. bussei* (p. 141)
   Leaves more distinctly pubescent, the upper more distinctly petiolate; mericarps not veined 14
14. Corolla mostly rather small with petals 7–13 mm long; mericarps 10–20; mericarps with short apical appendage 1–2 mm long or rarely distinctly acute 15
   Corolla usually larger with petals 12–38 mm long; mericarps 9–40; mericarps with longer apical appendage 3–5 mm long 16
15. Stems without long hairs except sometimes on basal thicker parts; leaves more triangular with lamina often longer than petiole; indumentum greyish and glaucous; inflorescences more subcapitate or pseudopaniculate, the pedicel often quite short; mericarps more greyish green or brown . . . . 15. *A. rehmannii* (p. 131)
   Stems mostly with long spreading hairs as well as shorter pubescence; leaves more ovate with lamina mostly shorter than petiole; indumentum not greyish and glaucous; inflorescence more open; mericarps often more blackish . . . . 16. *A. bidentatum* (p. 132)

16. Leaves slightly bullate in life, the venation slightly but distinctly impressed when dry; upper surface rather rough; staminal column often but not always glabrous; [calyx with lobes joined rather high up so that base is cup-shaped; lobes often 3-ribbed, the tips exceeding the mericarps; ripe mericarps dark brown with inner margin very densely and persistently hairy] . . . . 18. *A. guineense* (p. 139)
    Leaves not bullate and venation not impressed; upper surface smooth or velvety; staminal column nearly always stellate-pubescent (save in some variants of *A. mauritianum*) 17
17. Leaves mostly quite coarsely toothed; calyx with lobes joined rather high up so that base is cup-shaped; calyx lobe tips usually exceeding mericarps; mericarps dark brownish, densely pilose . . . . 19. *A. grandiflorum* (p. 140)
    Leaves much more finely serrate or crenate or almost entire; calyx with lobes joined lower down with base shallower; calyx lobe tips shorter or longer than mericarps; mericarps brown to quite black, hairy but eventually often much less so or even quite glabrous 18
18. Mericarps 9–16, the apical appendages more abrupt, almost subulate; calyx lobes much shorter than mericarps; **T** 6, 8 . . . . 14. *A. lauraster* (p. 130)
    Mericarps 17–40, nearly always 20–30, the apical appendages narrowly triangular; calyx lobes shorter, equalling or slightly longer than mericarps; widespread throughout Flora area . 17. *A. mauritianum* (p. 133)

1. **Abutilon longicuspe** *A. Rich.*, Tent. Fl. Abyss. 1: 69 (1847); Mast. in F.T.A. 1: 184 (1868); P.O.A. C: 265 (1895); Brenan, T.T.C.L.: 298 (1949) & in Mem. N.Y. Bot. Gard. 8: 223 (1953); Exell & Mendonça, C.F.A. 1: 373 (1951); Meeuse in F.Z. 1: 489, t. 93/4 (1961); Cribb & Leedal, Mt. Fl. S. Tanz.: 91, t. 6a (1982); Kabuye in U.K.W.F. ed. 2: 101, t. 29 (1994); Vollesen in Fl. Eth. & Eritr. 2 (2): 241, fig. 82.21.1–5 (1995). Types: Ethiopia, Memsach, Genniam, *Schimper* 258 (P, syn.; BM!, FT, K! isosyn.) & near Adua, Mt Scholoda, *Schimper* 1511 (P, syn.; BM!, FT, K!, isosyn.)

Shrubby herb to softly woody shrub (even described as a small tree) 1–4.5(–5.5) m tall, sometimes clump-forming. Young stems and most of plant densely velvety stellate-tomentellous, sometimes slightly angular; older parts of stem with typical malvaceous pattern of longitudinal impressions. Leaves somewhat discolorous; lamina ovate, 2.5–22 × 1.6–16 cm, acute to more usually acuminate to long-acuminate at the apex, shallowly crenate or serrate, the fine stellate indumentum above not obscuring the green leaf surface but fine greyish tomentum covering the undersurface or with long hairs in some variants; petiole 1.5–15(–19) cm long; stipules linear, 6 mm long. Flowers in terminal or lateral panicles up to 23 × 13 cm; pedicel 0.5–3(–4) cm long, the upper part beyond the joint 4 mm long; calyx 6–9(–12) mm long, brownish stellate-pubescent and sometimes with long hairs, lobed for $^{1}/_{3}$ to $^{1}/_{2}$ into oblong-ovate to triangular lobes 6–8 × 4–5 mm with midrib often raised. Corolla white, pink, lilac or mauve often with a dark purple to inky blue centre (never yellow); petals obovate, 0.8–1.4(–2) × ± 1 cm; staminal column often dark purple with characteristic transparent simple to 2–3-fid trichomes above (often flattened in dry state) but glabrous on the widened part though there may be long hairs at basal edge of petals. Fruit depressed oblate, ± 9 mm long, 1.3–1.5 cm diameter, densely stellate-hairy when young. Mericarps (13–)23–27, semicircular-reniform, 6–10 × 4–7 mm, rounded above without a dorsal angle, 1-seeded. Seed oblong-cordiform, 2 mm long, glabrous.

1. Calyx densely velvety stellate-tomentose; upper leaf-surfaces finely stellate-tomentose usually without long simple hairs ... a. var. *longicuspe*
   Calyx and/or upper leaf-surface pilose with long hairs as well as stellate-tomentose ... 2
2. Leaves finely stellate-pubescent; calyx with long simple hairs ... b. var. *pilosicalyx*
   Leaves with long simple hairs above as well as stellate pubescence; calyx tomentose ... c. var. *cecilii*

a. var. **longicuspe**

Leaf-blades finely stellate-tomentose above with few if any simple hairs*; calyx densely velvety stellate-tomentose, usually without long hairs.

UGANDA. Acholi District: Chua, Mt Rom, Dec. 1935, *Eggeling* 2374!; Kigezi District: Kachwekano Farm, July 1949, *Purseglove* 2977!; Mbale District: Bugishu, Elgon, Suam Ridge, 23 Dec. 1954, *Leggat* 100!

KENYA. Turkana District: Muruanysigar peak, 21 Sep. 1963, *Paulo* 966!; Kiambu District: Limuru, Sep. 1950, *Bally* 7995!; Masai District: Olekaitorror Escarpment, 21 Sep. 1962, *Glover & Wateridge* 3277!

TANZANIA. Ngara District: Rusumo, 23 June 1960, *Tanner* 505!; Masai District: Embagai (Empakazi) Crater, 17 July 1973, *Frame* 217!; Lushoto District: Lushoto, 15 Sep. 1960, *Semsei* 3098!

DISTR. **U** 1–3; **K** 1–7; **T** 1–5, 7; Sudan, Eritrea, Ethiopia, Zambia, Malawi, Mozambique, Zimbabwe; Yemen

HAB. Open *Podocarpus*, *Juniperus*, *Ekebergia* etc. forest, forest edges, riverine forest and woodland, hillside thicket; 1200–2400 m

SYN. *A. usambarense* Gürke *nom. nud.* (see Harms, P.O.A. C Index: 1 (1895) – this name has been used on labels

A. endlichii Ulbr. in E.J. 48: 367 (1912); T.T.C.L.: 298 (1949). Type: Tanzania, Moshi District: Engare Nairobi (North), *Endlich* 520 (B†, holo.)

A. crassinervium Mattei in Boll. Ort. Soc. Palermo NS 1: 100 (1915). Type: Ethiopia, without locality, *Schimper in Hohenacker* 169 (PAL, holo.) fide Vollesen

NOTE. The characteristic indumentum at the top of the staminal column varies in the number length and width of the hairs (0.5–2 mm long) which compose it. I am convinced *A. endlichii* belongs here despite a mention of 'flowers yellowish' (almost certainly not based on field data). E.G. Baker saw the type (rough sketch, BM!) and stated it was near *A. usambarense*.

b. var. **pilosicalyx** Verdc. **var. nov.** a var. *longicuspi* calyce longe piloso et stellato-tomentoso differt. Type: Kenya, Naivasha District: Kinangop, *Allan Turner* in CM 359 (K!, holo.; EA, iso.)

Leaves finely stellate-pubescent; calyx with long simple hairs as well as fine stellate tomentum.

KENYA. Naivasha District: Kinangop, Sep. 1931, *Albrechtsen* 9!; Kiambu District: Kikuyu, Sep. 1929, *Dale* 362!; Masai District: Ngong Hills to Magadi road, just below top of Ngong Hills, 28 July 1956, *Verdcourt* 1524!

DISTR. **K** 3, 4, 6; not known elsewhere

HAB. Grassland, hillside bushland, dry forest edges and other secondary vegetation; 1700–2400 m

* Long hairs are often present on under-surface.

c. var. **cecilii** (*N.E.Br.*) *Verdc.* **comb. nov.** Type: Zimbabwe, Manika, Inyanga, *Cecil* 196 (K!, holo.)

Leaves with upper surface with more simple hairs than stellate and densely tomentose beneath together with long hairs; calyx tomentose.

KENYA. Northern Frontier District: Leroghi range, Karissia Forest, July 1960, *Kerfoot* 2112!; NE Elgon, Nov. 1951, *Tweedie* 1069!; Narok District: Entasekera, Keshemoruo, 13 July 1961, *Glover* et al. 2152!

TANZANIA. Ufipa District: Musi, May 1955, *Lawton* 199! & Mbisi Forest, 20 July 1962, *Richards* 16822! & 16831!

DISTR. **K** 1, 3, 6; **T** 4; Ethiopia, East Zimbabwe and North Malawi (see note)

HAB. Forest and forest edges, bushland; 2100–2300 m

SYN. *A. cecilii* N.E.Br. in KB. 1906: 99 (1906); Vollesen in Fl. Eth. & Eritr. 2 (2): 242 (1995)
A. smenospermum Pic. Serm. in Miss. Stud. Lago Tana 7, Ricerche Bot. Pt. 1: 92, t. 16 (1951). Type: Ethiopia, forest of Chiesa di Gumbàt Uddus Michael, *Pichi Sermolli* 2323 (FT, holo.)

NOTE. I am not convinced that *cecilii* deserves specific rank. From northern Malawi on the Nyika Plateau there are several specimens with leaves very densely pilose with long hairs above and hardly any stellate ones and a mixture of coarse stellate and simple hairs beneath but other material from the same area lacks long hairs. Vollesen states there are differences in the shape of the mericarps at least in Ethiopia; he states the mericarps are 1-seeded in both *cecilii* and *longecuspe* but Pichi-Sermolli gives 3-seeded for his *smenospermum* and mentions the seeds are distinctly alveolate.

There are two other specimens of this variety which need mention. *Goujet* 139 (Kenya, Nakuru District: Elburgon, 9 Jan. 1976 at 2400 m) differs from var. *cecilii* in having echinulate fruits, the margins of the dehisced mericarps appearing serrated. This may be an abnormality due to disease etc. and until more material is discovered I leave it as a form of var. *cecilii*. The other specimen *Nattras* s.n. (Kenya, Scott Agricultural Laboratory, Nairobi, Oct. 1953) appears to be var. *cecilii* but it is not clear if it was growing wild on part of the original cover occurring within the boundary of the laboratory (at that date) or whether it was cultivated material originally from some other source. Assessments of wild fibre producing plants were carried out at this laboratory.

2. **Abutilon eggelingii** *Verdc.* **sp. nov.**, affinis *A. longicuspis* A. Rich. sed transitione inter unguem petali et basem columnae staminalis densissime longe pilosa, columna staminali aliter glabra, mericarpiis dorsaliter apice aristata. Typus: Ethiopia, Kaffa, 10 km SW of Agaro on Gera road, *F.G. Meyer* 8848 (K!, holo.; US, iso.)

Shrubby herb with woody stems to 4 m tall; stems with dense short pubescence and some longer hairs. Leaf-lamina rounded-ovate, 3.5–23 × 2.5–14 cm, ± 3-cusped at the apex, the apical one long and narrow, cordate at base, stellate-pubescent above and pilose with ± appressed longer hairs, more densely stellate-pubescent beneath and pilose on veins; petiole 1.5–26 cm long; stipules ? ± 2 mm long, soon falling. Flowers solitary, axillary; peduncle 3–9.5 cm long with indumentum as in stem; calyx 1–1.5 cm long, densely stellate-pubescent and pilose; lobes oblong-ovate, 6–7 × 5 mm, acute, 3-veined. Corolla lilac or pink with purple centre; petals obovate-oblong, 12–24 × 9 mm with ridge of dense long hairs at top of claw and sparse hairs further; staminal column without the characteristic hairs of *A. longicuspe*. Mericarps 13–16, elliptic, 15 × 8 mm, 1-seeded, with sparse short and long hairs on and near margin, awned for 2 mm at apex and with reflexed tooth on inner margin; seeds eccentric-reniform, 3 mm long, smooth and glabrous.

UGANDA. Karamoja District: Mt Debasien, Wuthei, Jan. 1936, *Eggeling* 2663!

DISTR. **U** 1; Ethiopia

HAB. ?Dry montane forest; ± 2100 m

SYN. "*A.* sp = *Meyer* 8848"; Vollesen in Fl. Eth. & Eritr. 2 (2): 242 (1995)

3. **Abutilon angulatum** (*Guill. & Perr.*) *Mast.* in F.T.A. 1: 183 (1868); Hochr. in Ann. Conserv. Jard. Bot. Genève 6: 12 (1902); Exell & Mendonça, C.F.A. 1 (1): 152 (1937) & 2: 373 (1951); T.T.C.L.: 297 (1949); Brenan in Mem. N.Y. Bot. Gard. 8 (3): 223 (1953); Keay, F.W.T.A. ed. 2, 1(2): 337 (1958); Meeuse in F.Z. 1: 488, t. 93, fig. 3 (1961); Vollesen in Fl. Eth. & Eritr. 2 (2): 242, fig. 82.22.5–7 (1995); Thulin, Fl. Somalia 2: 76, fig. 46F (1999). Types: Senegal, Walo, Dagana, near Saint-Louis (Four-à-Chaux) and Cayor, Laybar, *Perrottet/Leprieur* s.n. (P, syn.)*

Shrubby herb or shrub 1–2(–3) m tall; stems pale, greenish but often purplish or maroon in our area, subtomentose with very short stellate hairs, usually distinctly angular with longitudinal ridges. Leaves somewhat discolorous; lamina ovate, 1–23(–30) × 1–19(–25) cm, usually up to about 8 × 7 cm, acuminate at the apex, cordate at base, very finely stellate-tomentose, finely crenate or serrate, petiole 0.3–10(–24) cm long; stipules linear to linear-oblong, 4 mm long. Flowers in terminal lateral panicles (actually reduced leafy branchlets); pedicel 1–3.5 cm long, the apical joint 0.7–1.7 cm long; calyx 0.6–1(–1.5) cm long, finely tomentose, lobes ± 3 × 4 mm, acute, with long hairs on margins. Petals yellow or orange-yellow, obovate-elliptic, 1–2.2 × 0.7–1.6 cm, pubescent outside; staminal column densely stellate-hairy at base. Fruits rounded subglobose, 10–12 × 8–9 mm, deeply umbilicate at apex, densely stellate-tomentose or floccose; mericarps 20–33, elliptic, 7–9 × 5–6 mm, rounded or with obtuse dorsal angle at apex, at first densely stellate-hairy, 1-seeded; seeds reniform, ± 2.5 mm long, smooth to finely papillose.

var. **angulatum**; Meeuse in F.Z. 1: 488 (1961); Vollesen in Op. Bot. 59: 35 (1980)

Plant ± glaucous. Flower-buds not angular; calyx lobes usually 6 mm long, triangular, not or inconspicuously veined.

UGANDA. Acholi District: Chua, Ukuti, Dec. 1935, *Eggeling* 2420!; Toro District: Ruwenzori, Hima, Dec. 1925, *Maitland* 970!; Teso District: Agu, 27 Feb. 1966, *M.R. Smith* 56!
KENYA. Kisumu, May 1961, *Tweedie* 2152! & July 1968, *Tweedie* 3555!
TANZANIA. Musoma District: E side of Lake Magadi, 11 Apr. 1962, *Greenway et al.* 10595!; Tabora District: Masenge Camp, near Mopilingo R., 4 May 1970, *Sanane* 1184!; Iringa District: Msembi, 30 Apr. 1970, *Greenway & Kanuri* 14436!
DISTR. **U** 1–4; **K** 5; **T** 1, 2, 4–8; Senegal, Gambia, Nigeria, Burundi, Eritrea, Ethiopia, Somalia, Angola, Zambia, Malawi, Mozambique, Zimbabwe, Botswana, Namibia, South Africa
HAB. Wooded grassland, bushland with scattered trees, woodland e.g. *Acacia xanthophloea–Chloris–Setaria–Panicum* etc.; *Acacia–Adansonia–Albizia–Commiphora* etc., also old cultivations and coarse grassland, often on hardpan soil and termite mounds; 100–1350 m

SYN. *Bastardia angulata* Guill. & Perr, Fl. Seneg. Tent. 1: 65 (1831)
Abutilon intermedium Garcke in Schweinf., Beitr. Fl. Aethiop.: 49 (1867); Baker f. in J.L.S. 40: 27 (1911); Ulbr. in E.J. 51: 20 (1913). Type: Ethiopia, Gurrsarfa, *Schimper* in *Hohenacher* 2330 (B†, holo.; BM!, K!, iso.)
A. holstii Gürke *nom. nud.* (see Harms, P.O.A. C Index: 1 (1895) (has been used on labels)

var. **macrophyllum** (*Baill.*) *Hochr.*, Fl. Madag. 129: 139 (1955); Meeuse in F.Z. 1: 489 (1961); Vollesen in Op. Bot. 59: 35 (1980). Types: Madagascar, near Antananarivo [Tananarive], *Bojer* s.n.; Port Leven, Sato Is., *Bernier* s.n.; Port Leven, *Vesco* 2, without locality, *Boivin* 2591 (Baker f. gives 259) (P, syn.)

Plant with more yellowish or brownish indumentum, flower-buds angular; calyx usually more than 7 mm long with more distinct midveins or ± keeled. Inflorescences usually shorter and narrower and often more condensed than in var. *angulatum*.

TANZANIA. Kilwa District: Selous Game Reserve, Kingupira Forest, 24 July 1975, *Vollesen* in MRC 2577!; Lindi District: Lake Lutamba, 9 Sep. 1934, *Schlieben* 5256!

* A specimen *Roger* 138 at K is from Richard-Tol and not strictly an isosyntype; *Perrottet* 73 at BM may be an isosyntype.

DISTR. **T** 8; Malawi, Mozambique, Zimbabwe, Namibia, South Africa and Madagascar
HAB. Ground water forest; ± 115 m

SYN. *Sida macrophylla* Baill. in Bull. Soc. Linn. Paris (63): 504 (1885)
Abutilon intermedium Garcke var. *macrophylla* (Baill.) Baker f. in J. Bot. 31: 72 (1893)
A. pseudangulatum Hochr. in Ann. Conserv. Jard. Bot. Genève 6: 13 (1902). Types as above (*non A. macrophyllum* St. Hil. & Naudin)

NOTE. I have followed Meeuse's acceptance of the two varieties but the length of the calyx-lobes is somewhat puzzling from his description – in the general description of the species the calyx is said to be 10–15 mm long with the lobes 3–4(–5) mm but the key gives lobes usually under 6 for one variety and over 7 for the other. The status of this variety needs confirming.

4. **Abutilon anglosomaliae** *Thulin* in K.B. 53: 1013 (1998); Cufodontis in B.J.B.B. 29, Suppl.: 533 (1959) (not validly published); Vollesen in Fl. Eth. & Eritr. 2 (2): 248, fig. 82.23.9–10 (1995); Thulin, Fl. Somalia 2: 79 (1999). Type: Somalia (N), below Golis range, *E. Cole & Lort Phillips* (K!, holo.)

Woody herb or subshrub 0.6–1.5(–2) m tall; young stems pale, tomentose and usually densely spreading hairy, the floriferous parts usually densely glandular; old stems woody, brown, glabrescent. Leaves scarcely discolorous; lamina round to broadly ovate, 2.5–11 × 2.5–12 cm, completely rounded to shortly acuminate at the apex, cordate at base, stellate-pubescent on both sides, velvety and dense on young leaves but less so in adults where surface is not at all obscured, margin crenate to distinctly dentate; venation prominent beneath; petiole 1.5–10.5 cm long; stipules linear-lanceolate, 0.5–2 cm long. Flowers in panicles, the axes with dense glandular and also longer simple hairs; stipuliform bracts present in very young inflorescences; pedicel 1–4.5 cm long; calyx 5–9 mm long, densely glandular-pubescent, the lobes ovate, 2.5–4 × 4 mm. Corolla yellow to orange; petals broadly ovate, 1.5–2.5 cm long and wide, pubescent outside; staminal column stellate-pubescent all over or towards apex. Mericarps 10–16, 10–13 × 5–6 mm, with apical angle or short awn up to 0.5 mm long, ± 3-seeded; seeds 2.5 mm long with short simple curled hairs.

KENYA. Northern Frontier District: 17 km NE of El Wak, 11 Dec. 1971, *Bally & Radcliffe-Smith* 14558! & 53 km SW of Mandera on El Wak road, 30 May 1952, *Gillett* 13399! & S of Merti, lower Uaso Nyiro flood plain, 24 Feb. 1953, *Underhill* 3! (in *Bally* 8945)
DISTR. **K** 1; Somalia, Ethiopia
HAB. *Acacia tortilis–Commiphora* bushland on red sandy soil; 300–390 m

SYN. *A. molle* Bak. in K.B. 1895: 212 (1895), *non* (Ort.) Sweet (1830)

5. **Abutilon hirtum** (*Lam.*) *Sweet*, Hort. Brit.: 53 (1826) ('1827'); Don, Gen. Syst. 1: 503 (1831); Mast. in F.T.A. 1: 187 (1868); Ulbr. in E.J. 51: 33 (1913); Exell & Mendonça, C.F.A. 1: 373 (1951); F.W.T.A ed. 2, 2(1): 337 (1958); Meeuse in F.Z. 1: 487 (1961); Borss. Waalk. in Blumea 14: 169 (1966); Kabuye in U.K.W.F. ed. 2: 101 (1994); Vollesen in Fl. Eth. & Eritr. 2 (2): 248, fig. 82.25.6–8 (1995); Wood, Handb. Yemen Fl.: 107 (1997); Thulin, Fl. Somalia 2: 79 (1999). Type: "L'Inde", *Sonnerat* s.n. (P syn., K, photo.!) & Rumphius, Herb. Amb. 4: 29, t. 10 (1743) (syn.)

Usually very sticky slightly aromatic shrubby herb or shrub 0.6–2.4 m tall, stems etc. with dense tomentum and usually long hairs up to 4 mm long, apart from the dense yellow to orange basally swollen glandular hairs. Leaf-lamina rounded to ovate, (5–)13–24 × (4.5–)9.5–19.5 cm, rounded to shortly acuminate at the apex, cordate at base, densely tomentose to velvety stellate-pilose, mostly coarsely dentate or with teeth of various sizes; venation ± raised beneath; petiole (1–)2–20 cm long; stipules linear-attenuate, 3–12 mm long. Flowers solitary in leaf axils running into short terminal clusters or less often narrow panicles; pedicel 1.5–7 cm long including the 0.5–1 cm long jointed part; calyx 0.9–1.8 cm long, tomentose; lobes ovate to narrowly triangular, 4–7 × 6 mm, keeled. Corolla yellow to orange usually with reddish, maroon or purple centre; petals obovate, 1.5–2.7 cm long, rounded; staminal column 5–7 mm long with transparent stellate hairs, the arms swollen at base; styles deep crimson. Fruit narrowed to a concave apex, 1–1.5 cm long, 1.2–2.5 cm diameter; mericarps 16–25, 8–12(–14) × 5–7(–10) mm, the dorsal corner slightly angled or with distinct minute point up to 0.5 (rarely 1) mm, with stellate-pilose margins, (1–)3-seeded; seeds irregularly rounded cordate, 2.5 mm long and wide, finely papillate and sometimes with simple or stellate hairs.

UGANDA. Karamoja District: Lochomon, Aug. 1958, *J. Wilson* 595!; Toro District: Hima, Dec. 1925, *Maitland* 1041! & Queen Elizabeth Park, Kasese, 25 Jan 1962, *Loveridge* 460!

KENYA. Northern Frontier District: Furroli, 17 Sep. 1952, *Gillett* 13927!; Baringo District: Chebloch, Jan. 1962, *Tweedie* 2282!; Meru District: Meru Game Reserve, Kinna, 23 June 1963, *Mathenge* 184!

TANZANIA. Masai District: foot of Longido Mt, 31 Dec. 1968, *Richards* 23556!; Moshi District: Kilimanjaro, Mkao to ?Lerangwa, 20 Jan 1994, *Grimshaw* 94/76!; Lushoto District: Mkomazi, 23 Apr. 1934, *Greenway* 3950!

DISTR. **U** 1, 2; **K** 1–4, 6, 7; **T** 2, 3; ± pantropical

HAB. *Acacia tortilis* bushland on red soil, grassland with scattered *Acacia*, grassland on black soil, desert lava plains, weed of irrigated land; 250–1700 m

SYN. *Sida hirta* Lam., Encycl. 1: 7 (1783)

S. graveolens Roxb., Hort. Bengal: 50 (1814), *nomen*; Hornem., Suppl. Hort. Bot. Hafn.: 77 (1819); DC., Prodr. 1: 473 (1824). Type: grown in Copenhagen botanic gardens from seed sent by Roxburgh, *Hornemann* s.n. (C!, holo.)

Abutilon graveolens (Hornem.) Wight & Arn., Prod. Fl. Pen. Ind. Or. 1: 56 (1834) (*non sensu* Vollesen, Thulin etc.)

NOTE. *Gillett & Newbould* 19185 (Kenya, Northern Frontier District: 1–3 km NE of Mado Gashi, 5 June 1970) was considered by Gillett to be a species near *A. hirtum* differing in its large fruits, but while I first accepted this conclusion, it became clear it is no more than a form of *A. hirtum*. Most authors emphasise the ocellate flowers but specimens do occur which undoubtedly have pure orange or yellow flowers. Collectors should make special mention of this since it has been noticed in this genus that even in species which undoubtedly have no visible eye (spot) in life this can develop on drying and is clearly a latent character.

A specimen from Kenya, Northern Frontier District: Ol Doinyo Mara, 19 July 1912, *Mearns* 138, annotated as a new species by Exell, is this species.

6. **Abutilon pannosum** (*Forst. f.*) *Schltdl.* in Bot. Zeitung (Berlin) 9: 828 (1851); Keay, F.W.T.A. ed. 2, 1(2): 1: 337 (1958); Abedin in Fl. W. Pak. 130: 72, fig. 13, C, D (1979); Collenette, Illustr. Guide Fl. Saudi Arabia: 359 (1985); Marais, Fl. Mascareignes, 51 Malvacées: 22, t. 6/4–7 (1987); Vollesen in Fl. Eth. & Eritr. 2 (2): 246, fig. 82.23.11–12 (1995); Sirarajan & Pradeep, Malv. S. Pen. India: 197 (1996); Philcox in Rev. Fl. Ceylon 11: 343 (1997); Wood, Handb. Yemen Fl.: 107 (1997); Thulin, Fl. Somalia 2: 78 (1999); Boulos, Fl. Egypt 2: 102, t. 26, fig. 3, fig. (p. 32) (2000). Type: Cape Verde Islands, São Tiago [Sancti Jacobi Insula], *Forster* s.n. (BM!, holo.)

Annual or perennial herb 0.4–2(–3) m tall; stems and other parts velvety tomentose but usually not glandular or few glands on youngest shoots, with or without additional long simple hairs. Leaves often discolorous; lamina rounded to ovate, 2–14.5(–18) × 1.5–11.5(–15) cm, acuminate at the apex, cordate at base, mostly rather coarsely toothed or denticulate; venation prominent beneath; petiole 1–14(–17) cm long; stipules (4–)6–10 × 1.3–1.8(–3) mm. Flowers solitary or in narrow panicles; pedicel 0.5–4(–5.5) cm long; calyx 5–15 mm long, tomentose but without long simple hairs; lobes ovate to triangular, 5–8 × 4 mm, apiculate. Corolla bright yellow, to orange with or without a red centre, 2–4.5 cm wide; petals obovate, 0.8–2 cm long, wide; staminal column stellate-pubescent except for glabrous base*. Fruit depressed subglobose, deeply involute at apex, not or scarcely angled on outer edge; mericarps 16–30(–40),

* One specimen from Saudi Arabia, Djeddah, *Collenette* 1062 has a glabrous staminal column combined with ocellate flowers; *Waring* 12 also has a glabrous staminal column.

elliptic-reniform, 6–9 × 4–7 mm, rounded on outer edge, completely rounded at apex or very slightly angled, tomentose, 1–3-seeded; seeds reniform, ± 2.5 × 2 mm, papillate and stellate-puberulous and often with longer simple hairs around the hilum.

1. Inflorescences usually becoming ± long-paniculate; stems with or without long spreading hairs; leaf-blades often distinctly acuminate; corolla often small and without dark centre .... b. var. *figarianum*
   Flowers solitary or on few-flowered axillary branchlets becoming aggregated at apex but not long-paniculate; stems finely tomentose to densely spreading-pubescent but without additional spreading hairs; leaf-blades more rounded or subacute; corolla up to 4.5 cm wide, sometimes with a dark centre 2
2. Stems finely tomentose .... a. var. *pannosum*
   Stems densely spreading rather roughly pubescent ...... c. var. *scabrum*

a. var. **pannosum**

Stems finely velvety tomentose or softly shortly pubescent, rarely with long hairs as well. Leaves mostly rounded to ovate, rounded to subacute at the apex, typically shortly velvety tomentose. Flowers solitary, axillary or in short axillary few-flowered branchlets sometimes aggregated at the apex but not in elongate narrow panicles.

DISTR. West Africa (Cape Verde Is. and Senegal to Nigeria, Mali); North Africa from Mauretania to Egypt, Arabia, Syria, Iran, Pakistan, Afghanistan, India and Sri Lanka

SYN. *Sida pannosa* Forst. f., Fasc. Pl. Magell. 52, no. 106 (1788) & in Comm. Soc. Reg. Sci. Gottingen ser. 2 9 (1789)
S. glauca Cav., Icon. Desc. Pl. Hisp. 1: 8, t. 11 (1791). Type: cultivated in Spain from Senegal material, collector not known (MA, holo.)
S. mutica Del., Fl. Aegypt. 68, No. 633 (1813) *nom. nud.* ex DC., Prodr. 1: 470 (1824). Type: Egypt, *Coquebert de Montbret* s.n. (G, syn.) & Egypt, cult. at Rosetta, *Delile* (?MPU, syn.)*
Abutilon glaucum (Cav.) Sweet, Hort. Brit.: 54 (1826); G. Don, Gen. Syst. 1: 504 (1831); Webb in Hook., Fl. Nigrit.: 109 (1849); Mast. in F.T.A. 1: 185 (1868)
A. muticum (DC.) Sweet, Hort. Brit., ed. 2: 65 (1830); Abedin in Fl. W. Pak. 130: 71, fig. 16, C, D (1979)

NOTE. The only specimen I have seen which is very close to Atlantic Island material is one from which the leaves have fallen, *Savile* 35 (Tanzania, Dodoma, Makutapora, May 1938). It had originally been determined *A. muticum* and redetermined as *A. hirtum* but there appear to be no glands and the stem indumentum is finely tomentose. *A. muticum* has frequently been kept distinct particularly for Arabian and Egyptian material on account of the larger more attenuate sepals but variation is continuous. There are specimens approaching *A. hirtum* in the coarser indumentum e.g. *Burton* s.n. from Arabia, South Midian.

b. var. **figarianum** (*Webb*) *Verdc.* **var. et stat. nov.** Type: Sudan, Kordofan, Arasch Cool, *Kotschy* 180 (FT, holo., K!, iso.)

Stems etc. typically velvety tomentose with additional long simple hairs but these may be absent, usually without glands but sometimes slightly glandular on young shoots. Leaf-blades ovate to rounded, typically more acuminate than in other varieties; stipules 6–10 × 1.3–3 mm. Flowers axillary but typically inflorescences becoming narrowly paniculate. Corolla yellow to orange, almost always without a red centre, typically 2 cm wide but often petals 0.8–3.4 cm long.

* On the lower right hand side of the De Candolle herbarium sheet is a somewhat illegible 'Coquebert'. This is clearly Antoine François Ernest Coquebert de Montbret (1781–1801) who collected in Egypt.

UGANDA. Busoga District: Serere, Agu Swamp, July 1926, *Maitland* 1301! & Teso, Agu, June 1965, *M.R. Smith* 16! & 7 Feb. 1965, *M.R. Smith* 17!

KENYA. Turkana District: Lorengipe, 1 Oct. 1963, *Bogdan* 5640!; Baringo District: 16 km S of Lake Baringo, Perkerra Irrigation Stream, 5 Jan. 1959, *Bogdan* 4736!; Masai District: Selengai, 8 Feb. 1964, *Verdcourt* 3969!

TANZANIA. Masai District: N of Ketumbeine Mt, Lisingita area, 9 Jan. 1969, *Richards* 23693!

DISTR. **U** 3; **K** 1, 3, 4, 6, 7 (intermediates); **T** 2; Sudan, Ethiopia, Somalia; also recorded from Arabia, Egypt and recorded for Pakistan by Abedin

HAB. *Acacia–Commiphora* thicket and bush in coarse grassland, often on black soil, grasslands adjoining swamps, swamp pans, cultivations on riverine alluvium; 0–1350 m

SYN. *A. figarianum* Webb, Fragm. Fl. Aeth.: 53 (1854); F.P.S. 2: 15, fig. 5 (1952); Abedin in Fl. W. Pak. 130: 64 (1979); Vollesen in Fl. Eth. & Eritr. 2 (2): 246, fig. 82.23.1–5 (1995); Thulin, Fl. Somalia 2: 78 (1999)

A. graveolens sensu Mast. in F.T.A. 1: 184 (1868) pro parte, *non* (Hornem.) Wight & Arn.

A. hirtum sensu Wood, Handb. Yemen Fl.: 107 (1997) pro parte

NOTE. Several specimens with glandular paniculate inflorescences, small calyces and only 16–20 carpels show some approach to *A. anglosomaliae* but the habit, ovate-acuminate leaves, and corolla probably all yellow suggest *figarianum* – Kenya, Turkana District: Lorengipe, 1 Oct. 1952, *Bogdan* 5640 & 56 km SW of Lodwar, 19 Aug., *Vessey & Gales* (?) 50; Baringo District: 16 km S of Baringo, Perkerra irrigation scheme, 5 Jan. 1959, *Bogdan* 4736. All three are from irrigation scheme cultivations at 390–1020 m. I have treated them as forms of var. *figarianum* until there is more evidence.

c. var. **scabrum** *Verdc.* **var. nov.**, a var. **pannoso** ramis ramulisque scabre pubescentibus (haud tantum subtiliter tomentosis) differt. Typus: Kenya, Tana River District: 48 km S of Garsen, *Polhill & Paulo* 528A (K!, holo.)*

Stems and branchlets densely roughly pubescent, completely hiding the short indumentum. Leaves densely roughly velvety pubescent. Flowers solitary, axillary or few on axillary branchlets, not in a long ± leafless panicle. Corolla up to 4.7 cm diameter, not ocellate.

KENYA. Kilifi District: N of Sokoke/Arabuko forest, Lake Jilore, 14. Feb. 1984, *B.L. Simpson* 283!; Tana River District: Bura, 4 Mar. 1963, *Thairu* 18!; Lamu District: Witu, *F. Thomas* 155!

TANZANIA. Morogoro District: Ulugurus, 24 Dec. 1934, *E.M. Bruce* 364!

DISTR. **K** 1?, 7; **T** 6; Sudan (Red Sea hills)

HAB. *Acacia–Dobera–Thespesia–Hyphaene–Cordia* bushland and thicket on black clay soil flood plains; scattered tree grassland with *Dobera, Salvadora, Terminalia* and *Commiphora*; 15–105 (–600, **T** 6) m

GENERAL NOTE on species 5 and 6. *A. hirtum* (Lam.) Sweet, *A. pannosum* (Forst. f.) Schltdl. and *A. figarianum* Webb form a difficult complex and various keys which purport to separate them certainly do not work throughout the range. All three can have long hairs on the stem and some glands; all can have flowers with dark centres.

A. hirtum described from a specimen collected by Sonnerat from "l'Inde" and from Rumphius Herb. Amb. 4: 29 t. 10 (1743) typically has long hairs on the stems, rather rounded leaves, only shortly acute and glands on young parts; a separate label on Sonnerat's specimen states 'les fleurs souvent son tachée d'un pourpre-noirâtre à leur base' and Rumphius also states and indicates the centre is dark; flowers often solitary, axillary with pedicel up to 4 cm long. The mericarps are very shortly but distinctly acute at the apex on the outer edge, the projection being about 0.5 mm long or rarely up to 1 mm. *A. pannosum* was described from the Cape Verde Island, São Tiago [Sancti Jacobi] from a specimen collected by Forster; typically the plant is softly tomentose without glands or long hairs, the leaves rounded or subacute; corolla probably unspotted; flowers on axillary 1–several-flowered branchlets; the mericarps are completely rounded or with only a faint angle most noticeable when young. Throughout Mauritania, Senegal, Mali, Niger and Nigeria, the extensive material closely resembles the type; no field notes mention ocellate flowers but some specimens from Mali and Nigeria have dried with petal bases dark. In Egypt much material resembles the type but the corolla is often ocellate and the calyx lobes are often larger and more acuminate

* This excludes a detached leaf and flower at top right hand corner of sheet; the leaf has obvious stellate indumentum and the flower a glabrous staminal column.

(separated by Webb and others as *A. muticum* (DC.) Sweet which was described from Rosetta). In Israel around the Dead Sea e.g. at Ein Gedi the stems usually have long hairs; throughout Arabia, Syria, Iran, Afghanistan etc. material is mostly typical but many specimens have long hairs on the stem and some specimens certainly have distinctly ocellate flowers (e.g. *Collenette* 1062 from Jeddah). *A. figarianum* described from the Sudan (Kordofan) typically has stems with dense long white hairs but no glands; the corolla is small and no field notes from Sudan or Ethiopia mention ocellate flowers. The inflorescences are more narrowly paniculate than in typical *A. pannosum* and the leaves more ovate with more acuminate apices. The Sudan populations are clearly homogeneous but some specimens do have some glands and in Ethiopia many specimens have no long hairs, the stems being only tomentose; the corollas are often 3 cm or more wide. Further South in East Africa the problems increase. Specimens exactly similar to *A. pannosum* in habit, leaves and inflorescences have stem indumentum quite unlike. Material usually named *A. figarianum* is variable and many sheets are not easily named. *A. hirtum* and *A. pannosum* are quite distinct over wide areas and it would be unwise to alter their status but I have resolved some difficulties in eastern Africa by treating *A. figarianum* as a variety and establishing another variety of *A. pannosum* mainly occurring on the coast. It is clear there is still much gene flow between populations.

7. **Abutilon** sp.

Probably subshrubby herb but habit details not noted by collector; stems tomentose and with dense soft spreading pubescence. Leaves rounded-ovate, discolorous, 9 × 8–9 cm, shortly acuminate at the apex, cordate at base with lobes overlapping, irregularly toothed, finely velvety pubescent above, whitish velvety tomentose beneath and with dense longer hairs; petiole ± 6 cm long; stipules erect, filiform, ± 1 cm long. Inflorescences axillary, 2–several-flowered, 3–4 cm long; pedicel 1.5 cm long; calyx ± 10 mm long with ovate lobes 7 × 4.5 mm, tomentose and with longer pubescence. Immature fruits depressed subglobose, ± 1.5 cm diameter; mericarps ± 25, rounded at apex, densely stellate-pubescent, 1-seeded; seeds black, 2 mm long with pale simple hairs mostly around the hilum.

TANZANIA. Lushoto District: Lushoto, Jaegertal, 8 Jan. 1967, *Archbold* 897A!
DISTR. **T** 3
HAB. Margin of camphor forest; ± 1590 m

NOTE. This has the characters of *A. pannosum* but the small calyx is distinctive. The collector has mixed two plants together, the information on flowers in the field note referring to 897B. This differs from 897A in the fine tomentum of the stem, larger more ovate leaves up to 11.5 × 8 cm with longer acuminate tips and excessively fine velvety tomentum, also more obscurely crenate margins; stipules curved downwards, 6 mm long; flowers axillary with pedicel ± 9 cm long; calyx 1.6 cm long with triangular lobes 10 × 6 mm. The corolla is yellow with a brownish pink centre, about 5.5 cm wide (corolla preserved is ± 4 cm). It is probably a form of *A. mauritianum* (Jacq.) Medik. var. *grandiflorum* which is known from Jaegertal but the very fine tomentum does not agree. It is surprising that there is so little material from such a well known locality.

8. **Abutilon** sp. A

Habit data lacking, presumably subshrubs; stems tomentose and with long hairs but scarcely glandular. Leaves ovate, 4.5 × 3.7 cm, subacute. Flowers solitary in the leaf axils; calyx cupular, 1–1.2 cm long, rounded at base; lobes triangular-ovate, 8 × 5 mm, acuminate, densely pubescent. Corolla yellow, apparently not ocellate, ± 3 cm wide. Mericarps 20–25 with a distinct upper apical point –1 mm long.

UGANDA. Teso District: Agu, Aug. 1965, *M.R. Smith* 5!
KENYA. South Kavirondo District: Lambwe Valley, 1965, *Makin* 93!
DISTR. **U** 3; **K** 5
HAB. No data

NOTE. These appear to be identical but both specimens are poor. The calyx and general appearance would fit the *A. hirtum–figarianum–pannosum* complex but the carpels are much too acutely awned at the apex, the acute lobe being ± 1 mm long; they are on the other hand too short for *A. mauritianum* and its allies. Possibly a hybrid but nothing is known of any hybrids in African *Abutilon*.

9. **Abutilon pilosicalyx** *Verdc.* **sp. nov.** affinis *A. hirti* (Lam.) Sweet affiniumque calyce dense longe piloso calycis lobis lanceolatis usque anguste lanceolatis differt. Type: Kenya, Northern Frontier District: Wajir, *Hemming* 471 (K!, holo.; EA!, iso.)

Erect annual (fide Stannard & Gilbert) herb 1–1.5 m tall; stems woody, densely velvety pubescent with short ± spreading hairs and numerous to dense long simple hairs as well, probably somewhat glandular. Leaves scarcely discolorous; lamina broadly ovate to round, 5–13 × 4.5–12 cm, rounded to shortly acuminate at the apex, deeply cordate at base, crenate-serrate, ± softly to rather roughly stellate-pubescent on both surfaces but surface not obscured except in very young leaves; petiole very variable, some shorter that the lamina, others much longer, 4–14.5 cm long; stipules linear-lanceolate, 12–15 × 1–1.5 mm. Flowers solitary in leaf axils or running into lateral and terminal panicles; simple peduncles 4–7 cm long with indumentum as on stem; calyx 1.3–2.3 cm long, the lobes narrowly triangular, 8–18 × 4–6 mm at base, acute at apex, densely shortly pubescent and with dense long hairs as well. Corolla deep golden yellow or bronzy with red, purple or deep crimson centre or sometimes entirely yellow (see note); petals broadly obovate, 2–2.5 × 2.5 cm; staminal column stellate-pubescent. Fruit ± 1.2 cm long, 1.5 cm diameter, shorter than the calyx, smoothly rounded above when young; mericarps 17–19, distinctly shortly laterally acute at apex.

KENYA. Northern Frontier District: Garissa–Mado Gashi road, about 3 km N of Garissa–Hagadera road, 26 Nov. 1978, *Brenan et al.* 14730!; 74 km S of Mado Gashi on road to Garissa, 13 Dec. 1977, *Stannard & Gilbert* 1014! & 18 km from Garissa on road to Hagadera, 29 May 1977, *Gillett* 21185!
DISTR. **K** 1 (see note)
HAB. Open bushland with scattered trees of *Delonix, Euphorbia robecchii, Acacia turnbullii, Commiphora* etc.; 225–305 m

SYN. *A. graveolens* sensu Vollesen in Fl. Eth. & Eritr. 2(2): 248 (1995) & Thulin, Fl. Somalia 2: 79 (1999), *non* (Roxb. ex Hornem.*) Wight & Arn. sensu lato (see note)

NOTE. The specimens treated as *A. graveolens* by Vollesen and Thulin have for the most part a smaller calyx and apparently uniformly yellow or orange flowers. There is no doubt that these Somalian and Ethiopian populations are no more than varietally or subspecifically distinct from the Kenya material described above but I have not combined the descriptions of the two. Olof Ryding kindly searched the herbarium at Copenhagen for original material of *Sida graveolens* Roxb. ex Hornem. and discovered two sheets which I am sure are syntypes – the calyx is not densely long-hairy. Roxburgh material from the Wight herbarium confirms this. A specimen at K from Kenya, without locality (*Okungu* s.n., 14 Dec. 1971) is undoubtedly conspecific with *A. pilosicalyx* and is annotated as sheet II but no sheet I has been found - which presumably bears the data.

Originally the material from Kenya had been tentatively referred to *A. agnesae* Borzi but the description of that indicates a tomentose calyx and does not mention long hairs. Thulin puts *A. agnesae* in the synonymy of *A. guineense* (Schumach.) Bak. & Exell.

10. **Abutilon subprostratum** *Verdc.* **sp. nov.** affinis *A. fruticosae* Guill. & Perr. sed habitu ± prostrato vel subscandente, calyce stellato tomentoso et pilis simplicibus longis piloso differt. Typus: Kenya, Northern Frontier District: Mandera, 40 km from El Wak on Wajir Road, *Gilbert & Thulin* 1209 (UPS!, holo.; K!, iso.)

* Based on *Sida graveolens* Roxb. ex Hornem., Suppl. Hort. Bot. Hafn.: 77 (1819).

Prostrate or scandent subshrubby herb with slender woody stems to 1.5 m long, finely short spreading-pilose when young, later glabrescent and with characteristic malvaceous impressed pattern. Leaves discolorous; lamina narrowly ovate, 0.8–3.3 × 0.6–2 cm, rounded to subacute at the apex, cordate at base, obscurely distantly crenate, finely stellate–pubescent above, more greyish white velvety tomentose beneath; petiole very slender, 0.3–2.3 cm long with short pubescence and spreading hairs; stipules filiform, 6 mm long. Flowers solitary in the leaf axils; peduncle very slender, 3.5–4.5 cm long including 3 mm jointed part, finely pubescent and with some longer hairs; calyx 5 mm long, greyish stellate-tomentose and with long simple hairs, the lobes oblong, 3 × 2 mm, acute to apiculate, densely spreading-pubescent inside. Corolla uniform apricot yellow, petals ± 8 mm long; staminal column with biramous, triramous or stellate hairs, the central arm much longer than the others; ovary with 7 carpels, each with 3 ovules. Fruit not seen; very young fruits densely white spreading-pubescent.

KENYA. Northern Frontier District: Mandera, 40 km from El Wak on Wajir road, 29 Apr. 1978, *Gilbert & Thulin* 1209!

DISTR. **K** 1; not known elsewhere

HAB. *Commiphora, Boswellia, Acacia, Ghikaea* bushland with sparse grass (mainly *Tetrachaete*) on limestone ridge; ± 460 m

NOTE. This had been determined as *A. fruticosum* by C. Kabuye but is, I think, adequately distinct.

11. **Abutilon fruticosum** *Guill. & Perr.*, Fl. Seneg. Tent. 1: 70 (1831); Mast. in F.T.A. 1: 187 (1868); Exell & Mendonça, C.F.A. 1 (1): 153 (1937) & 2: 373 (1951); T.T.C.L.: 298 (1949); F.P.S. 2: 15, fig. 6 (1952); Keay, F.W.T.A. ed. 2, 1(2): 337 (1958); Meeuse in F.Z. 1: 489, t. 93, fig. 4 (1961); Abedin in Fl. W. Pak. 130: 55 (1979); Kabuye in U.K.W.F. ed. 2: 101 (1994) (sphalm (Cav.) Guill. & Perr.); Vollesen in Fl. Eth. & Eritr. 2 (2): 244, fig. 82.22.1–4 (1995); Wood, Handb. Yemen Fl.: 107 (1997); Thulin, Fl. Somalia 2: 76, fig. 46 A–C (1999). Type: Senegal, R. Senegal, Saffal island, near Saint Louis, *Perrottet* 80 (P, holo.; BM!, iso.)*

Perennial or sometimes annual erect branched herb, usually ± woody, 0.5–1.2 m tall with all parts velvety with very fine greyish stellate tomentum but long hairs usually absent. Leaves with lamina ovate, 0.8–6(–12) × 0.8–4(–9.5) cm, rounded to acute at apex, cordate at base, shallowly crenate-dentate; petiole 0.4–4.5(–9) cm long; stipules filiform, 8 mm long. Flowers axillary; pedicel 1–4(–8) cm long, the part above the joint 2–4 mm long; calyx 4–6.5(–10) mm long, lobed for $^1/_3$–$^2/_3$ its length into ovate-acuminate lobes 2–5(–7) × 1.8–3.5(–6) mm. Corolla yellow or orange; petals obovate, 5–12(–17) × 5–11 mm; staminal column hairy above. Fruit squarish, 7–10 mm long, subtruncate at apex, densely stellate-pubescent; mericarps 9–10, 5–9 × 3–4 mm, with dorsal angle subacute to acute but not awned, 2–3-seeded; seeds 1.5–2 mm long, papillose.

a. var. **fruticosum**

Calyx 4–6.5 mm long with lobes 2.5 × 1.8–3.5 mm with no long hairs. Petals mostly 5–10 × 5–7 mm.

KENYA. Northern Frontier District: Kerio R., Lokori, 19 July 1969, *Mwangangi* 1360!; Baringo District: shore of Lake Baringo, Feb. 1962, *Tweedie* 2300!; Masai District: 1.8 km S of Ol Orgesaile, 22 Apr. 1960, *Verdcourt* 2750!

TANZANIA. Masai District: Longido–Moshi track, 31 Dec. 1968, *Richards* 23541! & Kitumbeine Mt, 1 Mar. 1969, *Richards* 24228!; Lushoto District: 13 km NW of Mkomazi on Moshi–Korogwe road, 1 Apr. 1969, *Lye* 2409!

DISTR. **K** 1–4, 6, 7; **T** 2, 3; Senegal, Burkina Faso, Niger, Sudan, Ethiopia, Somalia, Angola, Zimbabwe, Botswana, Namibia; Egypt, Socotra, Arabia, Pakistan, India

* A 'Roger dedit' sheet from Herb Gay at K is probably an isotype.

HAB. Grassland, various bushland habitats, *Acacia–Salvadora*; *Acacia–Commiphora–Delonix*; *Acacia–Commiphora–Euphorbia–Cordia* etc.; riparian woodland *Hyphaene–Acacia elatior–Salvadora* etc., often near water; 30–1300 m

SYN. *Sida denticulata* Fres. in Mus. Senck.: 182 (1834). Type: Egypt, Thal Hebran, *Rüppell* s.n. (F, holo.)

Abutilon microphyllum A. Rich., Tent. Fl. Abyss. 1: 70 (1847) & Atlas, t. 15 (1851). Type: Ethiopia, Choho, *Petit* s.n. (P, holo.)

A. denticulatum (Fres.) Webb, Frag. Fl. Aethiop.-Aegypt.: 51 (1854)

A. kotschyi Webb, Frag. Fl. Aethiop.-Aegypt.: 52 (1854). Types: Sudan, Kordofan, base of Mt Arasch-Cool, *Kotschy* 380 (FT, syn.; K!, isosyn.) & bordering same area, under Mt Kohn, *Kotschy* 234 (FT, syn.; K!, isosyn.)

A. dubium Mattei in Boll. Ort. Bot. Palermo NS 1: 91 (1915). Type: Ethiopia, Tigré, without precise locality, *Figari* s.n. (FT, holo.)

A. fruticosum Guill. & Perr. var. *microphyllum* (A. Rich.) Abedin in Fl. W. Pak. 130: 58, f. 11, D (1979)

NOTE. Abedin recognises three varieties, var. *fruticosum* and var. *microphyllum* differing in leaf shape and var. *saidae* (published without a Latin description) differing by very thin thread-like pedicel. Meeuse mentions this is one of the few *Abutilon* to flower in the morning but several labels mention flowering in the afternoon. Annual plants under 20 cm tall with large leaves e.g. *Mathew* 6252 (Kenya, Northern Frontier Province, South Turkana, Lokori, 16 May 1970) are probably young plants growing under wetter conditions and flowering early; all material seen has been from ± the same place.

b. var. **hepperi** *Verdc.* **var. nov**. a var. *fruticoso* floribus majoribus differt. Typus: Kenya, Northern Frontier District: lower slopes of Mt Nyiro, *Hepper & Jaeger* 6816 (K!, holo.)

Calyx 10 mm long with lobes 6–7 × 4.5–6 mm, with some long hairs on margins and midvein outside. Petals 12–14 × 9–11 mm.

KENYA. Northern Frontier District: lower slopes of Mt Nyiro, South Horr, about 8 km N near R. Korungwe, 14 Nov. 1978, *Hepper & Jaeger* 6816!

DISTR. **K** 1; Ethiopia, Somalia

HAB. Rocky hillside; 800 m

SYN. *A.* sp. = *Gilbert* 2100; Vollesen in Fl. Eth. & Eritr. 2 (2): 244 (1995)

NOTE. Vollesen has annotated *Hepper & Jaeger* 6816 as *A.* sp.? nov. aff. *fruticosum.* Thulin in Fl. Somalia has pointed that flower size varies and treats *A.* sp. = *Gilbert* 2100 as simply a large-flowered form. The Kenya specimen does seem distinctive.

12. **Abutilon rotundifolium** *Mattei* in Boll. Ort. Bot. Palermo 7: 182 (1908) & in F.R. 9: 347 (1911) & in Boll. Ort. Bot. Palermo NS 1: 82 (1915); Thulin, Fl. Somalia 2: 78 (1999). Type: Somalia, Mogadishu, *Macaluso* 38 (PAL, holo.; K!, photo).

Prostrate or shrubby much-branched herb with pale finely tomentose stems 40–50 cm long but mostly only ± 15 cm tall; some long hairs sometimes present on young parts. Leaves shortly to quite long-petiolate often crowded on the stems; lamina round to rounded ovate, 0.5–6 × 0.5–7 cm, rounded to slightly apiculate at the apex, cordate at base, with rather spaced thickened crenae at the ends of the veins, very finely grey-green velvety tomentose, the surface obscured; venation raised beneath; petiole 0.2–8 cm long; stipules filiform, 2–4.5 mm long, deciduous. Flowers solitary in the leaf-axils or in axillary cymes; pedicel 0.4–3.8 cm long, the upper part beyond the joint 1.5–4 mm long; calyx finely velvety tomentose, without white hairs, 5–7 mm long, the lobes ovate, 3–5 × 3–4.5 mm, apiculate. Corolla yellow; petals 7–10 × 6 mm. Mericarps 9–10(–12), 7–10 × 4–6 mm, mucronate or with short awn 0.8–1.6 mm long, coarsely stellate-hairy and with fine stellate pubescence as well, 2–3-seeded; seeds subreniform, ± 2 mm long, ± smooth or very shortly papillate.

KENYA. Lamu District: Kui I., Sep. 1956, *Rawlins* 117! & E side of Simambaya I., 25 Oct. 1957, *Greenway & Rawlins* 9417! & Suthie I., 80 km NE Lamu, 2 July 1961, *Gillespie* 115!
DISTR. **K** 7; Somalia
HAB. On coral with *Commiphora* in *Capparis galeata–Salvadora persica* scrub, dunes and calcareous sandstone; 0–5 m

NOTE. This is closely allied to *A. fruticosum* and is part of the characteristic Mogadishu dune flora which extends down the coast into Kenya. The sparse Kenya material is rather diverse with *Gillespie* 115 having a spreading rather than prostrate habit and the leaves with long petioles rather than subsessile as in the other cited specimens. Study in the field may resolve the variation in this species.

13. **Abutilon wituense** *Baker f.* in J.B. 74: 193 (1936). Type: Kenya, Lamu District: Witu, *Thomas* 10 (BM!, holo.; K!, iso.)

Annual branched woody herb 1–1.5 m tall; stems with short hairs and long spreading hairs about 3 mm long but surface not obscured. Leaves rounded-ovate, distinctly but not deeply 3-lobed, 2–10 × 2.3–9 cm, acuminate at the apex, cordate, coarsely crenate-toothed, shortly pubescent above and more velvety so beneath but surface not obscured; petiole 1.5–11 cm long with same indumentum as stem; stipules filiform to linear-lanceolate, 5–8 mm long. Flowers few in the upper axils; peduncle ± 3 cm long; calyx 0.9–1.5 cm long with ovate lobes, 6–7 × 4 mm, cuspidate, shortly pubescent and with long hairs. Corolla yellow, sometimes with purple centre; petals obovate, 1.3 × 9 mm wide. Mericarps rather spreading in mature fruit, 9–11(–17 fide *Luke & Robertson* 2303), narrowly oblong-elliptic, 1.1(–1.4) cm long including a wing-like beak or awn which is 2–5 mm long, variable in the same capsule and venose; also short hook present on inner face; carpels coarsely stellate-hairy, later ± glabrous; seeds 2 per mericarp, ± reniform, 2.1 × 2 mm, compressed, densely dotted with small warts or minutely tuberculate.

KENYA. Kwale District: Miongoni forest, near Umba R., 25 May 1990, *Luke & Robertson* 2303!; Tana River District: Tana River National Primate Reserve, Mchelelo Forest, 11 Mar. 1990, *Kabuye et al.* TPR 66!; Lamu District: Witu, *Thomas* 10!
TANZANIA. Kilosa District: Ilonga, May 1970, *C.M. Wood* 8!; Rufiji District: Selous Game Reserve, Kibambawe R., S bank, 3 Aug. 1993, *Luke* 3642!; Kilwa District: Kingupira Forest, 24 July 1975, *Vollesen* in MRC 2578!
DISTR. **K** 7; **T** 6, 8; not known elsewhere
HAB. Riverine gallery forest, *Sterculia–Pachystela–Diospyros–Ficus* etc., thicket, dry sand rivers, dry waterholes on mud and as a weed in newly cleared shambas; 30–115 m

SYN. *A. bidentatum* sensu Vollesen in Op. Bot. 59: 35 (1980) pro parte, *non* A. Rich.

NOTE. Extremely similar to *A. lauraster* Hochr. in the shape of the ripe capsules with spreading follicles but quite different in foliage.

14. **Abutilon lauraster** *Hochr.* in Ann. Conserv. Jard. Bot. Genève 6: 24 (1902) & in Fl. Mad. 129, Malvaceae: 142, fig. 34/7–9 (1955); Meeuse in F.Z. 1: 494, t. 93, fig. 11 (1961); Vollesen in Opera Bot. 59: 35 (1980). Type: Madagascar, Lingvatt Bay, *Goudot* s.n. (GE, holo.)

Annual or biennial somewhat woody herb or subshrub 0.5–2 m tall; stems tomentose and often with long sometimes dense spreading hairs, glabrescent, sometimes glandular towards the apex. Leaves somewhat discolourous with lamina broadly ovate, 2.5–24 × 2–18 cm, acuminate at the apex, cordate at base, ± densely stellate-pubescent above but surface not obscured, more densely so beneath and often velvety tomentose and with long appressed hairs on venation, almost entire or very shallowly crenate; petiole 2.5–14 cm long; stipules filiform, 8 mm long. Flowers borne in axils of the upper leaves forming ± narrow pseudopanicles; pedicel 1–6 cm long, articulated at up to 10 mm from the apex, pubescent; calyx 1–1.2 cm long, pubescent and with long hairs; lobes ± ovate, 5–9 × 2.5–6.5 mm, acuminate, ciliate. Petals yellow, 1.2–2 cm long; staminal column stellate-pubescent above. Fruit with 9–16 ± spreading blackish mericarps, each elliptic-oblong, 10–14 × 3–5 mm, stellate-pubescent but ultimately ± glabrous, with subulate awn 3–4 mm long, 3-seeded usually with short ribs at base of awn; seeds reniform, ± 1.5 mm long and wide, smooth, sometimes with tuft of hair near the hilum.

TANZANIA. Uzaramo District: Kibaha, 35 km W of Dar es Salaam, 2 km W of Soga road, 21 May 1971, *Flock* 52!; Kilwa District: Kingupira, 8 May 1975, *Vollesen* in MRC 2275!
DISTR. **T** 6, 8; Zambia, Malawi, Mozambique, Zimbabwe, South Africa; Madagascar
HAB. Ground water forest, riverine forest and thicket, moist base of termite mounds; ± 150 m

SYN. *A. zanzibaricum* Mast. in F.T.A. 1: 186 (1818) pro parte excl. lecto.
A. indicum sensu Mast. in F.T.A. 1: 186 (1868) pro parte, *non* (L.) Sweet.

15. **Abutilon rehmannii** *Baker f.* in J. Bot. 31: 217 (1893); Ulbr. in E.J. 51: 30 (1913); Meeuse in F.Z. 1: 497, t. 93, fig. 14 (1961); Kabuye in U.K.W.F. ed. 2: 101 (1994). Types: South Africa, Transvaal, *Rehmann* 5221 (K! syn.) & Maadji Mt, *Burchell* 2372 (K! syn.)

Somewhat woody shrubby or sometimes annual herb (0.1–)0.9–1.5(–2) m tall, densely short stellate-tomentose and shortly spreading-pubescent but usually without long hairs; young stems sometimes slightly angled, old ones with characteristic malvaceous impressed pattern. Leaves ± discolorous, narrowly triangular-ovate to ovate, 2–10(–16) × 1–7(–9) cm, attenuate acute to caudate at the apex, cordate at base, usually distinctly serrate or doubly serrate-dentate, finely densely greyish velvety tomentose on both surfaces but paler beneath; petiole 0.5–5(–8) cm long; stipules lanceolate, 3–5 × 1 mm. Flowers axillary usually in upper axils and appearing subcapitate or pseudo-paniculate due to inconspicuous upper leaves; pedicel 0.5–3(–5) cm long with 3–5 mm long jointed apex; calyx 6–9(–11) mm long, velvety tomentose and pubescent, the lobes ovate to narrowly triangular, 4–7 × 4 mm, acute to apiculate, the median vein often visible. Corolla yellow; petals obovate, 7–13 mm long and wide; staminal column stellate-pubescent. Fruit subcylindrical-globose, 0.8–1 cm long, (1–)1.2–1.4 cm diameter, the sides convex, concave at apex, densely stellate-pilose; mericarps grey-green, 10–20, usually ± 13–15, ± oblong, 6.5–9 × 4–5 mm, acute to shortly (or up to 2 mm) awned at dorsal apical angle and ventrally toothed; seeds irregularly reniform, 1.5–2 mm long and wide, finely papillose.

UGANDA. Ankole District: Mbarara, slopes of Gayaza, Oct. 1925, *Maitland* 857!
KENYA. Naivasha District: shores of Lake Naivasha, 2 July 1963, *E. Polhill* 3!; Masai District: 40 km Magadi to Nairobi, 31 May 1958, *Verdcourt et al.* 2142!; Teita District: Voi, Mzinga Hill, 11 Jan. 1964, *Verdcourt* 3890!
TANZANIA. Musoma District: Ikoma, Banagi, 28 Nov. 1953, *Tanner* 1858!; Masai District: about 8 km N of old Banagi and Seronera road with Olduvai–Loliondo road, 2 Apr. 1964, *Verdcourt* 4029!; Pare District: Nyumba ya Mungu, 15 Nov. 1970, *Batty* 1127!
DISTR. **U** 2; **K** 1, 3, 4, 6, 7; **T** 1–3, ?6; Zimbabwe, Namibia, South Africa
HAB. *Acacia* and *Acacia-Delonix* etc. dry woodland and bushland, rocky bushland with *Melia, Sesamothamnus, Commiphora, Sansevieria* etc., grassland with scattered *Acacia*, riverside thicket of *Maerua, Croton, Grewia* etc., stony slopes near lakeside with *Acacia xanthophloea* etc. also submontane scrub; 450–1900 m

SYN. *A. braunii* Baker f. in J. B. 74: 193 (1936). Type: Tanzania, Lushoto District: Mombo steppe, *Braun* 1927 (EA, holo.)
A. sp. A, U.K.W.F. ed. 1: 205 (1974)

NOTE. I have followed C. Kabuye's identification of this taxon but am not sure what her interpretation covers since *A. bidentatum* and *A. engleranum* are not mentioned in her account in U.K.W.F. ed. 2. Although long hairs are usually absent some specimens do have them on the lower thickened part of the stem. The position is still unsatisfactory and some specimens are very close to *A. bidentatum* and the true relationships to South African material are not certain. For this reason I initially sank the two together but eventually had an instinctive feeling this was not correct. Field observations are necessary over a wide area.

16. **Abutilon bidentatum** (*Hochst.*) *A. Rich.*, Tent. Fl. Abyss. 1: 68 (1847); Mast. in F.T.A. 1: 186 (1868); P.O.A. C: 265 (1895); T.T.C.L.: 298 (1949); Hauman in F.C.B. 10: 61 (1963); Abedin in Fl. W. Pak. 130: 63 (1979); Vollesen in Op. Bot. 59: 35 (1980) & in Fl. Eth. & Eritr. 2 (2): 246, fig. 82.22.8–9 (1995); Wood, Handb. Yemen Flora: 107 (1997); Thulin, Fl. Somalia. 2: 78, fig. 46G (1999); Boulos, Fl. Egypt 2: 104 (2000). Type: Ethiopia, Tigray, Agau, *Schimper* 1003 (P, lecto.; FI-W, K!, iso.)

Shrubby herb 0.5–1.5 m tall; stems pubescent to tomentose and also often long-pilose with simple hairs. Leaves discolorous; lamina ovate to almost round, 1–12(–17) × 1–10(–14) cm, acuminate at the apex, cordate at base, serrate, dentate or crenate, often strongly, with alternately large and small teeth, finely stellate-tomentose; petiole 1–10(–13) cm long; stipules linear, 4.5 mm long. Flowers in leaf-axils or on axillary branches or sometimes merging into distinct panicles; pedicel 1.5–7(–9) cm long; calyx 6–10(–12) mm long with ovate lobes 7 mm long (including narrow cusp 2 mm long), 5 mm wide, tomentose and long-pilose. Corolla yellow or orange; petals obovate, 7–13(–?20) × 6 mm; staminal column usually stellate-pubescent all over. Fruit 8–17 mm long, 10–14 mm diameter; mericarps 12–17(–20), ultimately black and stellately spreading, oblong-elliptic, 8–10 × 3–5 mm, the outer edge rounded, the dorsal corner with 1–2 mm long awn and, the ventral edge with projection, sometimes persistently attached to receptacle by funicle, 2(–3)-seeded, pilose around the margins; seeds rounded reniform, 1.5–2 mm long and wide, minutely warted near hilum or all over.

UGANDA. Karamoja District: Moroto township, Sep. 1956, *J. Wilson* 257!; Busoga District: Butaleja, July 1926, *Maitland* 1155! & Busaba, July 1926, *Maitland* 1137!

KENYA. West Suk District: near Lokwien, below Turkwell Gorge, 10 Aug. 1978, *Lye* 9057!; Laikipia District: Kisima Farm, 40 km N of Rumuruti, Uaso Narok R., 13 Nov. 1977, *Carter & Stannard* 319!; South Nyeri District: Tebere, 17 Dec. 1971, *Okungu* s.n.!

TANZANIA. Musoma District: Serengeti Research Institute, 20 June 1970, *Harris* 4780!; Mbulu District: Lake Manyara National Park, plot 19, 24 Sep. 1968; *Vesey-FitzGerald* 5962!; Dodoma District: Chenenie, 19 May 1978, *Sigara* 215!

DISTR. **U** 1–3; **K** 1–7; **T** 1, 2, 5, 7; NE Congo-Kinshasa, Rwanda, Burundi, Ethiopia, Somalia; also Egypt, NW India and Pakistan and if *A. englerianum* is truly synonymous, Angola, Zambia, Mozambique, Zimbabwe, Botswana, Caprivi strip, Namibia, South Africa; Arabia

HAB. *Acacia* woodland, grassland with *A. tortilis*, grassland by forest edges, in dense vegetation among granite boulders by river, stony river gorges, rocky bushland with *Melia, Commiphora, Sansevieria, Sesamothamnus* etc.; 450–1900 m

SYN. *Sida bidentata* Hochst. in sched. sub num. *Schimper* 1003 (1842)

Abutilon englerianum Ulbr. in E.J. 51: 30 (1913); Meeuse in F.Z. 1: 497 (1961). Type: Namibia, Hereroland, Grootfontein, *Dinter* 905 (B, syn. †) (see note)

A. microcarpum Mattei in Boll. Ort. Soc. Palermo N.S. 1: 90 (1915). Type: Eritrea, Adi Ugri, *Cufino* 11 (PAL, holo.) (fide Vollesen and Thulin but E.G. Baker on his drawing (BM) records he saw this type at Florence)

NOTE. Much material at Kew has had the name *A. englerianum* pencilled on the sheets often with a query and there is no doubt the two are very close. I have examined a topotype from Grootfontein, *Dinter* 7385 (K) and there are small glands on the pedicel, calyces etc. but there are certainly small glands on the jointed part of the pedicel in *M.R. Smith* 53 from Uganda, Teso District: Agu. Neither *A. bidentatum* nor *A. englerianum* is mentioned in U.K.W.F. ed. 2. The species has frequently been misidentified as *A. mauritianum*. *Mathew* 6722 (Kenya, Northern Frontier Province, South Turkana, Lokori, 11.2 km S of Kangetet, 10 June 1970, 585 m) and *Mathew* 6726 (same area, R. Kerio, 12.8 km downstream from Lokori, 10 June 1970, 579 m) appear to be a distinct variety of *A. bidentatum* with a glabrous staminal column and annual habit. I have not been able to see enough material with open flowers to assess how unusual this is but all I have seen apart from these two have distinct indumentum on the column.

17. **Abutilon mauritianum** (*Jacq.*) *Medik.*, Künstl. Geschlecht. Malv.-Fam.: 28 (1787); Exell & Mendonça, C.F.A. 1 (1): 153 (1937); Brenan, T.T.C.L.: 298 (1949); U.O.P.Z.: 99 (1949); F.P.S. 2: 15 (1952); Keay, F.W.T.A. ed. 2, 1(2): 337 (1958); Meeuse in F.Z. 1: 493, t. 93, fig. 10 (1961); Hauman in F.C.B. 10: 159 (1963); Blundell, Wild Flow. E. Afr.: 75, t. 290 (1987); Kabuye in U.K.W.F. ed. 2: 101, t. 29 (1994); Vollesen in Fl. Eth. & Eritr. 2 (2): 246, fig. 82.22.10, 11 (1995); Thulin, Fl. Somalia 2: 78, fig. 46H (1999). Type: plant grown at Schönbrunn, Vienna from seed said to have been from Mauritius, Jacq., Ic. Pl. Rar. 1, t. 137 (1782) (lecto.!) (see general note)

Annual or perennial herb, subshrub or small shrub 0.4–2.7(–3.5) m tall; stems often purplish, ± woody, very finely tomentose, with or without additional long hairs. Leaves ± discolorous; lamina ovate, sometimes with a trace of 3-lobing, 1.4–18(–25) × 1.5–16(–18) cm, acuminate at the apex, cordate at base, the emargination often deep, denticulate to shallowly crenate-serrate or often ± entire, finely densely slightly roughly stellate-pubescent above but surface not obscured, very finely velvety tomentose beneath, the surface completely covered or often with long pubescence as well; petiole 0.5–20 cm long, either pubescent with longer hairs for entire length or finely tomentose except for a tuft of longer hairs ± 1 mm long near junction with lamina; stipules linear-subulate, 4–9 mm long, 0.5 mm wide, sometimes curved upwards, persistent or deciduous. Flowers solitary, axillary or on short axillary branches; peduncle 1.5–10.5(–12) cm long including jointed part 1–1.5 cm long, finely tomentose and often with long hairs as well all over or only on jointed part; calyx 0.6–2.2 cm long, finely tomentose or sometimes with few long hairs; lobes lanceolate to triangular or ovate-lanceolate, 4–18 × 4–8 mm, acute to attenuate, often keeled. Corolla yellow to orange, sometimes with a reddish centre (whole corolla often drying purplish-red); petals obovate, 1.2–2.5(–3.8) × 1–1.5(–2.5) cm; staminal tube with expanded part sparsely to densely stellate-pubescent but occasionally glabrous. Mericarps (17–)20–30(–40 fide Meeuse), ultimately black when mature, 10–16 × 4–6.5 mm with dorsal apical edge with long triangular projection 3–5 mm long, usually at least $^1/_3$ total length of mericarp, at first densely stellate-hairy later stellate-pubescent only on dorsal margin; seeds 2–3 per mericarp, purplish brown, rounded-reniform, compressed, 2–2.5 mm long and wide, very finely reticulate and with small papillae, some with projecting trichomes; funicle near hilum with few coarse stellate hairs.

KEY TO INFRASPECIFIC VARIANTS

1. Leaves and petiole with very fine dense stellate tomentum but no longer pubescence except for an isolated tuft of long pubescence at junction of petiole and lamina; stems without long hairs; 0–400 m ..... f. subsp. **zanzibaricum**
   Leaves and petiole often with longer hairs all over but without an isolated tuft; stems often with long hairs; widespread up to 2200 m .... (subsp. **mauritianum**) 2
2. Staminal column completely glabrous or virtually so; corolla yellow with crimson or purple eye; indumentum finely tomentose and no long hairs on the stem: Uganda .... c. var. **cabrae**
   Staminal column hairy at least in part; corolla yellow with or mostly without a dark eye; indumentum various with or without long hairs on the stem 3
3. Petals yellow with reddish purple base, 3–3.8 × 2.5 cm; staminal column 10–13 mm long; **K** 7 (Teita), **T** 3 (West Usambaras) .... e. var. **grandiflorum**
   Petals and staminal column not so large 3

4. Stem leaves etc. mostly finely tomentose with no long hairs; calyx about 7 mm long with lobes 4–6 mm long, usually acute not attenuate . . . . d. var. **brevicalyx**
   Stem leaves with or without long hairs; calyx exceeding 7 mm long with lobes over 6 mm long, acute to narrowly attenuate 5
5. Plant mostly with dense long hairs on stems, leaves etc. as well as short pubescence or velvety tomentum . . . . a. var. **mauritianum**
   Plant velvety tomentose or shortly pubescent but without long hairs or only on the very young parts . . . . b. var. **epilosum**

subsp. **mauritianum**

Stems with or without long hairs; leaves and petiole with very fine or longer pubescence but without a distinct isolated tuft at junction of petiole and lamina. Fig. 23, p. 135.

a. var. **mauritianum**

Plant with ± dense long hairs on stem leaves etc. as well as short tomentum or pubescence; calyx lobes 7–16 mm long, often long-attenuate.

KENYA. Northern Frontier District: Leroghi range, Karisia Forest, July 1960, *Kerfoot* 2113!; Nairobi District: Nairobi R., 23 Feb. 1968, *Thairu* 8!; Masai District: about 9 km NE of Entasakera, Emuuguroilline R., 8 June 1977, *Fayad* 54!

TANZANIA. Masai District: Ngorongoro Crater, 12 May 1952, *Tanner* 873!; Kilosa District: Ukaguru Mts, track from Mandege to Uponela, 1 Aug. 1972, *Mabberley* 1322! & Mikumi National Park, 30 Apr. 1968, *Renvoize & Abdallah* 1828!

DISTR. **K** 1, 3–7 (see note); **T** 1–3, 6; ?Ethiopia, Somalia

HAB. Upland and lowland grasslands, bushland, *Acacia xanthophloea* etc. woodland, forest edges, roadsides and waste places, riverine vegetation; also in cultivations; 1150–2200 m (see note)

SYN. *Sida mauritiana* Jacq., Misc. Austr. 2: 352 (1781) & Ic. Pl. Rar. 1, t. 137 (1782)
?*S. patens* Andr., Bot. Rep. 9, t. 571 (1809). Type: Ethiopia, specimens grown by Viscount Valentia (G. Annersley) in 1806 at Arley Castle, Staffordshire, no herbarium specimens, t. 571 (lecto.!)
Abutilon indicum sensu Mast. in F.T.A. 1: 186 (1868) pro parte, *non* (L.) Sweet
"*A. asiaticum* (vel *indicum*)" sensu Oliv. in Trans Linn. Soc. ser. 2, 2: 329 (1887), *non* (L.) Don (*nec* (L.) Sweet)
A. longicuspe A. Rich. var. *hildebrandtii* Baker f. in J.B. 31: 75 (1893). Type: Kenya, Teita District: N'di, *Hildebrandt* 2633 (BM!, holo.; K!, iso.)
?*Pavonia patens* (Andr.) Chiov. in Ann. Bot. Roma 13: 409 (1915), *non* sensu Chiov. *nec* auctt. mult. (see general note at end of species)

NOTE. *Holst* 8936 from Tanzania, West Usambaras, Kwa Mshuza, July 1893 bears the name *A. holstii* Gürke but the name does not appear to have been published; it has sometimes been used on herbarium sheets. *Holst* 8936 belongs to the variety dealt with above.

Humbert & Swingle 4302 labelled 'Monbasa (sic), bord de la mer, June 1928' is the only specimen of this variety and indeed of subsp. *mauritianum* I have seen from sea-level. It has a particularly densely hairy stem and very broad calyx lobes. It might be wrongly labelled or introduced from elsewhere in Kenya.

b. var. **epilosum** *Verdc.* **var. nov**. a var. *mauritiano* caulibus sine pilis longis differt. Type: Tanzania, Arusha District: Ngurdoto National Park, Big Momela Lake, 5 Nov. 1965, *Greenway & Kanuri* 12284 (K!, holo.)

Calyx lobes 7–16 mm long, often long-attenuate; plant finely tomentose or shortly pubescent but without long hairs or sometimes on young parts (many intermediates occur).

UGANDA. Acholi District: Agoro, Apr. 1943, *Purseglove* 1378!; Kigezi District: Queen Elizabeth National Park, Ishasha R. Camp, 12 May 1961, *Symes* 707!; Busoga District: 16 km S of Bugiri, Igwe, 14 Aug. 1952, *G.H.S. Wood* 328!

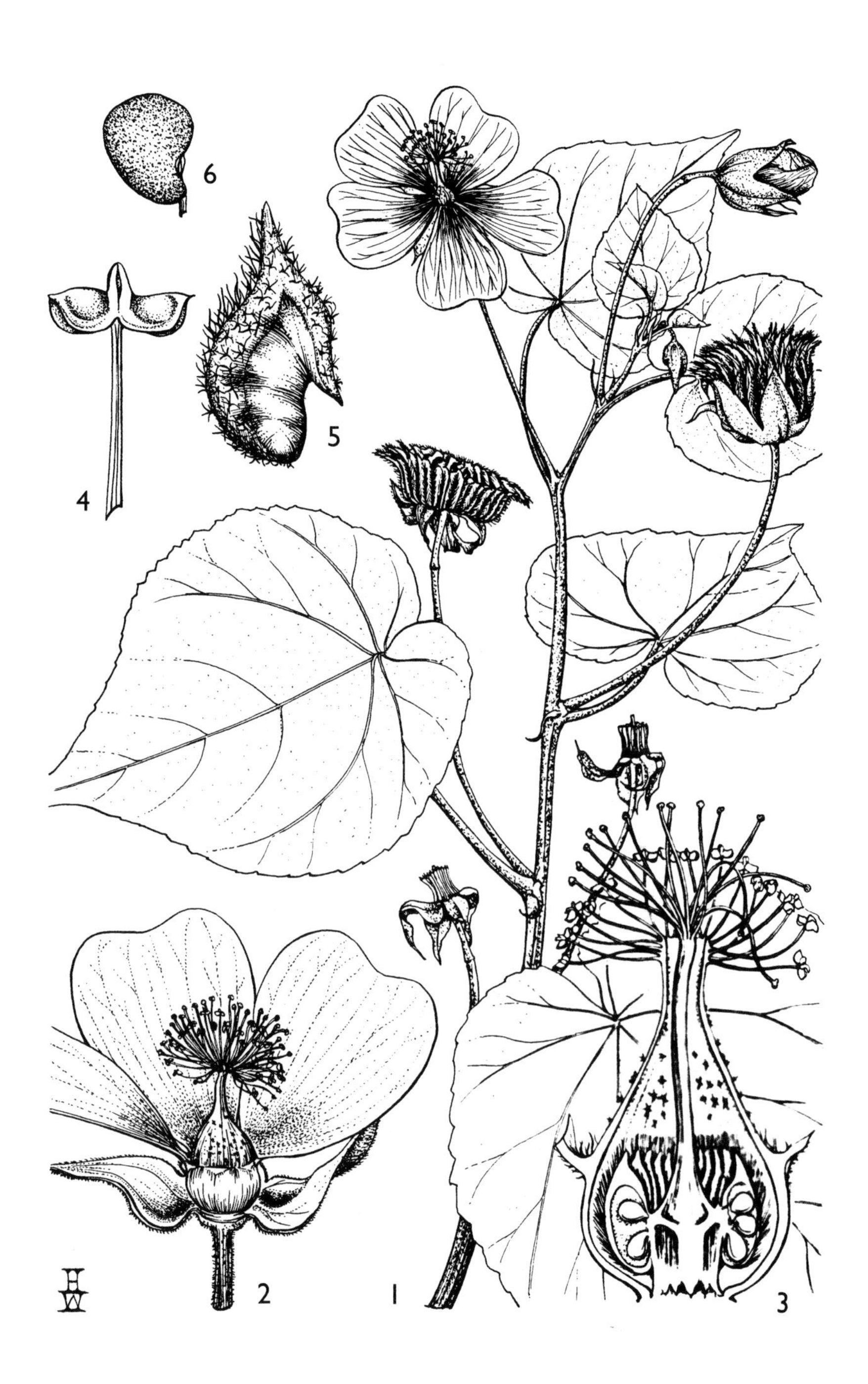

Fig. 25. *ABUTILON MAURITIANUM* — **1**, habit, × $^{2}/_{3}$; **2**, flower (half of calyx and petals removed), × $1^{1}/_{2}$; **3**, longitudinal section of androecium and gynoecium, × 4; **4**, dehisced anther, × 10; **5**, mericarp, × 3; **6**, seed, × 5. 1–4 from *Gillett* 12909; 5–6 from *Kokwaro* 274. Drawn by Heather Wood.

KENYA. Northern Frontier District: Mt Kulal, 9 Oct. 1947, *Bally* 5539!; Naivasha District: Lake Naivasha, 10 Apr. 1969, *E. Polhill* 108!; Meru District: Meru National Park, N bank of the Kindani R., Kindani Camp, 8 May 1972, *Ament & Magogo* 135!
TANZANIA. Ngara District: Bushubi, Keza, 18 Nov. 1959, *Tanner* 4500!; Arusha District: Ngurdoto National Park, Momela Lake, 4 Dec. 1966, *Richards* 21649!; Lushoto District: Korogwe area, Magunga Estate, 10 May 1952, *Faulkner* 960!
DISTR. **U** 1–4; **K** 1, 3, 4, 6; **T** 1–4, 6, 7; not known elsewhere
HAB. Grassland, *Acacia–Commiphora–Cordia–Croton* bushland and thicket, grassland with scattered trees, *Celtis–Albizia* etc. forest edges and clearings; 200–2000 m

SYN. *A. indicum* sensu Mast. in F.T.A. 1: 186 (1868) pro parte quoad *Grant* s.n. from Tanzania (**T** 6), Mbwiga, *non* (L.) Sweet

NOTE. *M.R. Smith* 518 (Uganda, Teso District: Serere, 12 Jan. 1968) is definitely described as having a yellow corolla with a crimson centre and the plant is very finely tomentose. There are several specimens which are ocellate when dry but no mention is made in the field notes so it may appear only on drying in some cases.

Tanner 5557 (Tanzania, Ngara District: Bugufi, Murgwanza, 29 Dec. 1960) bears two separate specimens, one with fine tomentum and no long hairs on stem or leaves, the other with long hairs on the stem and much longer pubescence on the leaves; the two varieties clearly grow together.

Luke et al. 5309 (Kenya, Teita District: Kasigau Mt, Rukanga route, 1050 m, 31 May 1998) appears to be this variety but the specimen lacks open flowers; the stipules ± point downwards and are up to 11 mm long.

c. var. **cabrae** (*De Wild. & T. Durand*) *Verdc.* **comb. et stat. nov.** Type: Congo-Kinshasa, Lower Congo-Kinshasa, 1897, *Capt. Cabra* 103 (BR!, holo.)

Indumentum of stems and foliage a very fine velvety tomentum with no long hairs; calyx ± 1.5 cm long. Corolla yellow with crimson or purple eye; staminal column glabrous. Mericarps with very long apical appendages.

UGANDA. Ankole District: Ishigiro, May 1939, *Purseglove* 705!; Mbale District: Tororo, Bukedi, Mar. 1916, *Snowden* 279!; Masaka District: Kyotera, Oct. 1925, *Maitland* 888!
DISTR. **U** 2–4; Congo-Kinshasa
HAB. Short grassland, uncleared land, sometimes near rock outcrops; 1050–1350 m

SYN. *A. cabrae* De Wild & T. Durand. in B.S.B.B. 38: 16 (1900); Robyns, F.P.N.A. 1: 582 (1948) pro parte

NOTE. Since much has been made of the importance of presence or absence of indumentum on the staminal column in this genus it is worth drawing attention to this variety and pointing out that it is not well marked. Many specimens which are almost identical e.g. *Tweedie* 177 (Kenya, slopes of Mt Elgon, June 1934), *Gillett* 12909 (Kenya, Moyale, 23 Apr. 1952) have varying amounts of indumentum on the staminal column. See also discussion on material grown at the Jardin des Plantes, Paris in the general note at the end of this species.

d. var. **brevicalyx** *Verdc.* **var. nov.** a varietatibus ceteris subspeciei *mauritiani* calyce breviore lobis 4–6 mm longis differt. Type: Kenya, SE slopes of Mt Elgon, *Padwa* 33 (K! holo.; EA, iso.)

Calyx lobes 4–6 mm long, mostly acute and not attenuate; plant tomentose with few or no long hairs.

UGANDA. Teso District: Serere, *M.R. Smith* S19!; Busoga District: Buraba and Molo, July 1926, *Maitland* s.n.!; Mengo District: 18.5 km, Wakyato to Ngoma, 1 Aug. 1956, *Langdale-Brown* 2285!
KENYA. SE Mt Elgon, 4 May 1953, *Padwa* 33!; Machakos District: Tana R., 7 Forks, slope W of Kindarumi Dam, 30 Apr. 1967, *Gillett & Faden* 18144!
TANZANIA. Mwanza, *R.L. Davies* 47/a!; Handeni District: Kwa Mkono, 27 June 1966, *Archbold* 742!; Kilosa District: Ilonga, 27 Aug. 1964, *Chambers* 4!
DISTR. **U** 3, 4; **K** 3–5; **T** 1, 3, 6; not known elsewhere
HAB. Grassland, *Acacia-Commiphora, Adansonia, Delonix, Erythrina* etc. bushland, *Euphorbia, Grewia, Rhus, Carissa, Harrisonia* etc. thicket, sometimes on termite mounds, rocky hills, also weed in cultivations; 450–1800 m

NOTE. Although extremes are very distinctive there are many intermediates with vars. a and b sometimes on one plant.

e. var. **grandiflorum** *Verdc.* **var. nov.** a var. *mauritiano* floribus majoribus, petalis usque 3.8 cm longis, 2.5 cm latis, columna staminali 10–13 mm longa differt. Type: Tanzania, Lushoto District: West Usambaras, 6.4 km NE of Lushoto, Mkuzi, *Drummond & Hemsley* 2077 (K!, holo.)

Shrubby to 3 m. Stems, petiole and peduncles with long hairs as well as fine pubescence; calyx 1.8 cm long. Petals yellow with reddish purple base, 3–3.8 × 2.5 cm; staminal column 10–13 mm long. Carpels up to ± 30.

KENYA. Teita District: Wundanyi Hill, Sep.-Oct. 1938, *Joanna* in CM 8965! & without precise locality, Oct. 1932, *M. Jex-Blake* in CM 2313! & Sep. 1932, *M. Jex-Blake* in CM 2295!
TANZANIA. Lushoto District: Mkuzi, 10 Apr. 1953, *Drummond & Hemsley* 2077! & Jaegertal, 17 Aug. 1932, *Geilinger* 1320! & Jaegertal, 23 Sep. 1983, *Kisena* 141!
DISTR. **K** 7; **T** 3; not known elsewhere
HAB. Forest near stream, secondary scrub and bushland; 1350–1650 m

NOTE. The label of 2295 states specimen grown in cultivation [in Nairobi] having originally been transplanted from Taita Hills; 2313 bears only locality Taita Hills although collected later. It is possibly what Jex-Blake (Gard. E. Afr. ed. 4: 102 (1957) calls *A. hildebrandtii* since E.G. Baker has made a pencil annotation on 2295 'confer *hildebrandtii* Baker f.'. *Gardner* in F.D. 2984 (Taita Hills, Mbololo Hill) has smaller flowers and mentions on the label 'Flowers yellow, large in sheltered spots but much smaller when exposed.' It is clear that varietal rank is all that can be given this striking plant but it needs further study in the field. It is known that strong links exist between the Taita Hills and the West Usambaras.
See also *Abutilon* sp. 7, page 126.

f. subsp. **zanzibaricum** (*Mast.*) *Verdc.* **subsp. et comb. nov.** Type: Tanzania, Zanzibar, *Bouton* s.n. (K!, lecto.) (chosen by Meeuse who did not realise sheet is a mixture)*

Leaves and petiole with very fine dense stellate tomentum and no longer pubescence except for an isolated tuft ± 3 mm long at junction of petiole and lamina. Stems without long hairs.

KENYA. Kwale District: Dzombo Mt, 8 April 1968, *Magogo & Glover* 798!; Mombasa District: Likoni Beach, 14 May 1962, *Wright* 13!; Kilifi District: Kibarani, 20 July 1945, *Jeffery* 267!
TANZANIA. Pangani District: Madanga, 10 Sep. 1956, *Tanner* 3113!; Bagamoyo District: 8 km S of Bagamoyo, 1 Nov. 1969, *Harris & Tadros* 3567!; Rufiji District: Mafia I., Kanga to Kinuni, 14 Aug. 1937, *Greenway* 5103!; Pemba, Chake Chake, 12 Oct. 1929, *Vaughan* 832!
DISTR. **K** 7; **T** 3, 6; **Z**; **P**; Comoro Is. (Moheli); Malawi**
HAB. Coastal thicket and bushland with *Dombeya, Markhamia, Hyphaene* etc., also riverine forest with *Sterculia, Diospyros, Ficus, Sorindeia* etc., sandy grassland and in banana and coconut plantations; 0–400 m

SYN. *A. zanzibaricum* Mast. in F.T.A. 1: 186 (1868) pro parte
A. mauritianum sensu U.O.P.Z.: 99 (1949), *non* (Jacq.) Medik. sensu stricto

NOTE. The hair tufts are ± missing in some specimens e.g. one sheet of *Faulkner* 587 but the very fine tomentum is characteristic.
Kassner 256 (Kenya, Mombasa District: Changamwe, 14 Mar. 1902) lacks the hair tufts and differs also in more distinctly toothed leaves, a less deeply lobed calyx with keeled lobes. Possible hybridisation with *A. guineensis* is not supported by non-impressed venation and glabrescent monocarps. The fragments have a slightly scandent look and the fieldnote gives the height as 15 ft which seems very improbable.

GENERAL NOTE. I have interpreted the Jacquin plate by means of a specimen at the BM from the herbarium of Peter Simon Pallas which bears a label *Sida mauritiana* in a contemporary hand. This was presumably sent as a specimen to Pallas by Jacquin rather than having been grown

* Only the bottom plant on the sheet is the lectotype of *Sida zanzibarica*; the top plant on the sheet is possibly *A. pannosum* but is in bud and nothing like it has been found in Zanzibar. The sheet is labelled '*Sida zanzibarica* Boj. in campis sylvestribus ins. Zanzibar leg. 1824'.
** *Pawek* 11309, Nkhata Bay District: 23 May 1976 from 465 m appears to be this.

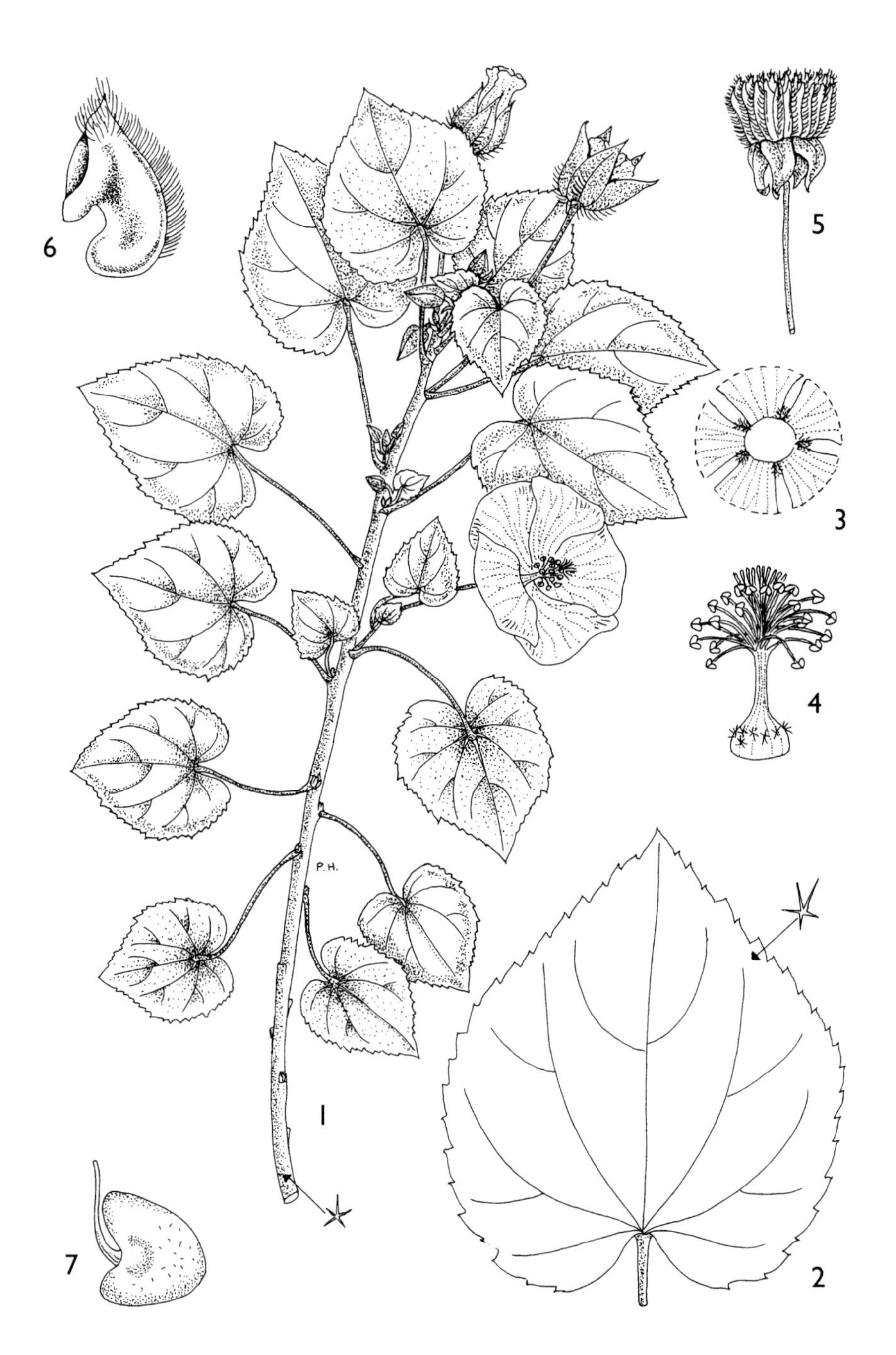

FIG. 24. *ABUTILON GUINEENSE* — **1**, habit, × $^2/_3$; **2**, mature leaf, × $^2/_3$; **3**, base of corolla with indument (diagrammatic); **4**, androecium and gynoecium, × 2; **5**, fruit, × $^2/_3$; **6**, mericarp, × 2; **7**, seed, × 9. 1, 3–4 from *Napier* 262; 2 from *Bally* 11617; 5–7 from unknown collector S13, Uganda. Drawn by Pat Halliday.

from seed at St. Petersburg. There is no specimen preserved at Vienna (fide Riedl, pers. comm.). Jacquin does not state from whence his seeds came merely that the species grows wild in Mauritius. Much material was sent to Vienna by Ceré who was Director of the Pamplemousse garden at the time. Marais, Fl. Mascar. 51 Malvaceae: 17–18 (1987) has explained that despite Jacquin's statement nothing resembling it has ever been found in the Mascarene Islands. He found there were five specimens (cibachromes now at K) of plants grown at the Jardin des Plantes, Paris in the Paris herbarium derived from the herbaria of Maire, Poiret and Richard all said to be originally Mauritian in origin. Lamarck also described *Sida planiflora* (Lam., Encycl. 1: 7 (1783)) from a plant grown at Paris from seed said to have been sent from Mauritius or 'L'Inde' by Commerson (P-LAM, ? holo. microfiche 83 II 4–this specimen is only in fruit and has a ticket stating *S. mauritiana* Jacq; seeds from Commerson but no locality given). Other figures are given by Cavanilles (Diss. 1: 32, t. 7, fig. 4 (1785) & t. 135, fig. 1a–f (1788)) and L'Héritier (Stirp. Nov.: 129, t. 62 (1789)). Marais states quite definitely that the African material always called *A. mauritianum* differs from the old specimens preserved at P in foliage and the staminal column having stellate hairs (not glabrous). The Pallas specimen mentioned above does, however, have dense stellate hairs on the staminal column and agrees with African material. It seems likely that African material was being cultivated in Mauritius but its origin is still not known. Moreover, variants with glabrous staminal columns do occur in East Africa.

Vollesen and others have put *Sida patens* Andr. (*Pavonia patens* (Andr.) Chiov.) in synonymy with *A. mauritianum*. The lack of an epicalyx rules out *Pavonia* but Andrews's figure shows a plant with long narrowly lanceolate sepals and the description speaks of five 2-beaked carpels which rules out *A. mauritianum* but the leaves and corolla certainly agree. It could possibly be a drastic abnormality due to cultivation but numerous carpels is a characteristic of the species.

18. **Abutilon guineense** (*Schumach. & Thonn.*) *Baker f. & Exell* in J.B. 74, Suppl. Polypet. Addenda: 22 (1936); Exell & Mendonça, C.F.A. 1: 154 (1937); Keay, F.W.T.A. ed. 2, 1(2): 337 (1958); Meeuse in F.Z. 1: 495, t. 93, fig. 12 (1961); Hepper, W. Afr. Herb. Isert & Thonning: 69 (1976); Vollesen in Op. Bot. 59: 35 (1980); Marais in Fl. Mascar. 51. Malvacées: 23, t. 5, fig. 9–11 (1987); Kabuye in U.K.W.F. ed. 2: 102 (1994); Thulin, Fl. Somalia 2: 79 (1999). Type: 'Guinea', *Thonning* 38 (C, S, syn.)

Woody branched herb or subshrub but sometimes annual or at least flowering in first year (0.15–)0.4–1 m tall; stems velvety pubescent and with long hairs. Leaves ovate, 7–11 × 4.5–11 cm, subacute to acuminate at the apex, cordate at base, toothed, ± roughly stellate-pubescent above but surface not obscured, more densely velvety beneath and somewhat discolorous, venation ± impressed above but raised reticulate beneath; petiole 2–7 cm long; stipules linear, ± 7 mm long. Flowers solitary, axillary; peduncle 3–9.5 cm long with same indumentum as the stems; calyx oblong-ellipsoid in bud becoming broadly cup-shaped at base when mature, at first ± 15 × 10 mm but in fruiting state up to 2 × 3 cm, densely roughly velvety pubescent, rather thick in consistency (actually mentioned as fleshy in one field note); lobes eventually broadly triangular, 11–13 × 6–11 mm, attenuate at the apex, often very distinctly 3-veined. Flowers up to 4.5 cm wide; petals yellow to apricot, 1.2–1.8 × 1.2–1.7 cm; staminal column glabrous to stellate-pilose. Fruits up to 2 × 3.5 cm; mericarps ± 20, 10–14 × (5–)6–6.5 mm, with apical tooth 3–5 (inclusive) mm long, very densely grey woolly pilose on outer margin, the hairs ± 3–4 mm long; seeds cordiform-reniform, 2.5 × 3.5 mm, smooth or minutely warted. Fig. 24, p. 138.

UGANDA. Karamoja District: between Kaabong and Kotido, July 1966, *M.R. Smith* S12!; Toro District: Hima, 20 Dec. 1925, *Maitland* 1044!; Teso District: Agu, Feb. 1966, *M.R. Smith* S13!

KENYA. Nairobi District: 32 km E of Nairobi, 19 Sep. 1951, *Bogdan* 3267!; Masai District: Athi Plains, near Kajiado, about 50 km from Nairobi on Namanga road, 21 Dec. 1963, *Verdcourt* 3847!; Tana River District: Tana River National Primate Reserve, 2.1 km N of Mchelelo, 13 Mar. 1990, *Kabuye et al.* TPR 322!

TANZANIA. Masai District: Ardai Plain, field 101, 27 June 1944, *Greenway* 6992!; Tanga District: Sawa, 10 Mar. 1965, *Faulkner* 3468!; Kondoa District: Kikore, 15 Apr. 1930, *B.D. Burtt* 2744!; Zanzibar, without locality, 1927, *Toms* 197!

DISTR. **U** 1–3; **K** 2, 4, 6, 7; **T** 1–3, 5, 6, 8; **Z**; Somalia, and from Ghana to Angola, Malawi, Mozambique, Zimbabwe, Natal, Swaziland, South Africa; Madagascar

HAB. Grassland, bushland and wooded (mainly *Themeda*) grassland with *Combretum, Cordia, Acacia, Terminalia* etc., edges of cultivations, swamp edges often on flood plains on black or grey heavy soil and sandy soil; recorded as a weed in sisal and wheat etc.; 0–1700 m

SYN. *Sida asiatica* L., Cent. Pl. II, 26 (1756) & Sp. Pl. ed. 2: 964 (1763) sensu descript. tantum (see note)

Abutilon asiaticum (L.) Sweet, Hort. Brit.: 53 (26) et auctt. mult. sensu descript. tantum (see note)

Sida guineensis Schumach. & Thonn., Beskr. Guin. Pl.: 307 (1827) & in Kongel. Dansk vid. Selsk. Naturvid. Math. Afr. 4: 21 (1829)

Abutilon agnesae Borzi in Boll. Ort. Bot. Palermo 10: 129, t. 10 (1911). Type: Somalia, Torda Goscia, *Macaluso* 117bis (PAL, holo.)

A. asperifolium Ulbr. in E.J. 51: 34 (1913). Type: Tanzania, Mwanza District: Ussukuma, *Conrads* 24 (B†, holo.)

A. densevillosum Mattei in Boll. Ort. Bot. Palermo N.S. 1: 91 (1915). Type: Somalia, Mogadishu (Mogadiscio), *Macaluso* 39 (PAL, holo.)

A. blepharocarpum Mattei in Boll. Ort. Bot. Palermo N.S. 1: 93 (1915). Type: Somalia, Merca, *Macaluso* 45bis (PAL, syn.), & Uebi Gof, Gamana, Giuba valley, *Scasselati* 132/15 (FT, syn.) (sphalm. Scasselari)

A. indicum (L.) Sweet subsp. *guineense* (Schumach. & Thonn.) Borss. Waalk. in Blumea 14: 175, fig. 19e (1966); Paul in Fl. India 3: 267 (1993), as *guineensis*; Sivarajan & Pradeep, Malv. S. Pen. India: 195, fig. 71 (1996); Philcox in Rev. Fl. Ceylon 11: 340 (1997)

NOTE. It is well known that there is a problem over the typification of *Sida asiatica* L. The description given by Linnaeus does not fit the two elements he cites nor the specimen in the Linnean Herbarium which agree with *S. indica* L. I have typified *S. asiatica* by one of these elements (Verdcourt in Taxon 52: 860 (2003)) so that it becomes a synonym of *S. indica*. I am fairly certain about the identity of *A. asperifolium* Ulbr. Its author mentions the slightly 3–5-lobed leaves and traces of 3-lobing can be found in several specimens of *A. guineensis*. E.G. Baker saw the type at B and *A. guineense* is widespread in Mwanza District, which is ± equivalent to Ussukuma.

19. **Abutilon grandiflorum** *Don*, Gen. Syst. 1: 504 (1831); Exell & Mendonça in C.F.A. 1 (1): 154 (1937); Exell, Cat. Vasc. Pl. S. Tomé: 116 (1944); Meeuse in F.Z. 1: 495, t. 93, fig. 13, 94 (1961); Vollesen in Opera Bot. 59: 35 (1980); Kabuye in U.K.W.F. ed. 2: 101 (1994). Type: 'native of Guinea' (fide Don), São Tomé, *Don* s.n. (BM, holo.)

Short-lived or annual sometimes shrubby herb 0.7–1.8 m tall with velvety tomentose stems, rarely with long spreading hairs, glabrescent and sometimes somewhat woody when older. Leaves discolorous; lamina ± round to ovate, 2.7–15 × 2–14 cm, usually long-acuminate at the apex, cordate at base, velvety-tomentose, coarsely serrate, often irregularly and unequally; venation prominent beneath; petiole 1–15 cm long; stipules linear-lanceolate, 0.5–1 cm long, deflexed. Flowers axillary; pedicel 3.5–9 cm long, the articulation ± 1 cm long; calyx 1.2–2 cm long; base cupular, 5-keeled; lobes ovate-triangular to ovate-lanceolate, 5.5–6 mm long, 4–10 mm wide, acute to cuspidate, mostly slightly keeled, usually exceeding the fruit (not in var. *iringense*). Corolla creamy-yellow to yellow or orange; petals obovate, 1.5–2 cm long and wide; staminal column stellate-pilose. Fruit 1.2–1.5 cm long, 1.5–2.5 cm diameter, ± overtopped by calyx lobe tips; mericarps over 25, eventually blackish-brown, 12–14 × 6–7 mm; apical triangular appendages up to 4–5 mm long, very coarsely stellate-pilose, 2–3-seeded; seeds blackish-purple, reniform, 2.5 × 2.2 mm, minutely lepidote, glabrous (or with tufts of hair near hilum, fide Meeuse), with minute spur from hilum formed by funicle remnant.

a. var. **grandiflorum**

Stems softly tomentose without spreading hairs or rarely with long spreading hairs. Leaf blades not trilobed or with only slight traces of lobing; calyx lobes up to 16 × 10 mm, the tips overtopping the mericarps.

UGANDA. Karamoja District: Bokora County, Iriri, June 1957, *Wilson* 362!; Teso District: near Labori (Lobori, Laboli), July 1926, *Maitland* s.n.!; ?District: Masaka–Mbarara road, Oct. 1925, *Maitland* 889!

KENYA. South Kavirondo District: Kanam, 30 Jan. 1935, *Allen Turner* in CM 3628!; Meru District: Meru Game Reserve, Muguongo, 14 Jan. 1963, *Mathenge* 146!; Masai District: about 50 km from Nairobi on Namanga road, Athi Plains, near Kajiado, 21 Dec. 1963, *Verdcourt* 3846!

TANZANIA. Musoma District: Seronera to Banagi and down the corridor and return km 58, 4 Apr. 1961, *Greenway* 9976!; Masai District: Lisingita area, 9 Jan. 1969, *Richards* 23686!; Morogoro District: Morogoro–Turiani road, *Semsei* in F.D. 1966! (error in flower colour)

DISTR. **U** 1, 3, ?4; **K** 3–7; **T** 1–3, 6–8; Sao Tomé, Congo-Kinshasa, Angola, Malawi, Mozambique, Zimbabwe, South Africa; Madagascar

HAB. Grassland on red laterite and black cotton clay soils, grassland with scattered *Acacia mellifera* or *A. tortilis*, *A. xanthophloea* woodland, *Acacia–Lannea–Commiphora* woodland, roadsides, riverine thicket and also recorded as a weed in wheat fields; 100–1550 m

SYN. *A. indicum* sensu Hook. f. in Niger Fl.: 230 (1849); Mast. in F.T.A. 1: 186 (1868), *non* (L.) Don (both authors refer to "*Sida grandiflora* Don" but Don described it under *Abutilon* not *Sida*)
A. asiaticum (L.) G. Don var. *lobulatum* Robyns & Lawalrée in B.J.B.B. 18: 273 (1947); Hauman in F.C.B. 10: 158 (1963). Type: Congo-Kinshasa, Lakes Edward & Kivu, Vitshumbi Bwera, *de Witte* 1005 (BR!, holo. & iso.)

NOTE. Meeuse states mericarps about 20 in his description but includes the species in part of the key under mericarps 16 or fewer. I collected *A. guineense* and *A. grandiflorum* in the same area (*Verdcourt* 3847 and *Verdcourt* 3846) and noted in my fieldnote that in the field *A. guineense* appears very different with leaves more bullate with coarser indumentum and fruits much hairier. Borssum Waalkes (Blumea 14: 175 (1966)) has said of *A. grandiflorum*–'this is in my opinion another form of *A. indicum*'.

b. var. **iringense** *Verdc.* **var. nov.** a var. *grandifloro* caulibus dense breve patenter pubescentibus, foliis leviter trilobatis, calycis lobis 5.5–9 mm longis, 4–8 mm latis quam fructu brevioribus differt. Typus: Tanzania, Iringa District: Msembe–Mdonya road, *Richards* 17360 (K!, holo.)

Stems with dense shortly spreading pubescence. Leaf blades slightly but distinctly trilobed towards the apex; calyx lobes 5.5–9 × 4–8 mm, not overtopping the mericarps.

TANZANIA. Mbeya District: Usangu Plain, near Utengule [Utencile], 29 Jan. 1963, *Richards* 17594!; Iringa District: Ipugulu, 27 km from Msembe via Kimirimatonge Circuit, 24 Apr. 1970, *Greenway & Kanuri* 14421! & Ruaha National Park, W of park H.Q. at Msembe, 16 May 1968, *Renvoize & Abdallah* 2185!

DISTR. **T** 7; not known elsewhere

HAB. Open *Acacia tortilis* wooded grassland, *Acacia* woodland and riverine thicket; 700–1050 m

20. **Abutilon bussei** *Ulbr.* in E.J. 48: 368 (1912); T.T.C.L.: 298 (1949). Type: Tanzania, Lindi District: Lindi, path to Pili-Pili source, *Busse* 2423 (B†, holo.; EA!, iso.)

Rather slender branched woody herb to 1.2 m, probably a short-lived perennial but rootstock of annual appearance; stems ± brown, slender, pubescent with short brownish stellate hairs except at base which is corky, ± furrowed and glabrous. Leaves thin; lamina ovate but becoming more narrowly so in upper leaves, 2–10 × 0.7–8.5 cm, the larger ones very narrowly acuminate at the apex, subcordate to cordate at base, entire or very obscurely crenate, with scattered small stellate hairs on both surfaces and margins but apearing ± glabrous; upper leaves and bracts subsessile or very shortly petiolate but lower petiole up to 10 cm long, ± densely stellate-pubescent; stipules filiform, 2–4 mm long, stellate-pubescent. Flowers several at the apices of branches and solitary or in cymes in axils of bract leaves lower down; pedicel up to 3 cm long including jointed part up to 8 mm long, stellate-pubescent; calyx 7–10 mm long, stellate-pubescent; lobes ovate, 5–8 × 4–6 mm, distinctly rather long-ciliate. Corolla orange, ± 3 cm wide; petals obovate, 1.4–1.7 × 1–1.5 cm; staminal column with few scattered stellate hairs; style arms 9–13. Fruit ± barrel-shaped in outline, 1.3–1.5 cm long, 1.2 cm diameter, distinctly narrowed at base; mericarps ± 13, pale

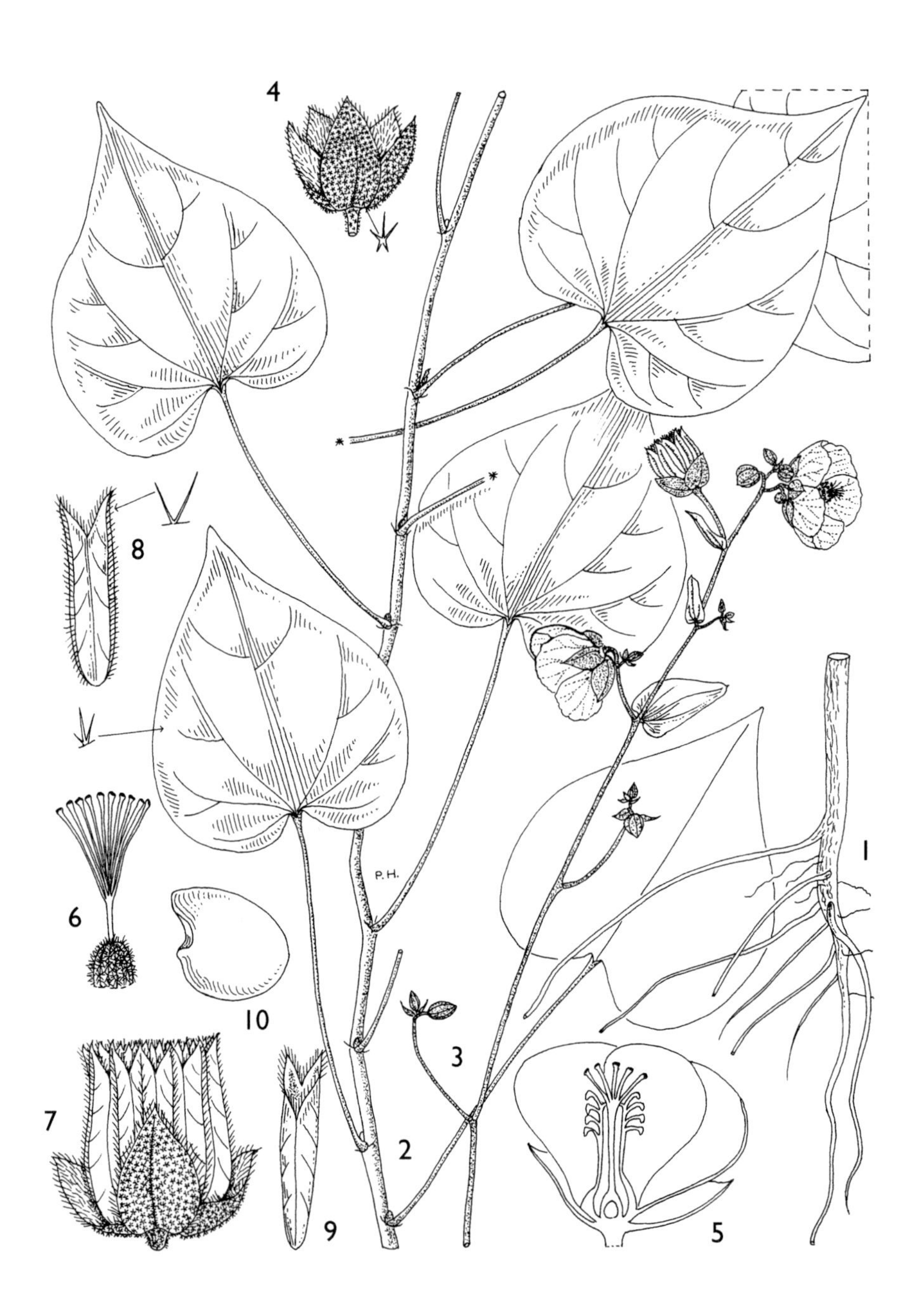

FIG. 25. *ABUTILON BUSSEI* — **1**, roots, × $^2/_3$; **2**, leafy twig, × $^2/_3$; **3**, inflorescence, × $^2/_3$; **4**, calyx; **5**, diagram of flower; **6**, gynoecium, × 2; **7**, fruit, × 2; **8**, single mericarp, × 2; **9**, single dehisced mericarp, × 2; **10**, seed, × 12. All from *Vollesen* in MRC 3488. Drawn by Pat Halliday.

green at first, never black except at tips, 3-seeded, membraneous, cohering concertina-like, ± oblong, 12 × 4.5 mm, the apical angular tip 2 × 1.5 mm, not diverging much, pubescent, the lateral surfaces distinctly obliquely veined, pubescent on outer angle; inner angular notch 5.5 mm from apical tips; seeds purplish brown, rounded reniform, compressed, ± 1 mm long and wide, glabrous but with pale elongate rugulations in the hilum. Fig. 25, p. 142.

TANZANIA. Kilwa District: Kingupira Forest, 18 Apr. 1976, *Vollesen* in MRC 3488!; Lindi District: Lindi, path to Pili-Pili source, 9 May 1903, *Busse* 2423!

DISTR. **T** 8; not known elsewhere

HAB. Light bushland (fide Busse); ground-water forest of *Celtis, Lepisanthes, Sorindeia, Trichilia* etc.; 0?–100 m

SYN. *A.* sp. nov. aff. *A. galpinii* Meeuse; Vollesen in Op. Bot. 59: 35 (1980)

NOTE. *A. wituense* Baker f. is the only other species I have seen in East Africa which has obviously veined mericarps.

21. **Abutilon ramosum** (*Cav.*) *Guill. & Perr.*, Fl. Senegamb. Tent. 1: 68 (1831); G. Don, Gen. Hist. 1: 501 (1831); Mast. in F.T.A. 1: 186 (1868); Ulbr. in E.J. 51: 27 (1913); Exell & Mendonça, C.F.A. 1 (1): 152 (1937); Keay, F.W.T.A. ed. 2, 1(2): 337 (1958); Meeuse in F.Z. 1: 498, t. 93, fig. 16 (1961); Hauman in F.C.B. 10: 161 (1963); Abedin in Fl. W. Pak. 130: 55, fig. 17 C–D (1979); Kabuye in U.K.W.F. ed. 2: 101 (1994); Vollesen in Fl. Eth. & Eritr. 2(2): 242, fig. 82.21.10–12 (1995); Thulin, Fl. Somalia 2: 76, fig. 46 E (1999). Type: Senegal, *Adanson* s.n. (MA, holo., KUH, photo.; P!, iso., microfiche)

Perennial branched shrubby herb, 0.45–1.8 m tall; stems pubescent or puberulous with small stellate hairs and usually pilose with long simple hairs as well. Leaves scarcely discolorous; lamina broadly ovate or almost round in outline, usually with some leaves shallowly 3-lobed, 2–19 × 1.5–17 cm, acute to shortly acuminate at the apex, cordate at base, crenate or serrate, stellate-pubescent on both surfaces with simple hairs but surface green and not obscured; petiole 2–11 cm long; stipules linear, 1–1.5 mm long, clearly univeined. Flowers in (1–)2–6-flowered axillary cymes, sometimes forming a terminal panicle; peduncles (actually truly branches since often with a small leaf at apex) 1.5–6 cm long; pedicel 0.2–2 cm long; calyx 4–7 mm long, stellate-pubescent, the lobes triangular, 3 mm long, acuminate. Petals yellow, 4–7 × 3 mm; staminal column with sparse simple or bifid hairs. Mericarps 6–8(–10), elliptic, 6–8 × 3 mm, 2–3-seeded, pubescent and sometimes glandular, with distinct spine-like usually curved awn 2–3 mm long; seeds 2.5 mm long, papillose-echinulate.

KENYA. Northern Frontier District: Mt Kulal, El Kajarta [Elgijada], 8 Sep. 1944, *J. Adamson* 109!; Baringo District: W side of Lake Baringo, Feb. 1962, *Tweedie* 2301!; Machakos District: Kiboko, Feb. 1970, *Tweedie* 3779!

TANZANIA. Mbulu District: Lake Manyara National Park, 26 Mar. 1962, *Dingle* 178!; Rufiji District: Selous Game Reserve, upriver from Kibembawe, 3 Aug. 1993, *Luke* 3643!; Iringa District: Mwagusi sand river, 12 May 1970, *Greenway & Kanuri* 14505!

DISTR. **K** 1, 3–5; **T** 2, 5–7; Senegal, Mali, Burkina Faso, Nigeria, Niger, Central African Republic, Cameroon, Congo-Kinshasa?, Sudan, Ethiopia, Somalia, Angola, Malawi, Mozambique, Zimbabwe, Namibia, Botswana; also Arabia, NW India and Pakistan

HAB. Lake shores in dry savanna, gravelly river beds, dry sand rivers and drifts with *Acacia* etc., riverine fringes with *Newtonia, Acacia, Kigelia, Tamarindus* etc.; (200–)750–1050 m

SYN. *Sida ramosa* Cav., Diss.: 28, t. 6, fig. 1 (1785) (wrongly quoted by Guill & Perr. as 34, t. 9, fig. 2 (which is *Sida obtusa* Vogel) but correctly cited on Jussieu herb. label)

Abutilon elaeocarpoides Webb., Fragm. Fl. Aethiop.: 53 (1854). Type: Sudan, Kordofan, Milbes, *Kotschy* 278 (FT, syn.; K!, isosyn.) & Ethiopia, Tacaze valley, *Schimper* 1679 (FT, syn.; K!, isosyn)

A. harmsianum Ulbr. in E.J. 51: 29 (1913). Type: Namibia, Hereroland, Omaruru, *Dinter* 1427 (B†, holo.)

NOTE. Fryxell (Lundellia 5: 104 (2002)) also points out that Guillemin & Perrottet cite the reference to Cavanilles wrongly but I think this a careless slip on their part and anyway at that date the fact that they associated the generic and specific names in a new combination is adequate.

IMPERFECTLY KNOWN SPECIES

Vollesen in Opera Bot. 59: 35 (1980) mentions a possible new species of *Abutilon* from ground-water forest in **T** 8 said to be a shrub to 2 m with red flowers; MRC 1138 – the sheet he saw – had no flowers and I have not seen any material.

16. **SIDA**

L; Sp. Pl.: 683 (1753) & Gen. Pl. ed. 5: 306 (1754)

Annual or perennial herbs or subshrubs, erect or prostrate; indumentum simple or mostly stellate. Leaves simple (rarely not so outside Africa), lobed or not, mostly distinctly toothed, cuneate to cordate at base, usually petiolate; stipules filiform to lanceolate. Flowers solitary or ± fasciculate in the leaf-axils, pedicellate or subsessile, often running into terminal clusters; pedicel often jointed. Epicalyx absent. Calyx cupular, often angular, 5-lobed. Corolla mostly white to yellow or orange, mostly small. Ovary of 5–15 carpels, each with a single pendulous ovule; style branches as many as carpels. Mericarps ± trigonous, ultimately separating from columella, dehiscent at apex or less often at base or indehiscent, smooth or variously sculptured, glabrous or pubescent, usually distinctly beaked, the beak often bearing 1–2 short to long awns. Seeds triangular-reniform with little or no endosperm.

About 200 species in tropics and subtropics of both hemispheres but mainly in America, some very widely distributed and weeds. A key to the sections of the genus is given by Fryxell in Sida 11: 63–64 (1985).

1. Mericarps with slender retrorsely barbed awns 3–6 mm long .... 5. *S. cordifolia* (p. 149)
   Mericarps either not awned or awns without retrorse barbs 2
2. Leaves with notched apex, the midrib projecting and forming a central tooth, otherwise entire or with 1–3 apical teeth only; small shrubby herbs with subsessile or shortly pedicellate flowers; fruit of 5 mericarps ('*cuneifolia*' group) 3
   Leaves not as above and without other characters combined 7
3. Mericarps smooth, not sculptured on back and sides, sides and/or basal part thin, breaking up to release seed 4
   Mericarps reticulately sculptured not breaking up to release seed 5
4. Leaves glabrous on upper surface, more than twice as long as wide; stem and calyx indumentum appressed; mericarps with lateral walls completely disintegrating; petals 7–10 mm long; widespread .... 6. *S. tenuicarpa* (p. 150)
   Leaves hairy on upper surface, less than twice as long as wide; stem and calyx indumentum spreading; mericarps with lower part of lateral walls and lower part of back disintegrating; petals 6–8 mm long; **K** 1, 7 .... 7. *S. tanaensis* (p. 152)

5. Leaves tomentose beneath with surface obscured; calyx densely pubescent to tomentose; mericarps densely pubescent . . . . 9. *S. massaica* (p. 153)
   Leaves ± pubescent beneath, the surface clearly visible; calyx pubescent or sparsely so; mericarps ± glabrous to pubescent 6
6. Petals 4–8 mm long; mericarps 2–2.5 mm long, back slightly laterally muricate but without horns (apart from awns); widespread . . . . 8. *S. schimperiana* (p. 153)
   Petals 8–9 mm long; mericarps 3–3.5 mm long, the back with a number of horns along edges with 2 stronger ones at base of beak (apart from awns of beak); **T** 1, 2 . . . . 10. *S. shinyangensis* (p. 154)
7. Leaves linear-oblong to linear-lanceolate, 2–12 × 0.3–1.5 cm, entire; flowers white or pink sometimes with red-purple eye; mericarps 5–7, scarcely beaked . . . . 1. *S. linifolia* (p. 146)
   Leaves not as above and other characters not correlating 8
8. Mericarps 5 9
   Mericarps 6–13 11
9. Leaves truncate to rounded at base . . . . 11. *S. alba* (p. 154)
   Leaves cordate at base 10
10. Trailing and rooting at nodes; plant pubescent; mericarps clearly birostrate with 2 distinct awns up to 1.5 mm long . . . . 2. *S. javensis* (p. 146)
    Erect or procumbent, not rooting at nodes; plant with dense seta-like hairs; mericarps scarcely beaked, the awns not separated . . . . 4. *S. urens* (p. 148)
11. Some or all leaves 3(–5)-lobed; plant scrambling, usually with white flowers; mericarps 8–13, the beak not divided . . . . 3. *S. ternata* (p. 148)
    Leaves not lobed 12
12. Leaves distichous; stipules of each pair often unequal, the larger 2–5-veined; plant glabrous or puberulous with simple hairs . . . . 12. *S. acuta* (p. 155)
    Leaves not distichous; stipules equal and 1-veined; indumentum usually denser and stellate in part 13
13. Base of calyx (or only bottom of ribs) yellow or pale yellow, distinctly 10-ribbed (may be obscure in some specimens); lobes glabrous inside; staminal column with stalked glands and simple hairs, mericarps with 0, 1 or 2 short to long awns . . . . 13. *S. rhombifolia* (p. 157)
    Base of calyx unicolorous with rest, distinctly 10-ribbed or not; lobes hairy inside; staminal column densely stellate-hairy, without glands 14
14. Prostrate herb of sand dunes etc. . . . . 14. *S. chrysantha* (p. 161)
    Erect herbs 15
15. Pedicels usually longer than petiole or leaves; mericarps with beak parts totally fused or ultimately divided, but actual awns small or absent . . . . 14. *S. chrysantha* (p. 161)
    Pedicels usually shorter than petiole; mericarps with connivent awns . . . . 15. *S. ovata* (p. 162)

1. **Sida linifolia** *Cav.*, Diss. 1: 14, t. 2, fig. 1 (1785); Mast. in F.T.A. 1: 129 (1868); Exell & Mendonça, C.F.A. 1: 148 (1937); Rodrigo in Rev. Mus. de La Plata, Bot. 6: 117, fig. 6, 2A–E, t. 12 (1944); Keay, F.W.T.A. ed. 2, 1(2): 339 (1958); Hauman in F.C.B. 10: 168 (1967); Sivarajan & Pradeep, Malv. S. Pen. India: 282, f. 106, 107 (1996). Type: Peru, *Joseph de Jussieu, Herb. A.L. de Jussieu* 12243A (P-JU, lecto.; microfiche!)

Strictly erect unbranched or branched annual herb 30–70 cm tall; stems setulose-pubescent. Leaves linear-oblong to linear-lanceolate, 20–85(–120) × 3–9(–15) mm, acute at the apex, narrowly rounded at base, ± appressed setulose, entire but with tufts of hair; petiole 3–8 mm long; stipules linear, 5–8 mm long. Flowers axillary, appearing close and terminal at apex of shoots; pedicel 5–13 mm long, jointed near apex; calyx 6 mm long, divided about half-way into triangular lobes, densely pubescent outside. Corolla white or pink (sometimes the petals red-purple at base), 1.3–2 cm wide. Mericarps 5–7, trigonous, 3 × 2 mm; outer curved face ± smooth, the inner flat faces slightly ribbed, hardly beaked. Fig. 26, p. 147.

UGANDA. Acholi District: Masindi–Gulu road, 3 June 1963, *Keithland* s.n.!; Busoga District: Busaba, July 1926, *Maitland* 1139!
TANZANIA. Mwanza District: probably Ukerewe I., 8 Apr. 1929, *Conrads* 353!; Kigoma District: Kigoma, Kibirizi, 18 Apr. 1994, *Bidgood & Vollesen* 3121!; Tabora District: about 5 km from Sikonge on Mbeya road, Mazinga, 21 Mar. 1991, *Kalema et al.* 73!
DISTR. **U** 1, 3; **T** 4; Mauritania and Senegal to Central African Republic, Congo-Kinshasa, Angola; widespread in tropical America and also recently recorded from India (Kerala)
HAB. Grassland by forest, wooded grassland, degraded *Julbernardia-Parinari* bushland, 850–1200 m

NOTE. Cavanilles's figure shows the leaves toothed in error; throughout its range they are entire but marginal tufts of hair do occur. Sivarajan & Pradeep mention it has been reported from Fiji but A.C. Smith (Fl. Vitiensis Nova 2: 436 (1951)) shows the record is erroneous. Its presence in India must be due to a recent introduction.
Rodrigo gives a photograph of a specimen 'Boldo legit in Havana' with a caption stating it is the type. However Cavanilles does not cite anything from Havana and this must, I think, be a specimen added to Cavanilles's collection later.

2. **Sida javensis** *Cav.*, Diss. 1: 10, t. 1. fig. 5 (1785); Borssum Waalkes in Blumea 14: 184 (1966); Vollesen in Fl. Eth. & Eritr. 2 (2): 251; fig. 82.24.6 (1995). Type: Java, *Commerson* s.n., *Herb. Jussieu* 12171 (P, P-JU, iso.)

Prostrate herb; stems branched, 0.2–1 m long, rooting at the nodes, ± glabrous, pubescent or rarely pilose. Leaves rounded ovate, 0.5–5 × 0.5–5.5 cm, unlobed to very shallowly 3-lobed, acute at the apex, cordate at base, distinctly serrate or biserrate, appressed pilose; petiole 0.5–5 cm long; stipules linear to lanceolate, 2–2.5 mm long. Flowers solitary; pedicel 0.5–2.5 cm long, jointed above the middle; calyx 5–7 mm long with ribbed tube, lobes 2–5 × 1.5–4 mm, acute, 3-veined. Corolla yellow or orange-yellow; petals 6–8 mm long; staminal column 1.5–3 mm long, hairy. Mericarps 5, 2.5 mm long (4–5 mm fide Borssum Waalkes), strigose, back with central ribs, sides thin-walled, partly disintegrating, lateral edges prominent; beak of two awns 0.5–1.5 mm long.

UGANDA. Toro District: Ruwenzori, Kirabowa, Dec. 1925, *Maitland* s.n.!; Mbale District: Bugishu, Bukonde, Nov. 1932, *Chandler* 1042!; Mengo District: Kipayu, May 1914, *Dummer* 789!
KENYA. South Kavirondo District: Kisii, Sep. 1933, *Napier* 2878 in CM 5276!
TANZANIA. Pare District: Gonja Maore, July 1955, *Mgaza* 53!; Buha District: 20 km NE of Kibondo, Keza Mission, 7 May 1994, *Bidgood & Vollesen* 3300!; Morogoro District: Mikese–Kisaki road, near Kimboza, 4 Sep. 1930, *Greenway* 2519!
DISTR. **U** 1–4; **K** 5; **T** 1, 3, 4, 6, 7; Sierra Leone to Central African Republic, Congo-Kinshasa and Ethiopia, Angola, Malawi, Mozambique; widespread in Asia
HAB. *Hyparrhenia-Sporobolus* grassland, *Terminalia–Combretum–Acacia* woodland and bushland, riverine forest, roadsides, on rocky soil and black soil; can be a weed in pastures and lawns; 250–1200 m

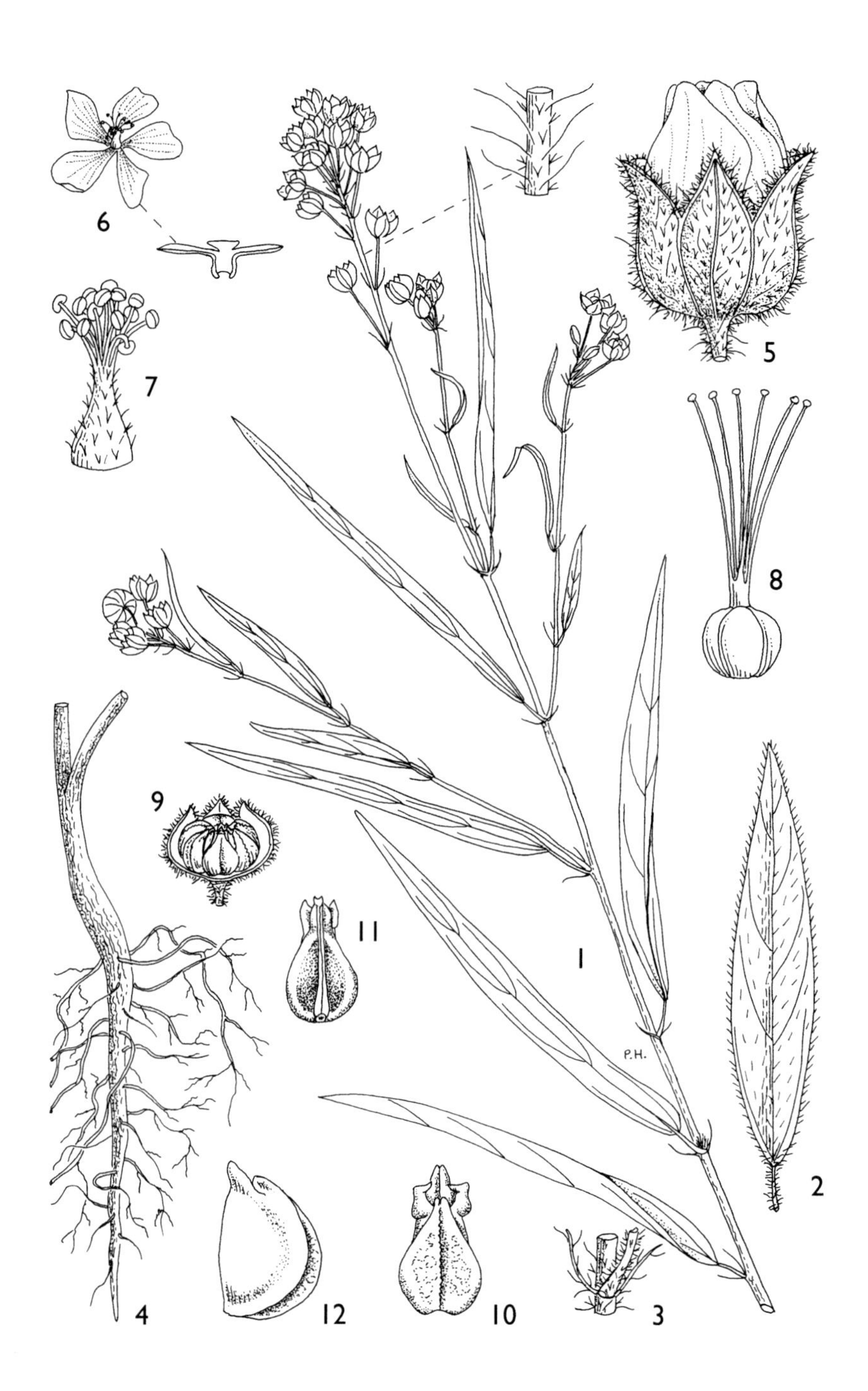

Fig. 26. *SIDA LINIFOLIA* — **1**, habit, × $^2/_3$; **2**, lower leaf, × $^2/_3$; **3**, stipules, × 2; **4**, root, × $^2/_3$; **5**, flower, × 4; **6**, open flower (diagrammatic); **7**, androecium, × 9; **8**, gynoecium, × 9; **9**, fruit (2 sepals removed), × 2; **10–12**, seed, dorsal and ventral view and profile, × 6. 1–3 from *Latilo* FH 63520; 4 from *Latilo* FH 63543; 5–12 from *Meikle* 898. Drawn by Pat Halliday.

SYN. *S. humilis* sensu Mast. in F.T.A. 1: 179 (1868), *non* Cav. (1788)
S. veronicifolia sensu Hiern, Cat. Afr. Pl. Welw. 1: 65 (1896); Exell & Mendonça, C.F.A. 1: 147 (1937); Keay, F.W.T.A. ed. 2, 1(2): 338 (1958); Meeuse in F.Z. 1: 476 (1961); Hauman in F.C.B. 10: 167 (1963); Haerdi in Acta Trop. Suppl. 8: 88 (1964); Vollesen, Opera Bot. 59: 36 (1980); U.K.W.F. ed. 2: 100 (1994), *non* Lam.

3. **Sida ternata** *L. f.*, Suppl. Pl.: 307 (1782); Meeuse in F.Z. 1: 475 (1961) (adnot.); Hauman in F.C.B. 10: 176 (1963); Troupin, Fl. Rwanda: 394, fig. 3 (1985); U.K.W.F. ed. 2: 100, t. 28 (1994); Vollesen in Fl. Eth. & Eritr. 2(2): 249, fig. 82.24.1–4 (1995). Type: South Africa, Cape of Good Hope, *Thunberg* s.n. (UPS, ? holo.)

Straggling, decumbent, subscandent or ± erect herb or undershrub 0.3–2(–4) m long from a woody rootstock; stems ± glabrous to sparsely pubescent. Leaves broadly ovate in outline, 2.5–10.5 × 2.5–8.5 cm, usually deeply 3-lobed (shallowly in some central Kenyan specimens), sparsely pubescent; lobes elliptic to ovate, 1–6 × 0.7–3.5 cm, acuminate, toothed or ± double-toothed, the outer lobes often with lobe on outer side; petiole 1.2–6.5(–12) cm long; stipules narrowly oblong, 5 × 1.2 mm. Flowers solitary, paired or in 2–3-flowered cymes; peduncle (1–)2.5–6(–7.5) cm long; pedicel 0.5–2 cm long; calyx tube ± 2 mm long, ± ribbed; lobes triangular-ovate, 5–9 × 2.5–4 mm, narrowly acuminate or cuspidate. Corolla usually white, sometimes purplish outside or pinkish in bud; petals elliptic, 6–10 × 3.5–4.5 mm, obtuse; staminal column 2–3 mm long, glandular hairy at base. Mericarps 8–13, semi-ellipsoid, ± 3.5 × 2 × 1.2 mm, compressed, glabrous, smooth or slightly warted; beak 1 mm long, not divided.

UGANDA. Toro District: Ruwenzori, foot of Msandama, Dec. 1925, *Maitland* 967!; Kigezi District: Kiswa, Dec. 1938, *Chandler & Hancock* 2610!; Mbale District: Sebei, Kyesoweri, R. Siti, 17 Oct. 1955, *Norman* 305!
KENYA. Northern Frontier Province: Maralal, Lorogi [Loroki] Plateau, 10 Nov. 1978, *Hepper & Jaeger* 6708!; Naivasha District: Kinangop, Dec. 1932, *Albrechtsen in Napier* in *CM* 2619!; Masai District: Enesambulai Valley, 12 Oct. 1969, *Greenway & Kanuri* 13831!
TANZANIA. Mbulu District: Mt Hanang, above Katesh, 3 May 1962, *Polhill & Paulo* 2305!; Lushoto District: West Usambaras, Shagai Forest, near Sunga, 24 May 1953, *Drummond & Hemsley* 2707!; Iringa District: Udzungwa Mts, Udekwa, Dec. 1981, *Rogers & Hall* 1477!
DISTR. **U** 2, 3; **K** 1, 3, 4, 6; **T** 2, 3, 7; Congo-Kinshasa (Kivu), Rwanda, Ethiopia, North Malawi, South Africa
HAB. Upland forest of bamboo, *Juniperus*, *Podocarpus* etc., riverine and ground-water forest, also drier *Olea* forest, hillside thickets, forest clearings with banana plantations; 1350–2500 m

SYN. *S. triloba* Cav., Diss. 1: 11, t. 1, fig. 11 (1785) & 5: t. 131, fig. 1 (1788). Type: South Africa, Cape of Good Hope, *Sonnerat* s.n. (MA?, P?, holo.) [name on several EA sheets]
S. permutata A. Rich., Tent. Fl. Aethiop.: 67 (1847). Type: Ethiopia, Mt Scholoda, *Schimper* III: 1911 (P!, holo.; F1, FT, K! iso.)

4. **Sida urens** *L.*, Syst. Nat. ed. 10, 2: 1145 (1759); Mast. in F.T.A. 1: 179 (1868); Exell & Mendonça, C.F.A. 1: 476 (1937); Rodrigo in Rev. Mus. de la Plata Bot. 6: 142, fig. 19, 2 A–E; 24 & t. 17 (1944); F.P.S. 2: 41 (1952); Keay, F.W.T.A. ed. 2, 1(2): 339 (1958); Meeuse in F.Z. 1: 476 (1961); Hauman in F.C.B. 10: 164 (1963); Marais, Fl. Mascar. 51: 6 (1987); U.K.W.F. ed. 2: 100 (1994); Vollesen in Fl. Eth. & Eritr. 2 (2): 249, fig. 82.24.5 (1995); Wood, Handb. Fl. Yemen: 106 (1997). Type: Jamaica, *Browne*, Herb. Linn. 866.20 (LINN, holo.)

Perennial herb or subshrub with procumbent or ± erect branches 13–120 cm tall, ± glandular-pubescent and with dense seta-like hairs. Leaves ovate, 2–10(–12) × 1–5.5(–7) cm, acute at apex, cordate at base, rather finely toothed or double-toothed, densely pubescent; petiole 0.3–5 cm long; stipules subulate, 3 mm long. Flowers solitary or up to 7 in axils, subfasciculate or often merging into short lateral or longer terminal raceme-like or subcapitate inflorescences; pedicel 1–10 mm long;

calyx 5–8 mm long, angular; lobes triangular, longer than the tube, sharply acute with concave margins, 3–5-veined, densely hairy. Corolla whitish or pale to deep yellow or apricot with often reddish, orange or purple centre or a pair of crimson spots at base of each petal; petals 5–8 mm long; staminal column ± 1 mm long, pubescent. Mericarps 5, subtrigonous, ± 2.5 × 1.2 mm, glabrous or slightly hairy, back with central, sides with weak sculpture and lateral edges rounded; beak not divided, ± 0.5 mm long.

UGANDA. Acholi District: near Chobi, 14 Aug. 1969, *Angus* 6144!; Bunyoro District: Hoima, Feb. 1943, *Purseglove* 1262!; Mengo District: Entebbe, Oct. 1924, *Maitland* 265!

KENYA. Meru District: Meru National Park, Campi ya Nyati, Jan. 2001, *Luke et al.* 7302!

TANZANIA. Moshi District: Lyamungu, 25 Oct. 1945, *Wallace* 1062!; Kigoma District: Bulimba, 25 May 1975, *Kahurananga et al.* 2790!; Ulanga District: Udzungwa Mts, Sanje, 6 Sep. 1984, *Bridson* 633!

DISTR. **U** 1–4; **K** 4, 5; **T** 2, 4, 6; Senegal to Ethiopia and south to Angola, Zimbabwe, Mozambique; Arabia, Réunion (introduced), Madagascar, New Caledonia (introduced) and widespread in tropical America and West Indies

HAB. *Hyparrhenia–Sporobolus* grassland, wooded grassland, *Terminalia–Combretum* etc. woodland, forest edges and clearings, weed of cultivations and disturbed ground, sometimes on black cotton soil; 800–1500 m

SYN. *S. densiflora* A. Rich., Tent. Fl. Aeth. 1: 66 (1847). Type: Ethiopia, Chire, *Quartin-Dillon & Petit* s.n. (P, holo.), *non* Hook. & Arn. (1833), *nom illegit.*

5. **Sida cordifolia** *L.*, Sp. Pl.: 684 (1753); Mast. in F.T.A. 1: 181 (1868); Exell & Mendonça, C.F.A. 1: 149 (1937); Rodrigo in Rev. Mus. de la Plata, Bot. 6: 184, fig. 37, 3 A–E, t. 31 (1944); T.T.C.L.: 306 (1949); Keay, F.W.T.A. ed. 2, 1(2): 339 (1958); Meeuse in F.Z. 1: 483 (1961); Hauman in F.C.B. 10: 169 (1963); Borssum-Waalkes in Blumea 14: 199 (1966); Abedin, Fl. W. Pak. 130: 83, fig. 19, C–E (1979); Blundell, Wild Fl. E. Afr.: 79, fig. 300 (1987); U.K.W.F. ed. 2: 100 (1994); Vollesen in Fl. Eth. & Eritr. 2 (2): 250, fig. 82.26.7 (1995); Sivarajan & Pradeep, Malv. S. Pen. India: 256, fig. 93 (1996); Philcox in Rev. Fl. Ceylon 11: 358 (1997); Thulin, Fl. Somalia 2: 80, fig. 47A (1999). Type: India, Linn. Herb. 866-12 (LINN, lecto.) (microfiche!)

Erect annual or perennial herb or shrublet 0.3–1.5 m tall; stems shortly stellate-pubescent, with or without long simple hairs, eventually glabrous. Leaves broadly ovate to ± round, 3–8 × 1–8 cm, acute to rounded at the apex, ± cordate to rounded at base, stellate-pubescent above, thinly to thickly velvety tomentose beneath; venation impressed above; petiole 1–5.5 cm long; stipules filiform, 2.5 mm long. Flowers fasciculate, mostly at tips of branches or forming panicles; lower ones sometimes solitary and axillary; pedicel 0.4–2 cm long, usually jointed near apex; calyx 4–8 mm long, lobed to about middle, densely tomentose. Corolla yellow, orange, buff-apricot or pinkish cream, sometimes darker in middle; petals 7–12 mm long; staminal column hairy. Mericarps 8–10(–12), 3–5 mm long, the lower part of outer face and sides reticulate, with two retrorsely barbed awns 3–6 mm long.

subsp. **cordifolia**; Marais in K.B. 38: 42 (1983) & in Fl. Mascar. 51: 14, t. 3, fig. 3–4 (1987)

Stems with long simple spreading hairs extending far beyond the fairly short mostly dense stellate pubescence.

KENYA. Kwale District: Mariakani, 10 Sep. 1953, *Drummond & Hemsley* 4236!; Mombasa District: Port Tudor, *MacNaughton* 93!; Lamu District: Kipini, Oct. 1956, *Rawlins* 175!

TANZANIA. Mwanza, *Conrads* 931!; Uzaramo District: Dar es Salaam, Ilala Boma, 12 Mar. 1969, *Harris* 2850!; Zanzibar, Fumba, 28 Jan. 1929, *Greenway* 1225!

DISTR. **K** 7; **T** 1, 5, 6, 8?; **Z**; scattered throughout tropical Africa (see note); Yemen, throughout India, Pakistan to China, Japan, Malaysia, N Australia (one old specimen seen); Annobon, St. Helena, Ascension, Seychelles, Madagascar and Mascarenes

HAB. Very mixed bushland and thicket at edge of *Cynometra–Brachystegia* forest, evergreen forest with *Adansonia*, common weed of cultivation; 0–450(–1150) m

subsp. **maculata** (*Cav.*) *Marais* in K.B. 38: 43 (1983) & in Fl. Mascar. 51 Malvacées: 15 (1983). Type: Santo Domingo, Herb. *Deletang* no 8 (P–JU 12266, holo., MA, iso.)

Stems with dense fairly short velvety stellate pubescence only; long hairs absent.

UGANDA. West Nile District: Terego, Aug. 1938, *Hazel* 664!; Ankole District: Ishigiro, May 1939, *Purseglove* 698!; Teso District: Serere, May 1965, *Nyiira* s.n.!

KENYA. Kiambu District: Ruiru Station, 12 July 1952, *Kirrika* 210!; Kisumu-Londiani District: Kisumu, *Carroll* H36!; Kwale District: Shimba Hills Development Scheme, Kidongo, Farm No 194, 25 Dec. 1968, *Mwangangi* 1308!

TANZANIA. Ngara District: Bugufi, Murgwanza, 6 Dec. 1960, *Tanner* 5421!; Pangani District: Bushiri, 21 Dec. 1950, *Faulkner* 748!; Ufipa District: Tanzania border crossing at Kalambo R., 9 April 1969, *Sanane* 567!; Zanzibar, Chukwani, 7 Feb. 1960, *Faulkner* 2495!

DISTR. **U** 1–4; **K** 4–7; **T** 1–8; **Z**; **P**; very widespread throughout tropical and subtropical Africa; Madagascar, Indian and Pacific Ocean Islands, throughout tropical and subtropical America from Florida to North Argentina, West Indies

HAB. Grassland, bushland, open and secondary woodland, clearings in dry evergreen forest, old cultivations, roadsides; 0–1800 m

SYN. *S. maculata* Cav., Diss. 1: 20, t. 3, fig. 7 (1785)

S. althaeifolia Sw., Prod. Veg. Ind. Occ.: 101 (1788) (as *althaeaefolia*). Type: Sloane, Hist. Jam. 1: 218, t. 136, fig. 2 (1707)

S. africana P. Beauv., Fl. Oware & Benin 2: 87, t. 116 (1807). Type: Nigeria, *P. Beauvois* s.n. (G, holo.)

Abutilon velutinum G. Don, Gen. Syst. 1: 504 (1831). Type: São Tomé [St. Thomas], *G. Don* s.n. (BM!, holo.)

Sida doniana (G. Don) D. Dietr., Syn. Pl. 4: 857 (1847). Note: the reference in I.K. to *Abutilon indicum* is quite erroneous.

NOTE. Varietal rank might be preferable for an indumentum difference but it is very likely the two were once geographically separated and that var. *cordifolia* was introduced into Africa. *Bouton* s.n., collected on Zanzibar I. in 1837, has 5 bits of var. *cordifolia* and one of var. *maculata*. Some material from **T** 1, 6 and 8 is var. *cordifolia* but with much fewer long hairs; hybridisation may occur between the two subspecies. The correlation with flower colour suggested by Marais does not work very well in the Flora area. Of hundreds of specimens examined from Asia and Malaysia all but about half a dozen have been var. *cordifolia*. Two from Sumatra which are subsp. *maculata* bear a varietal name signifying epilose, attributed to Borssum Waalkes but this is not mentioned in his 1966 paper and was presumably not published. With the exception of a very old specimen all material from New Guinea and Australia is subsp. *maculata*. Of the very extensive material available from Africa 95% are subsp. *maculata* but scattered specimens from Ivory Coast, Ghana, Togo, Nigeria, Central African Republic, Cameroon, Congo-Kinshasa, Sudan, Angola, Zambia, Mozambique, Zimbabwe, Botswana, Namibia and South Africa (particularly the last three) are subsp. *cordifolia* or more frequently hybrids; some with only traces of *cordifolia* characters are further hybrids. I suspect relatively few introductions of var. *cordifolia* have been diluted in this way; significantly in Somalia material is entirely subsp. *cordifolia* exactly similar to that from India and obviously directly introduced from there. There are a few specimens with glabrescent stems in Namibia, China and America which are aberrant forms of either subspecies.

6. **Sida tenuicarpa** *Vollesen* in K.B. 41: 92, fig. 1 B (1986); Blundell, Wild Fl. E. Africa: 79 (1987); U.K.W.F. ed. 2: 100 (1994); Vollesen in Fl. Eth. & Eritr. 2 (2): 251, fig. 82.25.6–11 (1995). Type: Ethiopia, Gonder, Bahar Dar to Debre Marcos, *Getachew & Gilbert* 998 (K!, holo.; ETH, iso.)

Erect to spreading much-branched shrublet 0.3–1 m tall; stems sparsely to densely appressed stellate-pubescent. Leaves cuneate to obovate or oblanceolate, 4–18(–26) × 1.5–9 mm, notched at apex, cuneate at base, entire or rarely with 1–2 teeth on either side, ± glabrous above, sparsely stellate-pubescent; petiole 1.5–5 mm long;

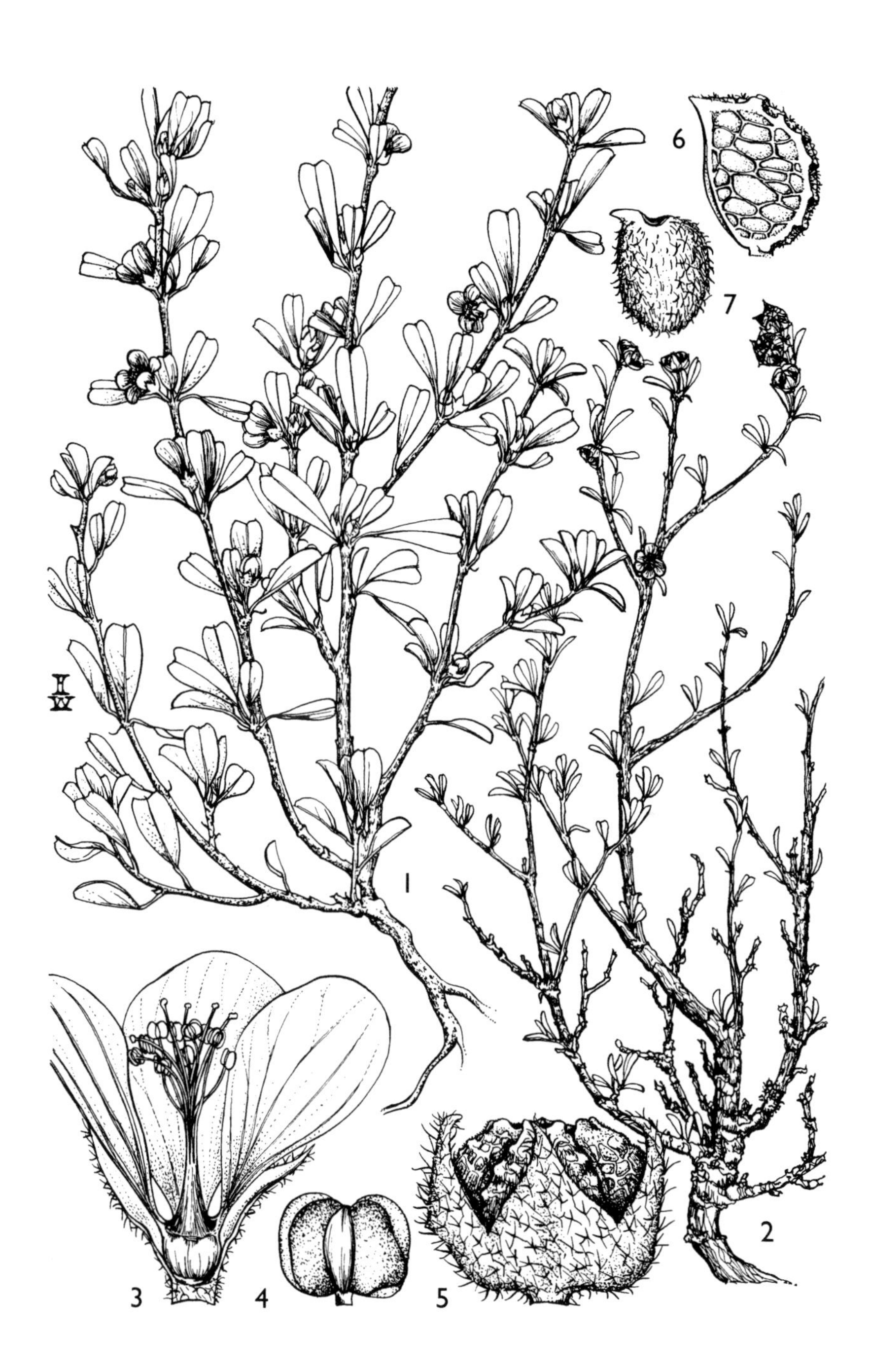

FIG. 27. *SIDA TENUICARPA* — **1**, habit (ungrazed), × 1; **2**, habit (grazed), × 1; **3**, flower (half of sepals and petals removed), × 5; **4**, anther, × 20; **5**, fruit (side view), × 7$^1/_2$; **6**, meriocarp (side view), × 7$^1/_2$; **7**, seed, × 7$^1/_2$. 1 from *Glover et al.* 2078; 2 from *Paulo* 753; 3–4 from *Battiscombe* 871; 5–7 from *Glover et al.* 1582. Drawn by Heather Wood.

stipules filiform to linear, 1–4 mm long. Flowers solitary and in up to 8-flowered terminal clusters at ends of branches; pedicel 1–3 mm long; calyx 3–4.5 mm long, appressed stellate-pubescent; lobes 2–3 mm long, 1 mm wide. Corolla yellow to orange-yellow, sometimes with reddish veins; petals 7–10 mm long with glabrous to ciliate claw; staminal column 2.5–3.5 mm long, glabrous to sparsely pubescent. Mericarps 5, 2–2.5 × 1.5–2 mm, with birostrate beak 0.5–1 mm long, hairy at apex, back smooth without sculpture, sides thin and papery soon completely disintegrating and releasing seed; seed 1.5–2 mm long, glabrous or with few hairs. Fig. 27, p. 151.

UGANDA. Toro District: Bukumbi, 22 Feb. 1932, *Hazel* 180!; Teso District: Serere, July 1932, *Chandler* 804!; Mengo District: Entebbe, Nov. 1932, *Eggeling* 716!

KENYA. Northern Frontier District: Mt Kulal, 3 km N of Gatab, 18 Nov. 1978, *Hepper & Jaeger* 6913!; Nairobi, 1919, *Battiscombe* in F.D. 871!; Masai District: Entasekera R., 11 July 1961, *Glover et al.* 2078!

TANZANIA. Musoma District: 16 km W of Klein's Camp, 12 Nov. 1953, *Tanner* 1824!; Kilimanjaro, Marangu, 16 Dec. 1963, *Archbold* 363!; Iringa District: Mufindi, Lugoda Tea Estate, 28 Jan. 1991, *Bidgood et al.* 1313!

DISTR. **U** 2–4; **K** 1, 3, 4–7 (see note); **T** 1, 2, 7; Congo-Kinshasa, Ethiopia

HAB. Dry short grassland, usually overgrazed, old cultivations, lakesides, riverine, forest glades, roadsides; (0 (see note)–)750–2150 m

SYN. *S. schimperiana* sensu Mast. in F.T.A. 1: 180 (1868); Oliv. in Trans Linn. Soc. Bot. ser. II, 2: 327 (1887); T.T.C.L.: 307 (1949) pro parte, *non* A. Rich.

S. cuneifolia sensu Cufod., E.P.A.: 543 (1959); Hauman in F.C.B. 10: 166 (1963); U.K.W.F.: 202 (1974); Blundell, Wild Fl. Kenya: 51 (1982) pro parte, *non* Roxb.

NOTE. *Tweedie* 2379 (Kenya, Kilifi District: near Sabaki R. ferry, June 1962) from riverine flats at ± sea level is the only specimen seen from **K** 7 or such a low altitude. Mrs. Tweedie mentions the flats were swept by floods in November. Vollesen suspected mislabelling but Mrs. Tweedie was extremely careful. Seed might have been carried a long way. The species has been cultivated for broom-making.

7. **Sida tanaensis** *Vollesen* in K.B. 41: 95, fig. 1C (1986); Blundell, Wild Fl. E. Afr.: 80 (1987); Thulin, Fl. Somalia 2: 80 (1999). Type: Kenya, Garissa District: Garissa–Jara Jila road 7 km from junction road Garissa/Hagadera, *Brenan, Gillett & Kanuri* 14746 (K!, holo.; EA, K!, iso.)

Low shrublet to 30 cm tall; stems with dense appressed to spreading stellate hairs. Leaves cuneate to obovate, 30–120 × 2–8 mm, entire or larger with 1–2 teeth on each side, stellate-hairy on both surfaces; petiole 1–3.5 mm long; stipules linear, 1.5–3 mm long. Flowers solitary or up to 4 in terminal clusters, subsessile or pedicel under 1 mm long; calyx 3–3.5 mm long, distinctly 10-ribbed, pubescent. Corolla yellow; petals 6–8 mm long with ciliate claw; staminal tube 1–2 mm long, glabrous or with scattered hairs at base. Mericarps 5, 2.5 mm long, 2 mm wide, ± glabrous and without sculpture, lower part of back and lower part of lateral walls thin and papery, breaking irregularly to release seed; beak under 0.5 mm long; seed ± 2 mm long, appressed crisped hairy.

KENYA. Northern Frontier District: Garissa–Jara Jila road 7 km from junction road Garissa/Hagadera, 26 Nov. 1978, *Brenan et al.* 14646! & Ijara, 30 Sep. 1957, *Greenway* 9253!; Tana River District: Tana River National Primate Reserve, middle road 0.5 km, 14 Mar. 1990, *Kabuye et al.* TPR 346!

DISTR. **K** 1, 7; Somalia

HAB. Open bushland, and scattered tree grassland with *Acacia–Commiphora* or more mixed *Combretum, Thespesia, Acacia, Commiphora* etc.; 30–250 m

8. **Sida schimperiana** *A. Rich.* in Tent. Fl. Abyss. 1: 66 (1847); Mast. in F.T.A. 1: 180 (1868) pro parte; P.O.A. C: 266 (1895); T.T.C.L.: 307 (1949) pro parte; Vollesen in K.B. 41: 95, fig. 1A (1986); Blundell, Wild Fl. E. Afr.: 79 (1987); U.K.W.F. ed. 2: 100 (1994); Vollesen in Fl. Eth. & Eritr. 2: 251, fig. 82.25.1–5 (1995). Type: Ethiopia, Tigré, between Maye Gouagoua [Mai-Gouagua] and Debra Sina, *Schimper* 1: 305, *Quartin Dillon & Petit* s.n. (P, syn.)

Intricately branched spreading or prostrate shrublet 10–50 cm tall; stems densely appressed stellate greyish hairy. Leaves narrowly cuneate, 3–10(–13) × 1–6 mm, notched at apex with a central tooth, otherwise entire, glabrous above, sparsely to densely appressed stellate-pubescent beneath; petiole 0.5–3.5 mm long; stipules linear to oblanceolate, 1–4 mm long. Flowers solitary or 2–4 congested at apices of shoots, subsessile or pedicel 1–2 mm long; calyx 2–4 mm long, sparsely to densely appressed-pubescent but tube often ± glabrous, lobes triangular, 2 × 1–2.5 mm. Corolla yellow to orange, sometimes with veins reddish; petals 4–8 × 3–4 mm, with glabrous claw; staminal tube 1–2.5 mm long, glabrous or sparsely pubescent. Mericarps 5, 2–2.5 × 1.5–2 mm, glabrous or with scattered appressed hairs, back reticulately sculptured, lateral angles muricate; sides reticulately sculptured disintegrating between the reticulations but not releasing seed, eventually opening slightly along ventral suture; beak birostrate, ± 0.5 mm long, the awns connivent; seed 2 mm long, appressed crisped hairy.

UGANDA. Mbale District: Mt Elgon, Kapchorwa, 9 Sep. 1954, *Lind* 292!

KENYA. Turkana District: Muruanysigar Peak, 24 Sep. 1963, *S. Paulo* 1022!; Naivasha District: SW of Lake Naivasha, near YMCA camp, 19 Apr. 1968, *Mwangangi* 767!; Masai District: about 17 km from Narok on Olokurto road, Orengitok, 17 May 1961, *Glover et al.* 1227!

TANZANIA. Musoma District: NE boundary 24 km from Naabi Hill, 10 May 1961, *Greenway et al.* 10155!; Masai District: East Serengeti, Lemuta, 20 July 1962, *Oteke* 195!; Lushoto District: West Usambaras, Kwai Valley, 25 Apr. 1953, *Drummond & Hemsley* 2246!

DISTR. **U** 3; **K** 1–7; **T** 1–3; Ethiopia

HAB. Grassland, grassland with *Tarchonanthus*, *Aloe* etc. also with *Acacia xanthophloea*, open *Olea-Acacia-Juniperus* woodland, sometimes in rocky places, overgrazed areas, roadsides; 1200–2700 m

SYN. *S. cuneifolia* sensu Cufod.: 543 (1959); Hauman in F.C.B. 10: 166 (1963) pro parte et auctt. mult., *non* Roxb.
S. sp. A sensu Agnew U.K.W.F.: 203 (1974) pro parte

NOTE. It is not possible to tell from some casual references which species is meant e.g. Jex-Blake (Gard. E. Afr. ed. 4: 126 (1957)) states *S. schimperiana* makes a good substitute for a 'Box edging'; Ivens (E. Afr. Weeds: 169, fig. 85) under *S. cuneifolia* is doubtless referring to both *S. tenuicarpa* and *S. schimperiana*.

9. **Sida massaica** *Vollesen* in K.B. 41: 97, fig. 1D (1986); U.K.W.F. ed. 2: 100 (1994). Type: Kenya, Nairobi, corner of Langata road and Uhuru Highway, *Kabuye & Ng'weno* 507 (K!, holo., EA, iso.)

Spreading subshrublet to 30 cm tall; stems whitish tomentellous. Leaves cuneate to broadly obovate, 3–12 × 2–8 mm, indented at apex with midrib projecting into emargination, entire or larger with 1–2 teeth per side, glabrous above, whitish tomentellous beneath; petiole 1.5–4 mm long; stipules filiform to linear, 1–3 mm long. Flowers solitary or up to 5 in terminal clusters; pedicel 1–2.5 mm long, pubescent; calyx 3–4.5 mm long, densely pubescent or tomentose. Corolla yellow; petals 7–9 mm long with claw sparsely ciliate; staminal tube 1.5–2 mm long, sparsely hairy at least at base. Mericarps 5, 3–3.5 × 2–2.5 mm, densely pubescent on back which is strongly reticulately sculptured and with two indistinct projections at transition from body to beak; lateral angles muricate; sides reticulate, thin and papery between ridges but not breaking, eventually opening slightly along ventral suture but not releasing seed; beak under 0.5 mm long; seed ± 2.5 mm long, appressed crisped hairy.

KENYA. Laikipia District: Rumuruti–Maralal road, 20 km S of Suguta Marmar, Ol Keju Losera dry river, 1 June 1979, *Gilbert et al.* 5423!; Nairobi District: 19 km E of Nairobi, Njiro Farm, 5 July 1951, *Bogdan* 3127!; Masai District: road from Keekorok gate to Narok, 16 Aug. 1971, *Kokwaro & Mathenge* 2727!

TANZANIA. Musoma District: Serengeti, Seronera, 23 Mar. 1961, *Greenway* 9883!; Maswa District: Serengeti, WNW of Lake Lgarya, 1 Oct. 1977, *Raynal* 19347!

DISTR. **K** 3, 4, 6; **T** 1; not known elsewhere

HAB. Especially grassland with *Acacia drepanolobium* etc. on black soil, *Acacia–Commiphora* bushland, dry stony riverbeds with *Pennisetum* etc.; 1500–1750(–1900) m

SYN. *S.* sp. A, Agnew in U.K.W.F. 203 (1974) pro parte

10. **Sida shinyangensis** *Vollesen* in K.B. 41: 97, fig. 1E (1986). Type: Tanzania, Shinyanga, near Ningwa R., *Burtt* 5610 (K!, holo., BM, iso.)

Low often spreading shrublet 7–30 cm tall; branchlets and leaves with dense appressed or slightly spreading stellate hairs. Leaves cuneate to obovate, 4–17 × 2–13 mm, glabrous above, the smaller entire, larger with 1–4 teeth on each side, the midrib produced beyond the apical emargination. Flowers solitary or in terminal heads of up to 5-flowers; pedicel 0–2 mm long, pubescent; calyx 3–4 mm long, pubescent especially on the lobes which also terminate with a few long hairs. Flowers yellow or white; petals 8–9 mm long with ciliate claw; staminal tube 2 mm long, glabrous near apex then with short hairs and longer hairs near base; filaments 1–2 mm long; style branches protruding 2–3 mm. Mericarps 5, 3–3.5 × 2.5 mm, ± glabrous to pubescent, back strongly reticulately sculptured and with a number of protuberances along the edges and two stronger ones near the base of the beak; lateral angles muricate; sides reticulate, thin between the ridges but not breaking, eventually opening slightly along ventral suture but not releasing the seed; beak birostrate, 0.5–1 mm long; seed ± 2.5 mm long, appressed crisped hairy.

TANZANIA. Mwanza District: Kalemera, 2 Apr. 1953, *Tanner* 1345!; Shinyanga District: near Shinyanga, *Bax* 93!; Mbulu District: Taringire National Park, road to Babati, 12 Feb. 1970, *Richards* 25364!

DISTR. **T** 1, 2; not known elsewhere

HAB. Short grassland on grey to black soil on hardpans; 1050–1350 m

SYN. *S. schimperiana* sensu Brenan, T.T.C.L.: 307 (1949) pro parte, *non* A. Rich.

11. **Sida alba** *L.*, Sp. Pl. ed. 2: 960 (1763); Exell & Mendonça, C.F.A. 1: 148 (1937); F.P.N.A. 1: 586 (1948); T.T.C.L.: 306 (1949); Keay, F.W.T.A. ed. 2, 1(2): 339 (1958); Meeuse in F.Z. 1: 477 (1961); Hauman in F.C.B. 10: 163 (1963); Abedin, Fl. W. Pak. 130 (Malvaceae): 81, fig. 19F (1979); Marais, Fl. Mascar. 51, Malvaceés: 11, t. 1, fig. 5 (1987); U.K.W.F. ed. 2: 101 (1994); Vollesen in Fl. Eth. & Eritr. 2 (2): 251, fig. 82.26.6 (1995); Wood, Handb. Yemen Fl.: 106 (1997); Thulin, Fl. Somalia 2: 80, fig. 47B (1999); Boulos, Fl. Egypt 2: 100, 321, t. 25, fig. 5 (2000). Type: Cult. Hort. Upsala, Linn. Herb. 866/2 (LINN, lecto.)

Annual, perennial or subshrubby herb 10–90 cm tall, often branched from base, the stems erect or slightly decumbent, densely stellate-pubescent, later glabrescent and often purplish brown. Leaves discolorous, elliptic or oblong, 0.8–5 × 0.5–3 cm, acute to rounded at the apex, truncate to rounded at base, regularly sharply serrate, dark green and thinly pubescent above, white or grey tomentose beneath; petiole 0.5–2.5(–4) cm long; stipules linear-subulate, 6–10 mm long, often with spurs on stem at their base. Flowers solitary in leaf axils or in 4–5-flowered clusters; pedicel 0.3–2.5 cm long; calyx 4–6 mm long, cupular, 10-ribbed, lobed to about the middle, pubescent, the lobes triangular, acute, apiculate. Corolla white, yellow or cream-buff; petals 4–6 mm long; staminal column 2 mm long, glabrous. Mericarps 5, 2–3 mm long excluding the birostrate beak with awns 2 mm long, puberulous; basal part of back thin, white and papery, breaking up to release the seed; sides thin walled, striate, the lateral edges distinct and straight.

UGANDA. Acholi District: Murchison Falls National Park, Chobi, 6 Sep. 1967, *Angus* 5871!; Toro District: Toro Game Reserve, 8 km SW of Sindikwa triangulation marker, 14 Dec. 1962, *Büchner* 17!; Busoga District: Kamuli, June 1926, *Maitland* s.n.!
KENYA. Northern Frontier District: Moyale, 10 Oct. 1952, *Gillett* 14029!; Baringo District: 16 km S of Lake Baringo, Perkerra Irrigation Scheme, 5 Jan. 1959, *Bogdan* 4734!; Fort Hall District: 16 km NE of Thika, 16 Aug. 1958, *Bogdan* 4606!
TANZANIA. Mwanza District: probably Ukerewe I., 5 May 1929, *Conrads* 927a!; Mbulu District: Lake Manyara National Park, Ndabash, 3 Mar. 1969, *Greenway & Kanuri* 11297!; Pangani District: Boza, 18 Jan. 1956, *Tanner* 2553!
DISTR. **U** 1–4; **K** 1–6; **T** 1–4, 6, 8; widespread in tropical and South Africa, Arabia, Mascarenes; also in India and America
HAB. Grassland, grassland with scattered trees and bushes, *Acacia* woodland, riverine flats, roadsides, weed in old cultivations, wheat etc., often on black cotton soil; 450–1900 m

SYN. *S. spinosa* L. var. ß L., Sp. Pl.: 684 (1753)
S. spinosa sensu Mast. in F.T.A. 1: 180 (1868) et auctt. mult., *non* L.

NOTE. I have followed Marais (1987) and Vollesen (1995) in keeping *S. alba* and *S. spinosa* separate but Sivarajan & Pradeep (Malv. S. Penins. India: 280 (1996)) and many others combine them.

12. **Sida acuta** *Burm. f.*, Fl. Ind.: 147 (1768); Rodrigo in Rev. Mus. de la Plata Bot. 6: 154 (1944); F.P.N.A. 1: 588 (1948); Keay, F.W.T.A. ed. 2, 1(2): 339 (1958); Meeuse in F.Z. 1: 477 (1961); Hauman in F.C.B. 10: 17 (1963); Haerdi in Acta Trop., Suppl. 8: 88 (1964); Borss. Waalk. in Blumea 14: 186 (1966); Marais, Fl. Mascar. 51 Malvacées: 11, t. 1, fig. 1 (1987); U.K.W.F. ed. 2: 101 (1994); Vollesen in Fl. Eth. & Eritr. 2 (2): 252, fig. 82.26.5 (1995); Sivarajan & Pradeep, Malv. S. Pen. India: 238 (1996); Thulin, Fl. Somalia 2: 80 (1999); Boulos, Fl. Egypt 2: 101, t. 25, fig. 5, 321 (2000); Krapovickas in Bonplandia 12: 87, fig. 1, 5c (2003). Type: Java, collector unknown, *Herb. Burman* s.n. (G, lecto.) (see Borss. Waalk.)

Perennial, annual or subshrubby herb 0.3–1(–1.5) (3 fide Meeuse) m tall; stems branched, puberulous. Leaves distichous, bright green, narrowly ovate or narrowly elliptic, 2–8.5(–10) × (0.5–)2–2.5(–3.8) cm, acute at the apex, obtuse to rounded at base, sharply toothed, ± glabrous or lower leaves sometimes pubescent; petiole 2.5–6 m long; stipules almost leafy, linear to lanceolate or falcate, up to 10 mm long, 1.5 mm wide, sometimes with those of a pair unequal, one linear and the other lanceolate (but not always in African material). Flowers solitary and axillary or less often few together in the axils; pedicel 2–12 mm long, puberulous; calyx 5–8 mm long, glabrous or ciliate, divided to middle into triangular acuminate lobes. Corolla creamy white, yellow, pale orange or apricot, sometimes darker orange at base; petals (4–)6–9 mm long; staminal column 1–2 mm long, pubescent and sometimes glandular. Mericarps (5–)6–8, 3–4 mm long, sparsely pubescent or ± glabrous, reticulate on back, striate on sides, breaking between the ridges in both areas; lateral edges sharp and muricate; awns rather stout, 1–1.5 mm long, diverging. Fig. 28, p. 156.

UGANDA. Teso District: Serere, Nov. 1932, *Chandler* 964!; Busoga District: 5 km W of Jinja, Buwenda, 15 Apr. 1987, *Rwaburindore* 2407!
KENYA. Nairobi, 10 Jan. 1950, *Bogdan* 2777!; Kisumu, 15 Dec. 1939, *Opiko* in *Bally* 693!; Kwale District: Mariakani, 10 Sep. 1953, *Drummond & Hemsley* 4237!
TANZANIA. Moshi District: Moshi–Arusha road, just after road to Machame, 15 Mar. 1955, *Huxley* 149!; Lushoto District: Korogwe, Magunga Estate, 13 June 1952, *Faulkner* 985!; Kilosa District: Mikumi National Park, Apr. 1967, *Procter* 3607!; Pemba: Chake–Chake, 12 Oct. 1929, *Vaughan* 782!

FIG. 28. *SIDA ACUTA* — **1**, habit, × 1; **2**, bud, × 4; **3**, sepals, × 4; **4**, flower (calyx removed), × 4; **5**, anther, × 32; **6**, pistil, × 4; **7**, **8**, seed head from above and side, × 4; **9**, **10**, fruit, different views, × 4; **11**, ovule, × 4; **12**, stipule, × 2; **13**, part of stem (older plant), × 1. 1–12 from *unknown collector* 1250; 13 from *Chandler* 964. Drawn by Dorothy R. Thompson.

DISTR. **U** 3, 4; **K** 4, 5, 7; **T** 2–6, 8; **Z**; **P**; pantropical
HAB. Grassland, bushland, forest edges, seasonally swampy areas on black cotton soil, riverine and lacustrine, stony sandy soil; weed of cultivation; 0–1650 m

SYN. *S. carpinifolia* sensu Mast. in F.T.A. 1: 180 (1868), *non* L. f.

NOTE. Borssum Waalkes treated *S. carpinifolia* L. f. as a subsp. of *S. acuta* and Rodrigo as a variety but it is a perfectly distinct species, formerly widespread in the Mascarene Is. Marais does not state that it occurs anywhere else but it is widely recorded in the tropics in the literature.

13. **Sida rhombifolia** *L.*, Sp. Pl.: 684 (1753); Mast. in F.T.A. 1: 181 (1868); K. Schum. in Fl. Brasil. 12 (3): 338 (1891); Fawcett & Rendle, Fl. Jamaica 5: 117, fig. 46 (1926); Exell & Mendonça C.F.A. 1: 149 (1937); Rodrigo in Rev. Mus. de la Plata, Bot. 6: 173, t. 28 (1944); T.T.C.L.: 307 (1949); Keay, F.W.T.A. ed 2, 1: 339 (1958); Meeuse in F.Z. 1: 480, t. 92A (1961); Hauman in F.A.C. 10: 172 (1963); Borssum Waalkes in Blumea 14: 193 (1966); Ugborogho in Bol. Soc. Bot. Ser. 2, 54: 65–85 (1980); Marais in K.B. 38: 41 (1983) pro parte and in Fl. Mascarenes 51 Malvacées: 15 (1987) pro parte; Vollesen in Fl. Eth. & Eritr. 2 (2): 254 (1995); Sivarajan & Pradeep, Malv. S. Pen. India: 245, figs. 82, 87, 109 A–B (1996); Boulos, Fl. Egypt 2: 101, t. 25, fig. 6 (2000); Verdcourt in K.B. 59: 233 (2004). Type: ? Jamaica, *Linn. Herb.* 886.3 (LINN!, lecto.) (left hand side fruiting specimen) (see note)

Perennial often subshrubby herb 0.3–2(–3) m tall; stems usually erect or ascending, stellate-pubescent or tomentose. Leaves oblanceolate to ovate, elliptic or rhombic (some Asiatic infraspecific taxa have obcuneate leaves with a retuse apex), 0.5–13 × 0.3–6 cm, rounded to acuminate at the apex, cuneate to rounded at base, finely stellate-pubescent, with simple hairs or almost glabrous above, usually more densely so beneath, the hairs often of different sizes (densely tomentose in some New World varieties), serrate except near base sometimes; petiole 0.2–2.5 cm long; stipules linear, 3–8(–12) mm long. Flowers solitary and/or in short axillary cymes sometimes aggregated at apex of shoots; pedicel 0.2–3.5(–6) cm long including up to 1.3 cm long joint; calyx 5–8 mm long, always 10-ribbed at base, the ribs distinctly yellowish, at least when dry, with scattered small and larger stellate hairs and sometimes long hairs, divided to middle into broadly ovate-triangular lobes, sometimes ciliate, glabrous inside. Petals white or pale yellow to apricot, sometimes reddish or purplish at base, 6–12 mm long; staminal column 1–3 mm long, with stalked glands and simple hairs. Mericarps (7–)8–14, (2–)3–3.5 mm long, ± smooth to slightly or strongly reticulate-ribbed on back and sides and the outer margins often ± serrate in some varieties, glabrous, glandular or puberulous, dehiscent or indehiscent, the beak part with 1–2 short or long awns or without awns but splitting and appearing birostrate but with nothing or scarcely anything projecting beyond the outline of the mericarp (in one South American variety the awns are very long, pubescent and flexuous).

1. Mericarps with a single awn, usually indehiscent .... c. var. **maderensis**
   Mericarps with two long or short awns or unawned but with beak part splitting into two; indehiscent to completely dehiscent 2
2. Mericarps without true awns projecting beyond the outline of mericarp but beak often splitting and easily mistaken for 2 awns 3
   Mericarps with 2 distinct short to quite long awns mostly 0.5–2 mm long 5
3. Mericarps ± smooth or with faint reticulation often totally dehiscing into two parts .... b. var. **serratifolia**
   Mericarps rugose, not splitting completely 4

4. Sepals with long simple hairs, mostly obvious on angles and margins .... e. var. **pethericki**
   Sepals without long simple hairs .... f. var. **afrorhomboidea**
5. Leaves finely stellate-puberulous above .... a. var. **riparia**
   Leaves more densely pubescent with simple hairs or mixture of stellate and ± simple (the simple hairs are reduced stellate hairs) .... d. var. **afroscabrida**

a. var. **riparia** *Burtt Davy* in Fl. Transvaal (2): 276, 277 (1932); Verdcourt in K.B. 59: 235, fig. 1a (2004). Type: South Africa, Transvaal, Rayton, *Rogers* 23562 (K!, lecto.) (see note)

Leaves finely stellate-puberulous above. Mericarps narrow with upper beak part often splitting but always bearing 2 short to quite long awns; back smooth or faintly reticulate, faces smooth or with faint sculpture particularly near margins.

UGANDA. Karamoja District: Pian, Lodoketeminit, 20 June 1962, *Kerfoot* 3814!; Kigezi District: Sabinio volcano, 12 Dec. 1930, *B.D. Burtt* 2974!; Mengo District: about 40 km SW of Jinja, Mariza Forest, near Nansagazi, 27 Nov. 1950, *Dawkins* 675!

KENYA. Northern Frontier District: Wamba, 1 Dec. 1958, *Newbould* 3000!; Nyeri District: 7 km on Nyeri–Kiganjo road, Zawadi Estate, Amboni R., 2 June 1974, *Faden et al.* 74/700!; Kakamega District: Iramba, Isukwa location, 12 July 1960, *Paulo* 542!

TANZANIA. Mbulu District: Marang Forest, Daudi, 4 May 1966, *Carmichael* 1199!; Lushoto District: Lushoto, 15 Aug. 1966, *Semsei* 4085!; Iringa District: Iringa, 19 Mar. 1932, *Lynes* 240!

DISTR. **U** 1, 2, 4; **K** 1, 3–6; **T** 1–3, 7; widespread in tropical and South Africa

HAB. Grassland, open woodland, forest glades; 900–2250 m

SYN. *S. rhombifolia* sensu auctt. afr. permult., *non* L. sensu stricto
S. grewioides sensu Oliv. in Trans. Linn. Soc. ser. 2, 2: 329 (1887), *non* Guill. & Perr.

NOTE. This is unlikely to be the first name at varietal rank for this taxon but there are many names with inadequate descriptions the types of which cannot be borrowed so it would be a lengthy expensive exercise.

Burtt Davy (Fl. Transvaal (1): 50 (1926)) mentions a new variety *S. rhombifolia* L. var. *riparia* Burtt Davy citing *S. riparia* Hochst. ex A. Rich., Tent. Fl. Abyss. 1: 65 (1847) as basionym but there it is only published in synonymy. Later (Fl. Transvaal (2): 276, 277 (1932)) he gives a brief descriptive phrase in the key and cites *Rogers* 22562 as well as referring back to p. 50. The lectotype is the Rogers Transvaal specimen. Actually the *Schimper* specimen cited by A. Richard is var. *maderensis*.

b. var. **serratifolia** (*Wilczek & Steyaert*) *Verdc.* in K.B. 59: 235, fig. 1/b–c (2004). Type: Congo-Kinshasa, Lower Shaba, Kaniama, *Mullenders* 199 (BR!, holo.)

Mericarps narrow with upper beak part without awns but splitting to form two ± parallel parts which scarcely project beyond the mericarp outline, often completely separating so that mericarp divides into thin halves and releases seed. Back and faces smooth or with very faint sculpture.

UGANDA. Kigezi District: Queen Elizabeth National Park, Ishasha R. camp, 15 May 1961, *Symes* 723!; Teso District: Serere, Aug. 1932, *Chandler* 861!; Masaka District: Kyotera, Oct. 1925, *Maitland* 1051!

KENYA. West Suk District: Kapenguria, 13 May 1932, *Napier* 1901!; Uasin Gishu District: Kipkarren, *Brodhurst Hill* 431!; Meru District: Tigania, Nov. 1955, *J. Adamson* 530!

TANZANIA. Kilosa District: Mugira Track, *Greenway & Kanuri* 15172!; Mbeya District: Mbozi, Mbimba Coffee Station, 22 May 1967, *Robertson* 577!; Songea District: about 32 km E of Songea, near Mkurira, *Milne-Redhead & Taylor* 8723!

DISTR. **U** ?1, 2–4; **K** 2–4; **T** 6, 7, 8; Congo-Kinshasa, Ethiopia, Malawi, Mozambique, Zimbabwe, Swaziland, South Africa

HAB. Grassland, open woodland with *Terminalia, Combretum, Julbernardia, Brachystegia* etc., sometimes in swampy places; 650–2100 m

SYN. *S. rhombifolia* sensu Oliv. in Trans Linn. Soc. 29: 34 (1872); U.K.W.F. ed. 2: 101 (1994) pro parte, *non* L.

S. serratifolia Wilczek & Steyaert in B.J.B.B. 22: 105 (1952); Meeuse in F.Z. 1: 481 (1961); Hauman in F.C.B. 10: 174 (1963); Vollesen in Fl. Eth. & Eritr. 2 (2): 254, fig. 82.27.8 (1995)
S. sp. B; Agnew, U.K.W. F. ed. 2: 101 (1994) pro parte quoad *Napier* 1901

NOTE. Some specimens intermediate with var. *riparia* do have minute projecting awns and those wishing to use specific rank for the taxa I am treating as varieties should note this. Hauman had reservations about *S. serratifolia* being distinct.

c. var. **maderensis** (*Lowe*) *Lowe*, Fl. Madeira: 69 (1868); K. Schum. in Fl. Brasil. 12 (3): 339 (1891); Verdc. in K.B. 59: 237, fig. 1d (2004). Type: Madeira, *Lowe* s.n. (K!, syn.)

Mericarps narrow with upper beak part bearing a single awn, indehiscent or beak part opening, the faces smooth or with some sculpturing.

KENYA. Northern Frontier District: Ndoto Mts, 8 June 1979, *Gilbert* et al. 5562! & 5602!
TANZANIA. Zanzibar: Ndagaa, 27 Aug. 1964, *Faulkner* 3423! & Kinyasini, 21 Jan. 1929, *Greenway* 1119!
DISTR. **K** 1; **Z**; Sudan, Ethiopia, Eritrea, eastern South Africa, Seychelles etc. and widespread throughout the tropics and in many places e.g. Java, Sumatra the commonest variant; it appears to be absent from mainland Australia and southern South America
HAB. Roadsides, plantation edges, thicket; 0–1300 m

SYN. *S. rhombifolia* sensu Willd., Sp. Pl. ed. 3: 740 (1800); Kunth, Nov. Gen. Sp. 5: 261 (1823); Griseb. in Ahh. Kon. Ges. Wiss. Göttingen 7: 179 (1857); K. Schum. in Fl. Brasil. 12 (3): 339 (1891) ('var. typica') pro parte; Rodrigo in Rev. Mus. de la Plata, Bot. 6: 176, fig. 38 (1944) ('var. typica'); Dupuy in Fl. Austr. 50: 154 (1993); Vollesen in Fl. Eth. & Eritr. 2 (2): 254 pro parte praesertim fig. 82. 27. 1–3 (1995)
S. maderensis Lowe in Trans. Cambr. Philos. Soc. 4: 35 (1831)
S. ostryifolia Webb, Fragm. Fl. Aeth. Aegypt.: 49 (1854), as *ostryaefolia*. Type: Sudan, Fazogli, *Figari* s.n. (FI–W, syn.), Ethiopia, Adua, *Schimper* 1105 (K!, syn.) & Cav., Diss. 5: 274, t. 134. fig. 1 (1788) (*carpinifolia* sensu Cav., *non* L.) (syn.)
S. rhombifolia L. var. *linnaeana* (i.e. typical variety); Griseb., Fl. Br. W. Indies: 74 (1859)
S. rhombifolia L. forma *aristata* Backer, Fl. Batavia 1: 101 (1907). Type: Java, *Backer* (BO, holo.)
S. unicornis Marais in K.B. 38: 42 (1983) & in Fl. Mascareignes 51, Malvacées: 15, t 2/9 (1987). Type: Mauritius, *Bouton* s.n. (K!, holo.)

NOTE. The Kenya material is much more pubescent than that from Zanzibar. I have examined material from throughout the range of the species and there is variation in the length of the awn, indumentum and whether the faces are smooth or rugose to reticulate. It is strange that Borssum Waalkes and Sivarajan and Pradeep do not mention 1-awned mericarps. Roxburgh drawing 1490 (K!) clearly shows 1-awned mericarps. Presumably *Malvinda unicornis* Dill., Hort. Elth. 216, t. 172 fig. 212 (1732), one of Linnaeus's syntypes of *S. rhombifolia*, belongs here.

d. var. **afroscabrida** *Verdc.* in K.B. 59: 237, fig. 1e (2004). Type: Kenya, Kiambu District: Mugutha R., *Kirrika* 194 (K!, holo., EA!, iso.)

Leaves ± densely hairy above with stellate and apparently simple hairs (actually probably always reduced stellate hairs). Mericarps as for var. *riparia*.

UGANDA. Toro District: South Kibale Forest, 9 Dec. 1938, *Loveridge* 203A!; Teso District: Serere, Sep. 1932, *Chandler* 908!; Masaka District: Kyotera, Nov. 1945, *Purseglove* 1846!
KENYA. Kiambu District: Mugutha R., 21 June 1952, *Kirrika* 194!; Nairobi District: Nairobi, Ainsworth Bridge, 16 Sep. 1971, *Mwangangi* 1803!; Masai District: Mara Masai Reserve, SSW of Narok, 17 Sep. 1947, *Bally* 5386!
TANZANIA. Musoma District: 24 km W of Klein's Camp, 12 Nov. 1953, *Tanner* 1798!; Lushoto District: Lushoto, 1 Sep. 1966, *Semsei* 4109!; Mpanda District: Kungwe-Mahali Peninsula, Kasoje, 16 July 1959, *Newbould & Harley* 4385!
DISTR. **U** 2–4; **K** 1–6; **T** 1–4, 6, 7; West Africa, Congo-Kinshasa, Burundi, Angola, Ethiopia, Malawi, Mozambique, Zimbabwe
HAB. Grassland, old forest clearings, often in marshy places or by rivers, alluvial black soil etc., abandoned cultivations and land resting fallow, roadsides also rocky places; 750–2400 m

SYN. *S. rhombifolia* L. var. γ; Keay, F.W.T.A. ed. 2, 1(2): 338, 339 (1958)
S. scabrida sensu annot. mult. exsicc. Herb. K, *non* Wight & Arn. nec *scabrida* sensu F.W.T.A. ed. 2, 1(2)

NOTE. Two covers of East African material labelled 'scabrida' had been sorted out from *S. rhombifolia* at Kew but these have different mericarps from those of *S. scabrida* Wight & Arn., Prodr. Fl. Pen. Ind. Or.: 57 (1834); Sivarajan & Pradeep, Malv. S. Pen. India: 250, fig. 89 (1996). Type: India, *Wight* 198 (K!, holo.). The Indian material has the sides and faces of the mericarps distinctly reticulately sculptured. Agnew has written on one of the covers for U.K.W.F. ed. 2 "included in *S. rhombifolia*".

S. scabrida sensu Keay in F.W.T.A. ed. 2, 1(2): 339 (1958) and Ugborogho in Bol. Soc. Brot. ser. 2a, 54: 100 (1980) is I think a different taxon from that of Wight & Arn.

e. var. **petherickii** *Verdc.* in K.B. 59: 238, fig. 1h (2004). Type: Uganda, West Nile District: Terego, *Hazel* 445 (K!, holo., KAW, iso.)

Calyx angular, mostly with lobe margins purplish and with long simple hairs on angles and margins. Mericarps small, rugose and reticulately sculptured on back and faces, indehiscent or with beak part splitting but with no true awns. Leaves ± glabrous above except near margins.

UGANDA. West Nile District: Terego, Mar. 1938, *Hazel* 445!; Busoga District: Busaba, July 1926, *Maitland* s.n.!
TANZANIA. Shinyanga, *Koritschoner* 2025A!
DISTR. **U** 1, 3; **T** 1; West Africa, widespread in Sudan, Ethiopia
HAB. Roadsides; 900–1300 m

SYN. *S. rhombifolia* sensu Andrews, F.P.S. 2: 41 (1952) pro parte, *non* L.
S. rhombifolia var. α; Keay, F.W.T.A. ed. 2, 1(2): 338, 339 (1958)
S. sp. 10; Vollesen in Fl. Eth. & Eritr. 2 (2): 252 (1995)

NOTE. Hazel records the flower colour as brownish red; other collectors state white with slight pinkish cream tint. This appears to have first been discovered by *Petherick* in the Sudan in Aug. 1862. Material of this taxon from Togo has been annotated *S. rhombifolia* subsp. *alnifolia* (L.) Ugborogho. This does have the same type of mericarp but very different foliage.

f. var. **afrorhomboidea** *Verdc.* in K.B. 59: 238, fig. 1/f–g (2004). Type: Tanzania, Lushoto District: Mnyussi, Gereza East, Manta, *Semsei* 1831 (K!, holo., EA, iso.)

Leaves almost glabrous above or very finely stellate-puberulous. Mericarps similar to var. *petherickii* but calyx lobes and angles without long simple hairs but fine stellate pubescence only.

KENYA. Kitui District: 5 km from Kitui on the Mutomo road, 17 Nov. 1979, *Gatheri et al.* 79/41!; Kwale District: Matuga, 25 Oct. 1982, *Robertson* 3439!; Mombasa township, 17 July 1953, *Templer* 6!
TANZANIA. Pangani District: Madanga, Kumbantoni, 27 Apr. 1956, *Tanner* 2804!; Kigoma, 20 Dec. 1971, *Harris* 6084!; Morogoro District: Turiani, Aug. 1964, *Shabani in Procter* 2776!
DISTR. **K** 4, 6; **T** 3, 4, 6; not known elsewhere
HAB. Waste ground, grassland, sandy lake beaches, abandoned cultivations, roadsides, weed in cassava etc; ± 0–1150(–2100) m

NOTE. Sivarajan & Pradeep (Malv. S. Pen. India: 246 (1996)) maintain *S. rhomboidea* Fleming as a distinct species and if this is followed then my varieties e and f would undoubtedly have to be transferred to it.

GENERAL NOTE. Borssum Waalkes chose Herb. *Clifford* 346/1 (BM) as the lectotype which he examined stating the mericarps had two short awns. Marais also gives this as the lectotype and also saw it. Rodrigo, however, gives a photograph of the Linnean specimen and captions it 'tipo'. This must be considered an earlier lectotypification, although he did not see it and totally misinterpreted it.

In order to try and understand this species I looked at every specimen preserved at Kew throughout its entire range–many hundreds of specimens. The pattern of varieties is much the same in most areas. I have examined the Linnean lectotype (which I suspect came from Jamaica) and it has somewhat rugose mericarps with two distinct awns but is too immature to show dehiscence. The third variety has narrower mericarps, mostly smooth and indehiscent and bearing a single usually stout rigid pungent awn which can be long or short; in some areas e.g. Java, Sumatra and other parts of Malesia it is the predominant variety and has on several occasions been described as a species but also frequently considered the typical variety. De Candolle (Prodr. 1: 462 (1824)) is one of the few early authors who associated 2-awned mericarps with *S. rhombifolia*. The second variety has the upper part dehiscent so that the two halves resemble beaks but do not actually protrude beyond the body of the mericarp. Forms of this variety have very short to short true awns protruding from near the ends of dehiscent edges and there are intermediates with the typical variety which is similarly dehiscent but with long awns. Individual plants are constant in the mericarp characters. It seems unlikely that the present very mixed distributions are due to introductions to and from areas which originally had only one variety. K. Schumann gives the most exhaustive account of the species with an extensive synonymy.

Borssum Waalkes has two subspecies *rhombifolia* and *retusa* L. (Borssum Waalkes) (type: India, *Linnean Herbarium* 866/7 (LINN, lecto.)) in which he includes *S. alnifolia* L. (type: Sri Lanka, *Hermann* 3: 4 (BM, lecto.). Marais recognised both *S. retusa* L. and *S. alnifolia* L. as separate species.

Ugborogho recognises both as subsp. of *S. rhombifolia* and records both from Nigeria but cites the first '*Sida rhombifolia* subsp. *retusa* (L.). Ugborogho comb. nov., *non* sensu Borssum Waalkes'. This of course is not possible; Borssum Waalkes's combination is the valid one and even if he has misunderstood the Linnean type his combination is based on it.

Several specimens from **U** 2, 3; **K** 2–4, 6; **T** 3 have been annotated *S. rhombifolia* sensu lato since no nutlets are available.

14. **Sida chrysantha** *Ulbr.* in E.J. 51: 46, fig. 2/K–T (1913); Burtt Davy, Man. Fl. Pl. Ferns Transvaal 2: 277 (1932); Meeuse in F.Z. 1: 481 (1961); Vollesen in Opera Bot. 59: 36 (1980). Type: Namibia, Hereroland, North foothills of Auas Mts, *Dinter* 1877 (B†, holo.)

Perennial spreading or prostrate herb or much branched subshrub 30–70 cm tall or long; stems branched, shortly mostly densely velvety pubescent. Leaves narrowly to broadly ovate-oblong or oblong, 1–4 × 0.5–2.5 cm, ± acute to rounded at the apex, cuneate to ± rounded at base, rather coarsely serrate, darker green above with sparse to ± dense stellate pubescence, usually much paler beneath, densely to very densely velvety stellate-pubescent and with prominent venation; petiole 2–10 mm long, stellate-pubescent; stipules ovate-lanceolate to lanceolate, 2–4 × 1 mm, pubescent. Flowers solitary, axillary; pedicel 1.3–6 cm long, much longer than subtending petiole, pubescent, jointed near apex; calyx obconic-campanulate, ± 8 mm long, densely stellate-tomentose outside, base strongly ribbed and drying yellowish, divided to ± middle into broadly triangular lobes, 5 mm wide at base, acute to acuminate at apex. Corolla yellow to orange or white; petals 10 × 8 mm; staminal tube with ± sparse stellate hairs. Mericarps 6–8(–9), 4 × 3 mm, subacute (the awns totally fused or obscure) and stellate-pubescent at apex, reticulately veined on back and flat sides, not splitting at base, the awn beaks totally fused or often separating into two divisions resembling thick awns and often each with a short true awn at end.

KENYA. Embu District: Gathigiriri, 9 Jan. 1968, *Kabuye* 142!; Tana River District: Karawa, June 1959, *Rawlins* 767!; Kwale District: Mariakani, 10 Sep. 1953, *Drummond & Hemsley* 4240!

TANZANIA. Lushoto District: Mkomazi Game Reserve, Kamakota Hill, 13 June 1996, *Abdallah et al.* 96/210!; Uzaramo District: 16 km NNW of Dar es Salaam, Kunduchi, 14 Sep. 1970, *Harris & Tadros* 5018!; Kilwa District: Kingupira, 10 June 1975, *Vollesen* MRC 2422!; Zanzibar, Sep. 1873, *Hildebrandt* 910!

DISTR. **K** ?1, 4, 7; **T** 3, 6, 8; **Z**; Zimbabwe, Botswana, Namibia and South Africa

HAB. Coastal grassland, wooded grassland and bushland with *Combretum*, *Xeroderris*, *Pteleopsis*, *Terminalia* etc. on rocky hills above *Commiphora–Acacia* bushland, roadsides, also on black cotton soil in old cultivations; 0–1400 m

NOTE. A distinctive totally prostrate plant (*Luke* 6144, Lamu District: Kiwayu KWS/WWF Camp, 2 Jan. 2000) which the collector could not match at EA and I first thought was a subspecies of *S. callantha* Thulin from S Somalia, is I think only a form of *S. chrysantha*. Rather similar prostrate forms occur in Tanzania.

The BM duplicate of *Hildebrandt* 910 bears a pencil annotation '*Sida longipes* E.Mey.? fide Garcke' but I have not discovered if this determination was published. *Sida longipes* E.Mey. ex Harv., *non* A. Gray is now called *S. dregei* Burtt Davy.

15. **Sida ovata** *Forssk.*, Fl. Aegypt.-Arab.: 124 (1775); T.T.C.L.: 306 (1949); Keay, F.W.T.A. ed. 2, 1(2): 339 (1958); Meeuse in F.Z. 1: 479 (1961); Hauman in F.C.B. 10: 75, t. 16 (1963); Ivens, E. Afr. Weeds: 169, 171 (1967); Abedin, Fl. W. Pak. 130 (Malvaceae): 86, fig. 21, C, D (1979); Hepper & Friis, Pl. Forssk. Fl. Aegypt.-Arab.: 199 (1994); Vollesen in Fl. Eth. & Eritr. 2 (2): 254, fig. 82.27.4–7 (1995); Sivarajan & Pradeep, Malv. S. Pen. India: 243, fig. 86 (1996); Wood, Handb. Fl. Yemen: 106 (1997); Thulin, Fl. Somalia 2: 81, fig. 47, O, G (1999); Boulos, Fl. Egypt 2: 101, fig. 7 (2000). Type: Yemen, Wadid Surdud, *Forsskål* 1728 (C, lecto.)

Erect or spreading rather woody herb or subshrub 0.2–1.2 m tall, finely to densely pubescent and sometimes with long simple hairs. Leaves usually ± round to elliptic, less often oblong, 1–5.5(–7.5) × 1–4(–5) cm wide, retuse to subacute at the apex, rounded at base, dentate; petiole 0.3–1.2 cm long; stipules linear-lanceolate, 3–5 mm long. Flowers solitary or sometimes paired; pedicel 0.2–1.5(–3) cm long, usually short, rarely some longer ones towards base; calyx 5–9 mm long, 10-ribbed at base; lobes triangular, 2.5–5 mm long, acute. Corolla white or pale yellow to orange; petals (5–)8–11 mm long; staminal column densely stellate-pubescent. Mericarps 7–8(–9), 3–3.5 mm long, puberulous to pubescent and glandular, dorsally rugose to ± reticulately sculptured, laterally weakly to distinctly sculptured, the margins distinct, muricate; awns 2, up to 1 mm long.

UGANDA. West Nile District: between Omugo Rest House and Oru R., 10 Aug. 1953, *Chancellor* 156!; Toro District: Kasese, Dec. 1926, *Maitland* 998!; Busoga, July 1926, *Maitland* 1083!

KENYA. West Suk District: 19 km N of Kacheliba, 7 Oct. 1964, *Leippert* 5042!; Machakos District: 5–6 km N of Nunguni, E side of mountain, Kithembe Village, 10 May 1968, *Mwangangi* 875!; Teita District: Tsavo National Park East, Voi Gate W to pipeline ± km 6, 11 Jan. 1967, *Greenway & Kanuri* 13022!

TANZANIA. Shinyanga District: Old Shinyanga, 11 June 1931, *B.D. Burtt* 2535!; Mbulu District: Lake Manyara National Park, Mbagaya to Ndabash, 2 Mar. 1964, *Greenway & Kanuri* 11285!; Iringa District: between Mdonya R. and Ibuguziwa track, 6 Apr. 1970, *Greenway & Kanuri* 14280!

DISTR. **U** 1–3; **K** 1–4, 6, 7; **T** 1–7; widespread in drier areas of tropical Africa and South Africa extending to Egypt, Arabia and India, Iran and Pakistan

HAB. Grassland (often overgrazed), grassland with scattered *Acacia, Olea* etc; *Acacia–Commiphora–Tarchonanthus–Salvadora* bushland and thicket on gravelly, sandy-alluvial, hard pan and black cotton soils; often in rocky places, wadi edges, also a weed in pasture and old cultivations; 5–2350 m

SYN. *S. grewioides* Guill. & Perr. in Fl. Seneg. Tent.: 71 (1830); Mast. in F.T.A. 1: 182 (1868); F.P.N.A. 1: 586, t. 58, fig. 28 (1948). Type: Senegal, Walo, *Perrottet* (P, holo., BM!, ?iso.)

NOTE. Vollesen includes in his synonymy *S. abyssinica* Hochst. ex Dietr. (Syn. Pl. 4: 359 (1847)) with *Schimper* III: 1453 (K, P) from Ethiopia, Djeladjeranne as type but Dietrich does not mention Hochstetter or Schimper but cites *S. patens* Andr. (see under *Abutilon* p. 139) of which *S. abyssinica* must be an illegitimate synonym.

17. SIDASTRUM

Baker f. in J. Bot. 30: 137 (1892); Fryxell in Brittonia 30: 449 (1978)

Erect ± shrubby herbs similar to *Sida* but differing in the ± reduced flowers often in extensive terminal panicles; pseudo-epicalyxs formed of stipules and leaf primordia present; true epicalyxs absent. Calyces ecostate. Androecium reduced with sometimes only 10 anthers (in one species with only 5 filaments with paired anthers). Stigmas and carpels 5–10. Mericarps thin-walled and 1-celled with little or no ornamentation.

FIG. 29. *SIDASTRUM PANICULATUM* — **1**, habit, × $^{2}/_{3}$; **2**, flower side view, × 5; **3**, styles, stigmas and stamens, × 8; **4**, back of anther and filament, × 20; **5**, fruit in calyx, × 5; **6**, carpels from above, × 5; **7**, carpel from side, × 6; **8**, seeds, × 6. All from *de Saeger* 1459. Drawn by Juliet Williamson.

A genus of 7 species in the New World from northern Argentine to southern U.S.A. and West Indies. One species occurs in tropical Africa and is also naturalised in Hawai.

Sidastrum paniculatum (*L.*) *Fryxell* in Brittonia 30: 453 (1978). Type: Jamaica, *P. Browne* Herb. Linn. 866. 17 (LINN, lecto.)

Perennial herb, woody at base, 0.9–2.7 m tall; stems and branchlets covered with short yellowish stellate hairs, ultimately glabrescent. Leaves lanceolate to ovate, 1.5–10 × 0.5–8.5 cm, ± acuminate at the apex, truncate, rounded or subcordate at base, unequally serrate, 5–7-veined, pubescent with stellate hairs; petiole 0.1–7 cm long; stipules filiform, 5–8 mm long. Flowers at first solitary on long slender peduncles in upper axils, followed generally by a flowering branch in each axil ultimately forming a large terminal leafy panicle up to 30 cm long; calyx 2.5–3 mm long, stellate-tomentose; lobes triangular. Petals crimson, purple or dark reddish brown, 5 × 1.5 mm wide; stamens ± 17; anthers dark-spotted. Carpels 5, 2.5–3.5 mm long, the valves acute at apex with 1–2 short beaks, stellate-pubescent; seeds purplish brown, 1.5 mm long with very minute scattered appressed hairs. Fig. 29, p. 163.

UGANDA. West Nile District: Koboko, July 1938, *Hazel* 653!
DISTR. **U** 1; Congo-Kinshasa, Sudan, Angola; tropical and subtropical Americas, W Indies
HAB. 'Scrub land'; ± 900 m

SYN. *Sida paniculata* L., Syst. Nat. ed. 10: 1145 (1759); Fawcett & Rendle, Fl. Jamaica 5: 114 (1926); Exell in C.F.A. 1: 373 (1951); F.P.S. 2: 42 (1952); Adams, Fl. Jamaica: 465 (1972); Friis & Vollesen, Vasc. Pl. Imatong Mts: 132 (1998)
S. schweinfurthii Baker f. in J.B. 30: 295 (1892). Type: Sudan, Mongolongbo, *Schweinfurth* 2916 (BM, holo., K!, iso.)

NOTE. The status of this plant in Africa is not completely certain. Fryxell does not mention it occurs in Africa. Exell says the Angolan record was an introduction. Schweinfurth's plant was collected in 1870 in an area where it was unlikely to be introduced. It certainly has not spread as most introduced plants have done. Vollesen considers it is native in Africa.

INDEX TO MALVACEAE

New names validated in this part

Abutilon eggelingii *Verdc.* **sp. nov.**
Abutilon fruticosum *Guill. & Perr.* var. **hepperi** *Verdc.* **var. nov.**
Abutilon grandiflorum *Don* var. **iringense** *Verdc.* **var. nov.**
Abutilon longicuspe *A. Rich.* var. **cecilii** (*N.E.Br.*) *Verdc.* **comb. nov.**
Abutilon longicuspe *A. Rich.* var. **pilosicalyx** *Verdc.* **var. nov.**
Abutilon mauritianum (*Jacq.*) *Medik.* var. **brevicalyx** *Verdc.* **var. nov.**
Abutilon mauritianum (*Jacq.*) *Medik.* var. **cabrae** (*De Wild. & T. Durand*) *Verdc.* **comb. et stat. nov.**
Abutilon mauritianum (*Jacq.*) *Medik.* var. **epilosum** *Verdc.* **var. nov.**
Abutilon pannosum (*Forst. f.*) *Schltdl.* var. **figarianum** (*Webb*) *Verdc.* **var. et stat. nov.**
Abutilon mauritianum (*Jacq.*) *Medik.* var. **grandiflorum** *Verdc.* **var. nov.**
Abutilon pannosum (*Forst. f.*) *Schltdl.* var. **scabrum** *Verdc.* **var. nov.**
Abutilon mauritianum (*Jacq.*) *Medik.* subsp. **zanzibaricum** (*Mast.*) *Verdc.* **comb. nov.**
Abutilon pilosicalyx *Verdc.* **sp. nov.**
Abutilon subprostratum *Verdc.* **sp. nov.**
Hibiscus fuscus *Garcke* subsp. **naivashense** *Mwachala* **subsp. nov.**
Hibiscus holstii *Mwachala* **sp. nov.**
Hibiscus kabuyeana *Mwachala* **sp. nov.**
Hibiscus masasiana *Mwachala* **sp. nov.**
Pavonia arenaria (*Murr.*) *Roth* var. **microphylla** (*Ulbr.*) *Verdc.* **comb. nov.**
Pavonia dimorphostemon *Verdc.* **sp. nov.**
Pavonia flavoferruginea (*Forssk.*) *Hepper & Wood* var. **microphylla** *Verdc.* **var. nov.**
Pavonia urens *Cav.* var. **hanangensis** *Verdc.* **var. nov.**
Pavonia urens *Cav.* var. **irakuensis** (*Ulbr.*) *Verdc.* **comb. nov.**
Roifia *Verdc.*, **gen. nov.**
Roifia dictyocarpa (*Webb*) *Verdc.* **comb. nov.**
Thespesia garckeana *F. Hoffm.* var. **schliebenii** *Verdc.* **var. nov.**

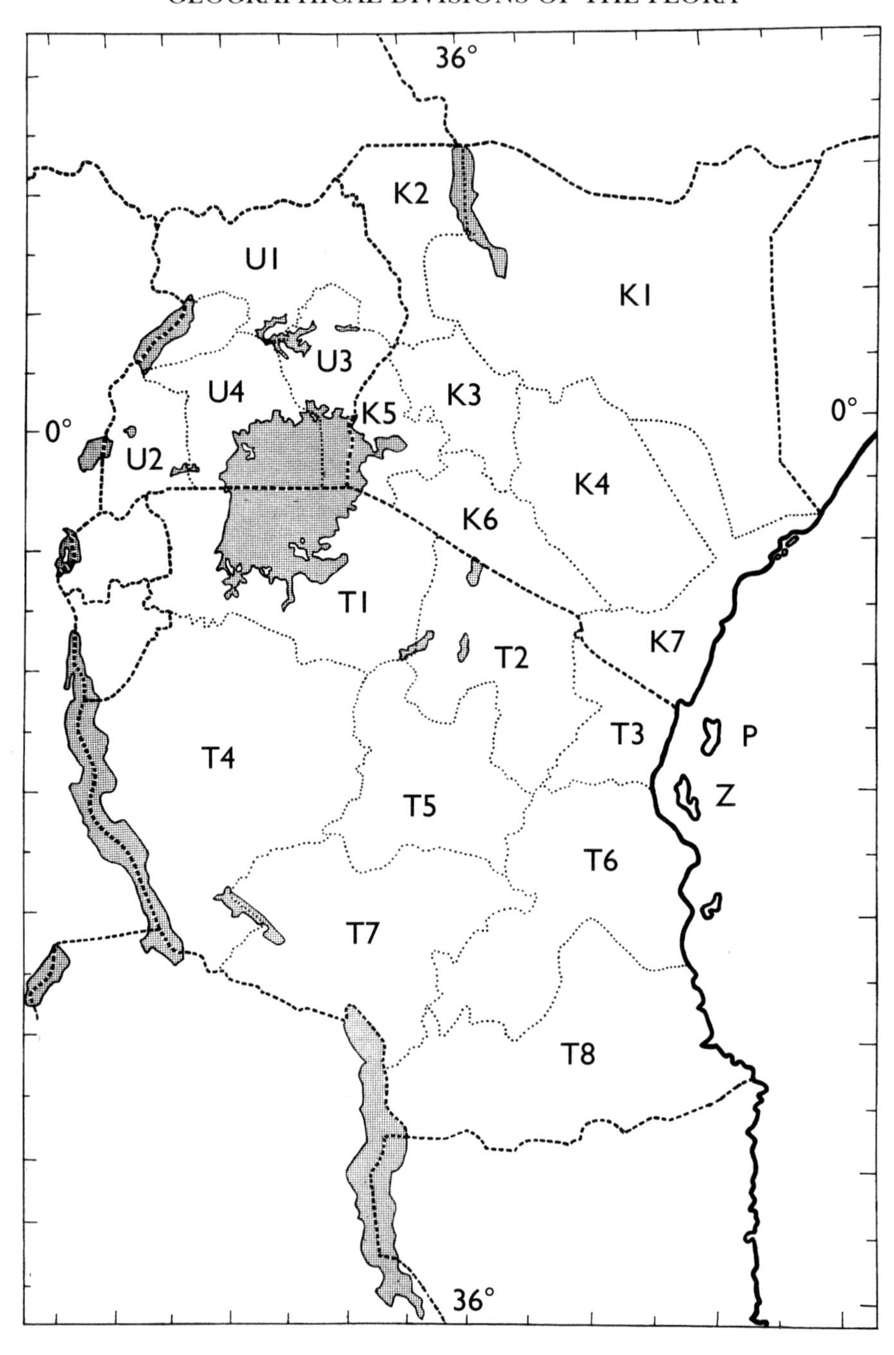
36°
K2
U1
K1
U3
U4
K3
K5
0°
0°
U2
K4
K6
T1
K7
T2
T3
P
T4
Z
T5
T6
T7
T8
36°

First published in 2009 by
Royal Botanic Gardens, Kew
Richmond, Surrey, TW9 3AB, UK
www.kew.org

ISBN 978 1 84246 189 1

British Library Cataloguing in Publication Data
A catalogue record for this book is available from the British Library

Design and typesetting by Margaret Newman,
Kew Publishing, Royal Botanic Gardens, Kew.

For information or to purchase all Kew titles please visit
www.kewbooks.com or email publishing@kew.org

All proceeds go to support Kew's work in saving the world's plants for life

LIST OF ABBREVIATIONS

A.V.P. = O. Hedberg, Afroalpine Vascular Plants; **B.J.B.B.** = Bulletin du Jardin Botanique de l'Etat, Bruxelles; Bulletin du Jardin Botanique Nationale de Belgique; **B.S.B.B.** = Bulletin de la Société Royale de Botanique de Belgique; **C.F.A.** = Conspectus Florae Angolensis; **E.J.** = A. Engler, Botanische Jahrbücher für Systematik, Pflanzengeschichte und Pflanzengeographie; **E.M.** = A. Engler, Monographieen Afrikanischer Pflanzen-Familien und Gattungen; **E.P.** = A. Engler, Das Pflanzenreich; **E.P.A.** = G. Cufodontis, Enumeratio Plantarum Aethiopiae Spermatophyta; in B.J.B.B. 23, Suppl. (1953) et seq.; **E. & P. Pf.** = A. Engler & K. Prantl, Die Natürlichen Pflanzenfamilien; **F.A.C.** = Flore d'Afrique Centrale (*formerly* F.C.B.); **F.C.B.** = Flore du Congo Belge et du Ruanda-Urundi; Flore du Congo, du Rwanda et du Burundi; **F.E.E.** = Flora of Ethiopia & Eritrea; **F.D.-O.A.** = A. Peter, Flora von Deutsch-Ostafrika; **F.F.N.R.** = F. White, Forest Flora of Northern Rhodesia; **F.P.N.A.** = W. Robyns, Flore des Spermatophytes du Parc National Albert; **F.P.S.** = F.W. Andrews, Flowering Plants of the Anglo-Egyptian Sudan *or* Flowering Plants of the Sudan; **F.P.U.** = E. Lind & A. Tallantire, Some Common Flowering Plants of Uganda; **F.R.** = F. Fedde, Repertorium Speciorum Novarum Regni Vegetabilis; **F.S.A.** = Flora of Southern Africa; **F.T.A.** = Flora of Tropical Africa; **F.W.T.A.** = Flora of West Tropical Africa; **F.Z.** = Flora Zambesiaca; **G.F.P.** = J. Hutchinson, The Genera of Flowering Plants; **G.P.** = G. Bentham & J.D. Hooker, Genera Plantarum; **G.T.** = D.M. Napper, Grasses of Tanganyika; **I.G.U.** = K.W. Harker & D.M. Napper, An Illustrated Guide to the Grasses of Uganda; **I.T.U.** = W.J. Eggeling, Indigenous Trees of the Uganda Protectorate; **J.B.** = Journal of Botany; **J.L.S.** = Journal of the Linnean Society of London, Botany; **K.B.** = Kew Bulletin, *or* Bulletin of Miscellaneous Information, Kew; **K.T.S.** = I. Dale & P.J. Greenway, Kenya Trees and Shrubs; **K.T.S.L.** = H.J. Beentje, Kenya Trees, Shrubs and Lianas; **L.T.A.** = E.G. Baker, Leguminosae of Tropical Africa; **N.B.G.B.** = Notizblatt des Botanischen Gartens und Museums zu Berlin-Dahlem; **P.O.A.** = A. Engler, Die Pflanzenwelt Ost-Afrikas und der Nachbargebiete; **R.K.G.** = A.V. Bogdan, A Revised List of Kenya Grasses; **T.S.K.** = E. Battiscombe, Trees and Shrubs of Kenya Colony; **T.T.C.L.** = J.P.M. Brenan, Check-lists of the Forest Trees and Shrubs of the British Empire no. 5, part II, Tanganyika Territory; **U.K.W.F.** = A.D.Q. Agnew (or for ed. 2, A.D.Q. Agnew & S. Agnew), Upland Kenya Wild Flowers; **U.O.P.Z.** = R.O. Williams, Useful and Ornamental Plants in Zanzibar and Pemba; **V.E.** = A. Engler & O. Drude, Die Vegetation der Erde, IX, Pflanzenwelt Afrikas; **W.F.K.** = A.J. Jex-Blake, Some Wild Flowers of Kenya; **Z.A.E.** = Wissenschaftliche Ergebnisse der Deutschen Zentral-Afrika-Expedition 1907–1908, 2 (Botanik).

FAMILIES OF VASCULAR PLANTS REPRESENTED IN THE FLORA OF TROPICAL EAST AFRICA

The family system used in the Flora has diverged in some respects from that now in use at Kew and the herbaria in East Africa. The accepted family name of a synonym or alternative is indicated by the word "see". Included family names are referred to the one used in the Flora by "in" if in accordance with the current system, and "as" if not. Where two families are included in one fascicle the subsidiary family is referred to the main family by "with".

PUBLISHED PARTS

Foreword and preface
*Glossary
Index of Collecting Localities

Acanthaceae
Part 1
*Actiniopteridaceae
*Adiantaceae
Aizoaceae
Alangiaceae
Alismataceae
*Alliaceae
*Aloaceae
*Amaranthaceae
*Amaryllidaceae
*Anacardiaceae
*Ancistrocladaceae
Anisophyllaceae — as Rhizophoraceae
Annonaceae
*Anthericaceae
Apiaceae — see Umbelliferae
Apocynaceae
*Part 1
*Aponogetonaceae
Aquifoliaceae
*Araceae
Araliaceae
Arecaceae — see Palmae
*Aristolochiaceae
Asparagaceae
*Asphodelaceae
Aspleniaceae
Asteraceae — see Compositae
Avicenniaceae — as Verbenaceae
*Azollaceae

*Balanitaceae
*Balanophoraceae
*Balsaminaceae
Basellaceae
Begoniaceae
Berberidaceae
Bignoniaceae
Bischofiaceae — in Euphorbiaceae
Bixaceae
Blechnaceae
*Bombacaceae
*Boraginaceae
Brassicaceae — see Cruciferae
Brexiaceae
Buddlejaceae — as Loganiaceae
*Burmanniaceae
*Burseraceae
Butomaceae
Buxaceae

Cabombaceae
Cactaceae
Caesalpiniaceae — in Leguminosae
*Callitrichaceae
Campanulaceae
Canellaceae
Cannabaceae
Cannaceae — with Musaceae
Capparaceae
Caprifoliaceae
Caricaceae
Caryophyllaceae
*Casuarinaceae
Cecropiaceae — with Moraceae
*Celastraceae
*Ceratophyllaceae
Chenopodiaceae
Chrysobalanaceae — as Rosaceae
Clusiaceae — see Guttiferae
Cobaeaceae — with Bignoniaceae
Cochlospermaceae

Papaveraceae
Papilionaceae — in Leguminosae
*Parkeriaceae
Passifloraceae
Pedaliaceae
Periplocaceae — see Apocynaceae (Part 2)
Phytolaccaceae
*Piperaceae
Pittosporaceae
Plantaginaceae
Plumbaginaceae
Poaceae — see Gramineae
Podocarpaceae
Podostemaceae
Polemoniaceae — see Cobaeaceae
Polygalaceae
Polygonaceae
*Polypodiaceae
Pontederiaceae
*Portulacaceae
Potamogetonaceae
Primulaceae
*Proteaceae
*Psilotaceae
*Ptaeroxylaceae
*Pteridaceae

*Rafflesiaceae
Ranunculaceae
Resedaceae
Restionaceae
Rhamnaceae
Rhizophoraceae
Rosaceae
Rubiaceae
Part 1
*Part 2
*Part 3
*Ruppiaceae
*Rutaceae

*Salicaceae
Salvadoraceae
*Salviniaceae
Santalaceae
*Sapindaceae
Sapotaceae
*Schizaeaceae
Scrophulariaceae
Scytopetalaceae
Selaginellaceae
Selaginaceae — in Scrophulariaceae
*Simaroubaceae
*Smilacaceae
Sonneratiaceae
Sphenocleaceae
Strychnaceae — in Loganiaceae
*Surianaceae
Sterculiaceae

Taccaceae
Tamaricaceae
Tecophilaeaceae
Ternstroemiaceae — in Theaceae
Tetragoniaceae — in Aizoaceae
Theaceae
Thelypteridaceae
Thismiaceae — in Burmanniaceae
Thymelaeaceae
*Tiliaceae
Trapaceae
Tribulaceae — in Zygophyllaceae
*Triuridaceae
Turneraceae
Typhaceae

Uapacaceae — in Euphorbiaceae
Ulmaceae
*Umbelliferae
*Urticaceae

Vacciniaceae — in Ericaceae
Valerianaceae
Velloziaceae
*Verbenaceae
*Violaceae
*Viscaceae
*Vitaceae
*Vittariaceae

*Woodsiaceae

*Xyridaceae

*Zannichelliaceae
*Zingiberaceae
*Zosteraceae
*Zygophyllaceae

FORTHCOMING PARTS

Acanthaceae
Part 2
Apocynaceae
Part 2
Asclepiadaceae — see Apocynaceae
Commelinaceae
Cyperaceae
Solanaceae

Editorial adviser, National Museums of Kenya: Quentin Luke
Adviser on Linnaean types: C. Jarvis

Parts of this Flora, unless otherwise indicated, are obtainable from:
Royal Botanic Gardens, Kew, Richmond, Surrey TW9 3AB, England. www.kew.org or www.kewbooks.com

*** only available through CRC Press at:**
UK and Rest of World (except North and South America):
CRS Press/ITPS,
Cheriton House, North Way, Andover, Hants SP10 5BE.
e: uk.tandf@thomsonpublishingservices. co.uk

North and South America:
CRC Press,
2000NW Corporate Blvd, Boco Raton, FL 33431-9868,
USA.
e: orders@crcpress.com

Information on current prices can be found at www.kewbooks.com or www.tandf.co.uk/books/